第二版

北京四合院建筑

The Architecture of the Quadrangle in Beijing (2nd Edition)

马炳坚　编著

图书在版编目（CIP）数据

北京四合院建筑／马炳坚编著.－2版.－天津：天津大学出版社，2020.4
ISBN 978-7-5618-6606-1

Ⅰ.①北… Ⅱ.①马… Ⅲ.①北京四合院－研究 Ⅳ.①TU241.5

中国版本图书馆 CIP 数据核字(2020)第 027395号

Beijing Siheyuan Jianzhu

出版发行	天津大学出版社
地　　址	天津市卫津路92号天津大学内
邮　　编	300072
电　　话	022-27403647
网　　址	www.tjupress.com.cn
印　　刷	北京雅昌艺术印刷有限公司
经　　销	全国各地新华书店
开　　本	210mm × 285mm
印　　张	19.75
字　　数	660千
版　　次	2020年4月第2版
印　　次	2020年4月第1次
定　　价	298.00元

谨以此书献给
第二十届世界建筑师大会

DEDICATED TO
THE XX UIA CONGRESS BEIJING 1999

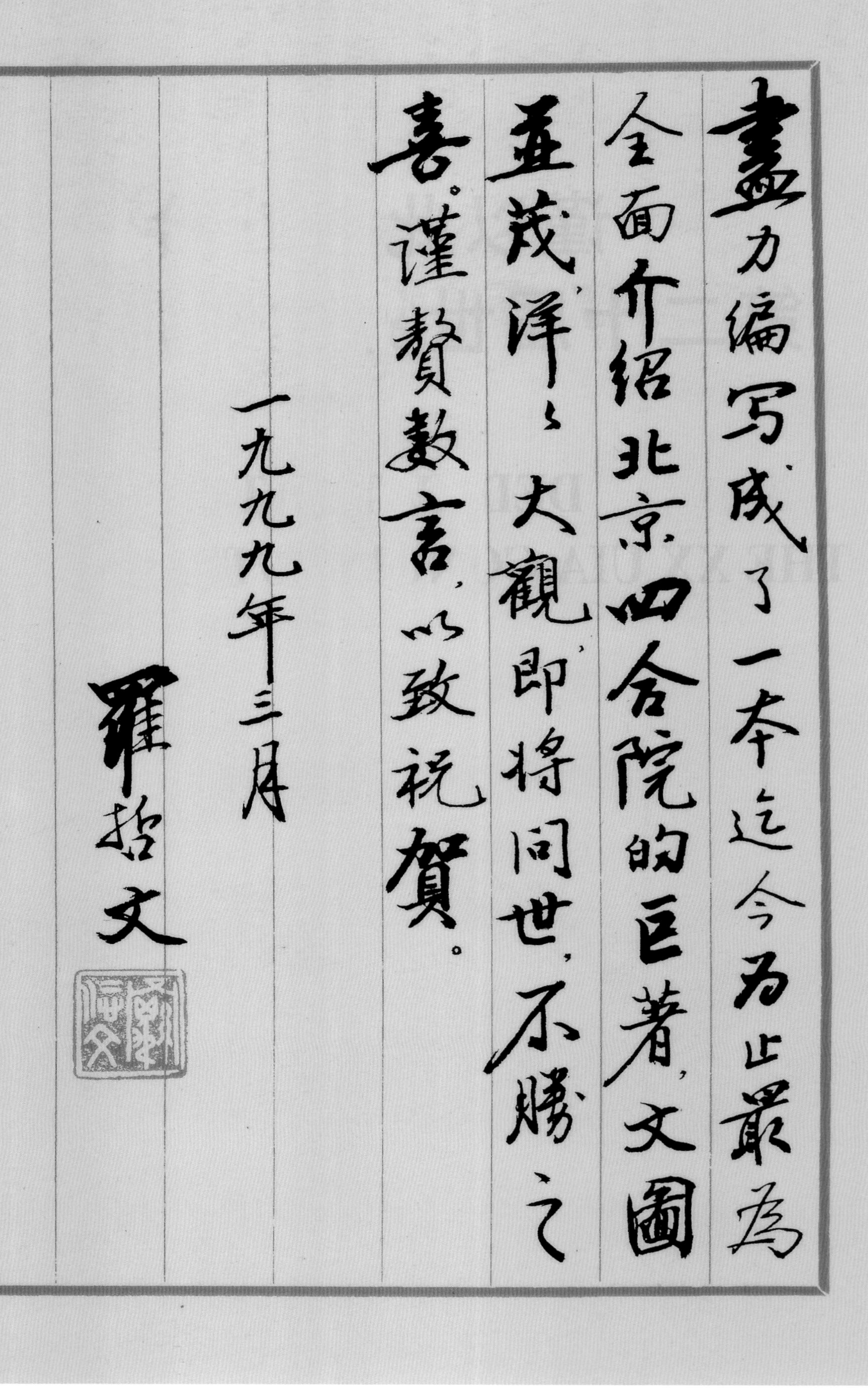

盡力編写成了一本迄今為止最為全面介绍北京四合院的巨著，文圖並茂，洋洋大觀，即將問世，不勝之喜。謹贅數言，以致祝賀。

一九九九年三月

羅哲文

四合院是中國傳統建築布局构成的基本形式，不仅有其發生發展的悠久歷史，而且有着丰富深邃的文化内涵，而北京四合院則堪稱為代表。如何保护研究這份珍貴的文化遺产，弘揚其优秀傳統，是當代建筑学人和熱心同道们十分关注的课題。茲有馬炳堅等同志費心

内容简介

《北京四合院建筑》一书是作者在多年从事北京四合院保护、研究、设计、施工的基础上写成的一部学术、技术专著。全书涉及四合院的历史、文化、格局、风水、空间、构造、装修、装饰、设计、施工、保护、修缮等全部内容，是迄今为止包含内容最全，技术、艺术信息含量最大，涉及方面最广的一部四合院建筑专著。本书不仅内容翔实，文字叙述流畅，通俗易懂，而且选用了大量照片和墨线图配合文字内容，使读者一目了然，使用方便。

本书第二版对基本内容进行了校订和补充，将最后一章进行了改写，使内容更加符合《北京城市总体规划(2016—2035)》精神。

本书的出版，对继承、弘扬中华传统建筑文化、保护古都风貌，对北京四合院及其他传统民居的保护、研究、开发利用均有重要意义。本书适合于从事古建筑文物保护、建筑历史研究、古建园林设计施工的广大科技人员及大专院校建筑系师生、中外建筑学工作者、房地产开发经营者、广大的古建筑爱好者阅读和参考。

Synopsis

THE ARCHITECTURE OF THE QUADRANGLE IN BEIJING is an academic and technical work based on the author' s work over the years on the research, design and construction of the quadrangle. It covers the history, culture, pattern, fengshui(localization), space, structure, design, fitting, decoration, protection and repair of the quadrangle. It is a monograph on the quadrangle, comprehensive in content,complete in the information about technology and art. Its statement is full and accurate, easy to read, and it is richly illustrated with photos and drawings.

In the revision, the basic content was revised and supplemented, and the last chapter was rewriten, to make the content more in line with *Beijing Urban Master Plan (2016–2035)*.

The appearance of this book will have great significance in the work on the inheritance,and dissemination of Chinese architectural culture, the protection of the style and feature of the ancient capital—Beijing, and the protection, research, development of the quadrangle and other forms of traditional folk dwellings.This book will be valuable to the departments in charge of the protection of architectural relics, study on history of architecture, design and construction of landscape gardens. Teachers and students of the architecture faculty in universities, researchers on architecture at home and abroad, and investors of real estate, and readers interested in ancient buildings, will benefit from this book as well.

作者简介

马炳坚，1947年生，高级工程师。从事中国传统建筑施工、设计、研究、教学、办刊等工作五十余年，业绩显赫，著作颇丰。

1967年步入传统建筑行业，曾经历天安门城楼重建，中山公园前区复建，北海、景山公园大修等工程。1983年与同仁共同发起创办《古建园林技术》杂志。1981年起，开始古建筑技术研究工作，并为创办北京房地产职工大学古建筑工程专业撰写木作专业课教材。该教材1991年定名为《中国古建筑木作营造技术》并由科学出版社正式出版，被海内外学者誉为“近代对中国古建筑最有分量的书”，并多次获奖。1987年开始涉足传统建筑设计工作，2003年任北京市古代建筑设计研究所所长，其间主持传统建筑及文物古建筑保护修缮设计项目数十个，并将设计成果编辑成《大壮营造录——北京市古代建筑设计研究所设计作品集》和《五十年传承与思考——马炳坚传统建筑设计作品集》。

曾任中国建筑学会建筑史学分会常务理事、中国文物学会传统建筑园林委员会常务理事、中国紫禁城学会常务理事、中国民族建筑研究会副会长。2014年发起创办中国勘察设计协会传统建筑分会并任首届会长。现为《古建园林技术》杂志主编、北京历史文化名城保护专家委员会专家、北京房地集团古建专家委员会顾问等。

About the Author

Born in 1947, Mr. Ma Bingjian is a senior engineer. He has been engaged for more than 50 years in design, construction, research, teaching and publication of ancient buildings with striking achievements and fruitful writings.

In 1967, he entered the profession of traditional architecture and has experienced reconstruction of the Tian' anmen Tower, reconstruction of the front area of Zhongshan Park, repairs of Beihai and Jingshan Park. In 1983 Mr. Ma and his colleagues started publishing a journal named *TRADITIONAL CHINESE ARCHITECTURE AND GARDENS*. From 1981, he started research work on the techniques of ancient buildings and wrote teaching material of timberwork for the major of ancient building engineering for the establishment of the corporate university of Beijing Real Estate. And the teaching material was named as *THE BUILDING TECHNOLOGY OF THE TIMBERWORK OF THE ANCIENT BUILDING IN CHINA* and published by Science Press. It is considered by Chinese and foreign scholars as the monumental work on Chinese ancient architecture and has been rewarded many times. From 1987, he started doing some work on design of traditional building and in 2003, he was appointed as the director of the Beijing Institute of Design & Research of Ancient Building. And after that, he has hosted dozens of conservation and repair design projects of traditional building and historic monument and he compiled all the design projects as two books published to the public, *RECORDING OF DESIGN & CONSTRUCTION PROCESS BASED ON TRADITIONAL ARCHITECTURAL CULTURE — PORTFOLIO OF BEIJING INSTITUTE OF DESIGN & RESEARCH OF ANCIENT BUILDING* and *INHERITANCE AND THINKING OF 50 YEARS — PORTFOLIO OF DESIGN WORKS OF TRADITIONAL BUILDING BY MA BINGJIAN.*

He had been the executive member of the Society of the History of Architecture under China Society of Architecture, the executive member of the standing committee of Traditional Architecture & Landscape under the Society of Culture Relics of China, the executive member of China Society of Forbidden City and the vice-president of National Architecture Institute of China. In 2014, he proposed and established the Society of Traditional Building under China Engineering & Consulting Association and was appointed as the president. Now he is the editor-in-chief of the journal *TRADITIONAL CHINESE ARCHITECTURE AND GARDENS*, the expert of the Expert Committee on Conservation of Beijing as a Historical & Cultural city and the advisor to the Expert Committee of Ancient Building of Beijing Real Estate Group.

目　　录

代序

《北京四合院建筑》读后[①]

按：本文是郑孝燮先生对《北京四合院建筑》一书的一篇书面评价。文中提出了关于保护北京古都风貌、保护四合院的重要意义；抒发了作者对传统民居建筑及其建筑艺术、环境艺术的深厚情感；表达了一位老学者对中国传统建筑及其保护现状的深切关注和对中青年学者从事传统建筑研究工作的热情支持，这一切都值得我们认真学习和思考。现全文发表，以飨读者。

马炳坚先生著《北京四合院建筑》是一部理论与实践相结合、历史文化与现代发展相结合、后来居上的居住建筑专著。这部书已走向世界，成为向北京举行的第20届世界建筑师大会献礼的专著。这件事实本身就等于对他作出了最好的评价。北京四合院民居很久以来就以规模之大、空间布局组合与建筑之精、庭院绿化与建筑装修之美闻名于世。新华社记者王军、刘劼在《寻找历史与现实的平衡》一文中说：现代主义建筑大师贝聿铭先生"对北京明清古城有着很高的评价"。贝大师认为"四合院不但是北京的代表建筑，还是中国的代表建筑。四合院应该保留，要一片一片地保留。不要这儿找一个王府，那儿找一个王府，孤零零地保护，这个是不行的"。他对"园林、四合院包蕴着的宜人尺度和变幻无穷的空间格局，投以更为关注的目光"（《瞭望新闻周刊》1999年10月25日，第43期，第47页）。

从事古建筑施工、设计、研究、教学、办刊三十余年的马炳坚先生，具有古建筑方面丰富的实践经验和专深的研究功力。如此两全的基础是许多专家学者不容易兼而有之的。实践出真知。这条朴素的真理，加上勤奋不懈的钻研努力，从而使作者情不自禁地在书中说出了"悟"字。"悟"其实是来自深化理论研究或实践经验的"知"的升华。先辈建筑大师梁思成教授对中国古建筑的研究极其精辟地"悟"出了"千篇一律，又千变文化"的至理。《北京四合院建筑》书中如"基本方位"与"基本格局"问题，如四合院有小、中、大型同类而不同型的问题，如一种风格、多种形式——四合院外形设计的原则的问题，以及院门等级、形制的同与不同等等，可以认为在相当程度上就是"千篇一律，又千变文化"这条至理的实物论证。

大科学家钱学森先生说"建筑是科学的艺术，或者艺术的科学"。《北京四合院建筑》也为建筑的这一综合性的本质特征作出了分解的实例辨证。例如，书中关于四合院建筑及其构造，关于四合院的营造与保护这两大部分均以重点突出建筑营造的工程技术为中心内容。又如关于外檐内檐装修、砖雕、木雕、石雕及油漆彩画等，则在于美术创作与工艺技术的紧密结合。再如家具与陈设、园林和庭院绿化等章节，则把环境艺术创造放在首位。

①原文刊登于《古建园林技术》2000年第1期。

尤其是园林与庭院绿化就格外要求匠心独运，追求人与自然和谐，以悦目赏心为首要目的。明清北京为五方杂处，全国首善之区。除皇家及王府园林外，达官贵人、富商大贾的四合院住宅，往往也都建有自己的花园——北方风格的私家园林，可惜早已毁坏殆尽。这种高、中档次的北京四合院的私家园林，同样讲求叠现人工山水，种植花木，追求自然情趣，甚至诗情画意。绿化居住环境，美化家居生活本来是我国居住文化的优良传统。一般普通的四合院民居十之八九无力自辟园林，不过他们却有自己的一套绿化院落、和谐自然、美化生活的做法。《北京四合院建筑》为此而借用了郑板桥小院落布置的一段描写："十笏茅斋，一方天井，修竹数竿，石笋数尺，其地无多。"而"风中雨中有声，日中月中有影，诗中酒中有情，闲中闷中有伴。非唯我爱竹石，即竹石亦爱我也"。"一室小景，有情有味。"仅仅如此简洁的茅屋小院，竹石点点，风雨日月，就能带给这位小院主人以丰富的心理感受。书到这里作者加了一道重笔："可见营造环境对人心理影响之大"，"前人经验，值得研究继承"。

《北京四合院建筑》对于我国四合院历史的追溯，颇有史可考，又有遗址和文物考古可证。不仅如此，更可贵的是作者还立足现实，为了有利于城区居住条件的改善和人口、交通等压力的减轻，进而落实古都风貌保护，提出了在保护维修原有四合院的同时，还需要重视新四合院的设计问题。主张发挥建筑新材料、新工艺、新技术、新内容的效用和传统的建筑形式风格特点相结合的设计。而且认为：一、这种新设计（不属于文物）以钢筋混凝土构架代替木构架已成为不可逆转的趋势。二、新四合院设计是一种仿建的建筑，以仿得真仿得像，达到以假乱真为原则。三、当今新生事物层出不穷，现代生活设施必须与四合院结缘，但这种结缘不应冲淡四合院本来特有的古朴、典雅的风格气韵。

《北京四合院建筑》不但文字内容好，选用的照片、图绘也精美可贵。看到书中有些四合院的图形，就会引人联想，不知有多少迄今还活着，又有多少已在推土机下消失成了"遗影"！书中前言背页引的那张雪中的北京四合院建筑全景照，把整个四合院及左邻右舍前后的邻居院落全部照了出来。那里是那样的安静、纯朴自然、美观实用，那样的"京味"十足，然而不知它现在是否还活着？不知它今后的命运又将如何？

总之，历史早已证明，人类的建筑文化始于民居。《北京四合院建筑》这部民居著作的出版，必将为丰富中国建筑史系，为抢救、保护古都风貌，为研究老北京的家居礼俗历史文化，为研究推陈出新设计新四合院民居，乃至为改革开放加强对外文化交流等方面，都会作出很大的贡献。

郑孝燮

1999 年 11 月 23 日于北京

After reading
THE ARCHITECTURE OF THE QUADRANGLE IN BEIJING

Note: This article is a book review written by Mr. Zheng Xiaoxie on *THE ARCHITECTURE OF THE QUADRANGLE IN BEIJING*. The article puts forward the significance of protecting historic urban landscape of Beijing and quadrangle dwellings in Beijing, and expresses the author's deep feelings about traditional residential buildings and their architectural, environmental art and expresses an old scholar's deep concern about the status of traditional Chinese architecture and its conservation and enthusiastic support for the traditional architectural research work done by the young and middle-aged scholars, and all these mentioned above deserve our serious study and thinking. Now it is published in full version for readers.

THE ARCHITECTURE OF THE QUADRANGLE IN BEIJING by Mr. Ma Bingjian is a monograph on residential buildings combining theory with practice, historical culture with contemporary development, which surpass the former relevant books. This book has spread globally and become a monograph dedicated to the 20th World Architects Conference held in Beijing. This fact is the best evaluation of him in itself. Quadrangle residential buildings in Beijing have long been famous for their large scale, the organization of spatial layout, architectural value, courtyard greening and architectural decoration. The reporters Wang Jun and Liu Jie from Xinhua News Agency said in the article *Finding the Balance between History and Reality*: Mr. Ieoh Ming Pei, a master of modernist architecture, "has a very high opinion of the historic urban landscape of Beijing in the Ming and Qing Dynasties". Master Pei believes that "quadrangle dwelling is not only a representative building of Beijing, but also a representative building of China. The quadrangle dwellings should be safeguarded, and it should be kept block by block. Don't conserve a royal palace here, and then conserve another there, and keep them isolated from each other, and it doesn't work". He paid more attention to "the pleasant scale and ever-changing spatial pattern contained in the garden and the quadrangle dwellings" (*Outlook News Weekly*, October 25, 1999, No. 43, p. 47).

Mr. Ma Bingjian has been engaged in the construction, design, research, teaching and publication of historical buildings for more than 30 years. He has rich practical experience and deep research skills in this field. It is rare that scholars and experts have such a basis. Real knowledge comes out of practice. This simple truth, coupled with diligent research efforts, led the author to utter "what he comprehend" in the book in spite of himself. "Comprehension" is actually the sublimation of "knowledge" from deepening theoretical research or practical experience. Our predecessor, Professor Liang Sicheng, a master of architecture, has brilliantly "comprehend" the truth of "sameness with infinite changes" to describe Chinese historical bindings based on his research on Chinese traditional architecture. In the book *THE ARCHITECTURE OF THE QUADRANGLE IN BEIJING*, such as "basic orientation" and "basic pattern", such as issues of the small, medium and large scale of quadrangle dwellings, such as one style and multiple forms—the design principles of quadrangle dwellings, as well as the similarities and differences of the level and the shape of the gate to the courtyard, etc., can be considered a physical argument of the truth "the sameness with infinite changes" to a considerable extent.

Mr. Qian Xuesen, a great scientist, said "the art of architect-science, or the science of art". *THE ARCHITECTURE OF THE QUADRANGLE IN BEIJING* also is a detailed illustration of this comprehensive essential feature of architecture. For example, the content about the building and its structure, the content about the construction and protection of quadrangle dwellings, these two parts focus on the engineering technology of building construction. And about the eave decoration inside and outside, brick carving, wood carving, stone carving and color painting, etc., it lies in the close combination of art creation and technology. And chapters on furniture and furnishings, gardens and courtyard greening put environmental art creation first. In particular, gardens and courtyard greening require ingenious design, the pursuit of harmony between man and nature and the primary purpose of pleasing the eye and the body. In the Ming and Qing Dynasties, Beijing was a place where people of all kinds live and it was the best place in the country. In addition to the royal gardens, the quadrangle dwellings of noble officials and wealthy businessmen often have their own gardens which show northern style. Unfortunately, they have already been destroyed. The private gardens of this high and medium quality of quadrangle dwellings in Beijing also emphasize the organizing and arranging artificial landscapes, planting flowers and trees, and pursuing natural taste and even poetry. Greening the living environment and beautifying the household life is originally a fine tradition of Chinese living culture. Generally, the ordinary residences of quadrangle dwellings are incapable of creating gardens by themselves, but they have their own methods to green courtyards, achieve harmony with nature, and beautify life. *THE ARCHITECTURE OF THE QUADRANGLE IN BEIJING* borrows a description of the layout of the small courtyard of Zheng Banqiao: "a thatched cottage in a quite small courtyard, with a well and some slime bamboos and also some stalagmites and there is not much land left." But "there is sound in the rain and the wind, a shadow in the sun and the moon, love in the wine and poetry, and companionship in the boredom and the leisure. Not only I love bamboos and stones, but also they love me" and "a small room with a beautiful scene, it' s affectionate and interesting". Just such a simple thatched cottage and courtyard, a little bit of bamboos and stones, the wind and the rain, can bring rich psychological feelings to the owner. Here the author spills more ink: "It is obvious that the environment has a great impact on people' s psychology" and "the previous experience is worth studying and inheriting."

THE ARCHITECTURE OF THE QUADRANGLE IN BEIJING traces the history of the quadrangle dwellings in China based on both considerable historical documentation and archeology of site and heritage site. In order to help improve the living conditions in the urban area and reduce the pressure on population and traffic, and furthermore implement the protection of historical urban landscape of the city, the author propose that while protecting and maintaining the original quadrangle dwellings, it is necessary to pay attention to the design issues of new quadrangle dwellings. He advocates a design that combines the utility of new building materials, new processes, new craftsmanship, and new contents with the traditional architectural form and style. Moreover, it is proposed that: first, this new design (not a cultural relic) has replaced the wooden frame with a reinforced concrete frame, which has become an irreversible trend; second, the design of the new quadrangle dwellings is a kind of imitation building, which shall be more real and appear identical with the original as much as possible; third, there are many new things emerging today, and contemporary living facilities must be associated with the dwellings, but this connection should not dilute the original and elegant style of the building.

THE ARCHITECTURE OF THE QUADRANGLE IN BEIJING has not only exquisite content, but also beautiful photos and drawings. When you see some graphics of a quadrangle dwelling in the book, it will make you wonder how many are still alive so far, and how many have disappeared under the bulldozer and become a "left-over shadow"! The panoramic view of the quadrangle dwelling of Beijing in the snow cited on the back of the foreword in the book photographed an entire quadrangle dwelling and its surrounding yards. It was so quiet, simple, natural, beautiful and practical, and full of "Beijing Flavor", but I wonder if it is still alive now? I wonder what will happen to its future?

In short, history has long proven that the architectural culture of mankind began in residential houses. The publication of the book *THE ARCHITECTURE OF THE QUADRANGLE IN BEIJING* is bound to enrich the history of Chinese architecture and make a contribution to saving and protecting historic urban landscape of Beijing and studying the history, the culture and customs of traditional Beijing, and helping design new quadrangle dwellings and strengthening foreign cultural exchanges in the course of reform and opening-up.

ZHENG Xiaoxie

In Beijing, 23 Nov, 1999.

再版前言

20 年前，我怀着对中华传统建筑文化及北京传统四合院建筑的挚爱之情，应天津大学出版社之邀，撰写了《北京四合院建筑》一书。这本书出版至今已经 20 年了。20 年来，该书印刷八次，印数达 2.2 万册。对于这样一本专业性极强，涉及面也不宽的专业学术技术专著来说，可以说是一个破记录的数字了。

这本书写成于第 20 届世界建筑师大会在北京召开之际，是一本向大会献礼的书，旨在使中华传统建筑文化重要组成部分的北京四合院建筑文化在中国、在世界得以传播和弘扬。今天看来，这个目的是达到了。

20 年来，这本书在北京传统四合院保护方面发挥了应有的作用。这 20 年，正是对北京老城胡同四合院是“拆”还是“保”两种主张争论最为激烈的年代。相关规划管理部门、研究部门、大专院校、设计单位中热爱传统建筑文化的专家、同仁在进行老北京四合院保护规划编制以及对破旧院落进行翻改建过程中，大量参考并采纳了《北京四合院建筑》一书中的相关内容，使该书在北京四合院保护和修建中发挥了重要作用。

20 年来，许多在改革开放中先富起来的成功人士出自对北京四合院文化的热爱，不惜重金，在北京老城区购置旧院，建造新宅。他们把拥有一所四合院看作身份和地位的象征。他们建造四合院的重要依据就是《北京四合院建筑》这本书。当年我在北京市古代建筑设计研究所工作时，每年都会接到数起这类高档四合院的设计委托。

20 年来，重视中华传统建筑文化的房地产开发企业也把开发（应为“翻改建”）高档四合院住宅作为企业追求的目标。尽管当时北京市在四合院翻改建方面有“小规模，微循环，渐进式”的政策约束，但还是有不少成片的老房子被翻改建成具有传统风格特点的现代四合院。这些房地产企业规划设计的重要参考依据也是《北京四合院建筑》这本书。

20 年来，一些移居到北京的外国人也在北京老城区购地置宅，当起了北京寓公。有位美国朋友在北京先后建了三个四合院。最早建的那个四合院就是看了这本书后找到我，由我亲手设计的。

20 年来，还有移民美国、加拿大、澳大利亚、俄罗斯等国的华人，来到北京与我探讨有关在当地建造北京传统四合院事宜。他们找到我时，往往都是带着《北京四合院建筑》这块“敲门砖”。

由于这本书通俗易懂，图文并茂，通篇充满北京传统文化内容，很多领导同志和相关部门常把这本书作为赠予外国友人的礼品，使这本书进一步走向世界。

……

近 20 年新建的四合院有许多变化，它们已经不再是明清或民国时期那种室内设施比较简陋的传统住宅，而是院落格局、建筑风貌（包括建筑形式、材料做法、装饰、色彩等）保持北京四合院的传统特色，室内现代设施应有尽有，生活十分方便舒适的现代高档次住

宅，是最符合中国人“天人合一”的宇宙观，最接地气、最健康舒适的理想居所，是传统四合院在当代的“涅槃和重生”。

较之20年前，北京老城区的胡同四合院（杂院）也发生着另外一些变化。配合北京亚运会、奥运会以及其他重要活动的举办，北京老城区胡同四合院也不止一次进行了整饰、包装和改造。由于执行者不够专业，每改造一次，都会使胡同四合院失去一分当年的风采。加上近些年来一些住在大杂院的原住民的迁出和其他务工人员的进驻，原有的大杂院愈加杂乱无序。生活在杂院里的居民的居住环境愈加恶化。为改善老百姓的居住环境，一些新悦派建筑师也不断在老城区“做试验”“搞样板”，甚至把外国的居住理念或模式搬到北京老城区，使已经为数不多的四合院面临着“转基因”的危险。

2016年，北京市出台了《北京城市总体规划（2016—2035）》。这是由党中央、国务院批复的规划，贯彻了习近平总书记两次视察北京的讲话精神，体现了党中央对首都功能的新定位。“规划”特别指出，对老城区“不能再拆”，“要应保尽保”，并制定了对胡同四合院进行“恢复性修建”的方针。这是新中国成立以来对老北京四合院保护力度最大，指导思想最明确的一轮规划。认真执行这个规划，对北京四合院的保护和传统建筑文化的传承，具有深远意义。

自2017年以来，我本人作为北京历史文化名城保护专家，深度参与了北京胡同四合院的保护整治工作。通过保护实践，不仅对四合院的文化内涵有了更加清晰的认识，对如何进行“恢复性修建”也形成了一套较为成型的想法。在这种背景下，对《北京四合院建筑》一书进行修订，更具有实践依据和现实意义。

本次修订，主要有以下内容。

一、将我国著名规划和建筑专家郑孝燮先生2000年为《北京四合院建筑》一书写的书评《〈北京四合院建筑〉读后》作为第二版序言，置于全书之首。

二、增加再版前言，回顾本书出版20年来对四合院建筑文化传承及修缮事业发展发挥的作用及作出的贡献。

三、对与四合院构成有关的宅门、影壁、墙体、建造技艺以及油饰、彩画等相关内容进行必要的补充、调整。

四、重写第九章。将原第九章“北京四合院前景展望”删除，改为“北京老城保护和四合院的恢复性修建”，重点谈在《北京城市总体规划（2016—2035）》指导下，对北京老城区胡同四合院进行整治、提升，“恢复性修建”的策略和做法。

修订后的《北京四合院建筑》，对原有的基本内容完全保留，对如何进行“恢复性修建”将进行较为系统、详尽的阐述，以期对当前的四合院保护修建工作发挥应有的作用。

20年，既漫长又短暂。

回顾20年前将《北京四合院建筑》一书奉献给第20届世界建筑师大会和广大读者的情景，感觉恍如昨日。而当年仅有52岁的我，如今已步入古稀之年。

回望20年，不免唏嘘感叹，同时，也为《北京四合院建筑》一书能对继承弘扬祖国传统建筑文化有所贡献而安慰。在此，我要对20年来酷爱四合院，并为传承和弘扬四合院建筑文化作出贡献的同仁好友表示衷心感谢！我们还要继续努力，并肩携手，为北京四合院的保护、传承与发展，为中华传统建筑文化的发扬光大不断作出新贡献！

馬炳坚

2019年5月1日于北京营宸斋

Preface to the second edition

20 years ago, with the love for Chinese traditional architectural culture and traditional quadrangle dwellings in Beijing, I was invited by Tianjin University Press to write a book named *THE ARCHITECTURE OF THE QUADRANGLE IN BEIJING*. It has been 20 years since the book was published. In the past 20 years, the book has been printed eight times and the number of copies has reached 22,000. It is a record-breaking number for such a professional, academic and technical monograph, which is extremely professional and has no extensive coverage.

The book was written on the occasion of the 20th World Architects Conference in Beijing. It is a book dedicated to the conference. It aims to make the architectural culture of quadrangle dwellings in Beijing, an important part of Chinese traditional architectural culture, spread in China and worldwide. It seems that this goal is achieved today.

For the past 20 years, the book has played its due role in the protection of traditional quadrangle dwellings in Beijing. These two decades, is an era when the debate is the most intense about whether to dismantle or protect Hutong and quadrangle dwellings of traditional Beijing. Many experts and colleagues interested in traditional architectural culture, have referred to and adopted the relevant content a lot in the course of making the master plan for the quadrangle dwellings and rebuilding the shabby courtyards of traditional Beijing. They come from relevant planning management departments, research departments, colleges and universities, and design institutes, and what they did makes the book play an important role in the protection and construction of quadrangle dwellings in Beijing.

In the past 20 years, many successful people who first got rich during the reform and opening-up bought old courtyards and built new houses in the old town of Beijing regardless of great expense, only because of the love for quadrangle dwellings in Beijing. They considered the possession of a quadrangle dwelling as a symbol of identity and status. The important basis for their construction is this book *THE ARCHITECTURE OF THE QUADRANGLE IN BEIJING*. When I was working in Beijing Institute of Design & Research of Ancient Building, I received several design commissions for such high-end quadrangle dwelling every year.

In the past 20 years, real estate development companies that value Chinese traditional architectural culture have also put the development (should be "reconstruction") of high-end quadrangle dwellings as a target. Although Beijing had "small-scale, micro-circulation and step-by-step" policy restraints in the renovation of the courtyard house at that time, many groups of old houses were transformed into modern quadrangle dwellings in traditional styles. The important reference is also the book *THE ARCHITECTURE OF THE QUADRANGLE IN*

BEIJING, for the planning and design finished by these real estate companies.

Over the past 20 years, some foreigners who had migrated to Beijing have also purchased land, bought houses in the old town of Beijing, and settled down in Beijing. An American friend built three quadrangle dwellings one after the other in Beijing. The reason why the earliest one was designed by me is that he read the book and came to me for help.

In the past 20 years, some Chinese who have immigrated to the United States, Canada, Australia, Russia and other countries came to Beijing to discuss with me about the construction of the quadrangle dwellings of traditional Beijing. When they came to me, they often brought *THE ARCHITECTURE OF THE QUADRANGLE IN BEIJING* with them as a stepping-stone.

The book is easy to understand, with lots of photos and texts being interrelated and exquisite, and full of Beijing' s traditional culture. Many leading comrades and related departments often use this book as a gift for foreign friends, making it go into the world further.

There have been many changes in the newly built quadrangle dwellings in the past 20 years. They are no longer traditional houses with relatively simple indoor facilities like in the Ming and Qing Dynasties or the Republic of China, but their courtyard pattern and architectural style (including architectural forms, material practices, decoration, colors, etc.) have kept the traditional characteristics of quadrangle dwellings in Beijing. Now they are modern high-end residences with all the modern facilities, which can provide quite convenient and comfortable living conditions. It is the most ideal place to live in harmony with nature, the most down-to-earth, the most healthy and comfortable. It is the nirvana and rebirth of a traditional quadrangle dwelling in contemporary China.

Compared with 20 years ago, Hutong and quadrangle dwellings in the old town of Beijing (miscellaneous courtyard) has undergone other changes. In conjunction with the hosting of the Asian Games, the Olympic Games and other important events, Hutong and quadrangle dwellings has been repaired, renovated and transformed more than once in the old town of Beijing. The performers are not professional enough, so every time they are intervened and thus they would appear less attractive than before. Coupled with the migration of some aborigines who have lived in the quadrangle dwellings in recent years and the arrival of other migrant workers, the original miscellaneous courtyards have become even more chaotic. The living conditions of residents living there is getting worse. In order to improve the living environment of ordinary people, some architects of *Xinyue* School (those trying designing new quadrangle dwellings with traditional characteristics) also "made experiments" and "set

up models" continually in the old town of Beijing, and even employ foreign living ideas or models in the old town of Beijing, and what they did makes the few quadrangle dwellings faced with the danger of "GMO".

In 2016, Beijing enacted *Beijing Urban Master Plan (2016–2035)*. This is a plan approved by the Party Central Committee and the State Council, which carries out the spirit of the speech of General Secretary Xi Jinping during his two inspections in Beijing and reflects a new orientation of the capital's function proposed by the Party Central Committee. The *Plan* specifically pointed out that demolition was no longer allowed and we should protect the old town as much as possible and in the *Plan* the policy of "restorative construction" of Hutong and quadrangle dwellings was formulated. This is a round of plan that has provided the greatest protection efforts to the quadrangle dwellings of traditional Beijing and the clearest guiding ideology since the founding of the People's Republic of China. It is of profound significance to implement the *Plan* carefully for the protection of the quadrangle dwellings in Beijing and the inheritance of traditional architectural culture.

Since 2017, as an expert in the protection of Beijing as a famous, historical and cultural city, I have been deeply engaged in the protection and renovation of Hutong and quadrangle dwellings in Beijing. Through the protection practice, I not only has a clearer understanding of the cultural connotation of quadrangle dwellings, but also form a set of ideas on how to carry out "restorative construction". Within this context, it will have more practical evidence and be of more realistic significance to revise the book *THE ARCHITECTURE OF THE QUADRANGLE IN BEIJING*.

The revision of this time mainly includes the following.

First, the book review *After reading THE ARCHITECTURE OF THE QUADRANGLE IN BEIJING* written by Mr. Zheng Xiaoxie, a well-known expert of architecture and planning in China, for the first edition of *THE ARCHITECTURE OF THE QUADRANGLE IN BEIJING* in 2000, is placed in front of all the content as the preface of the second edition.

Second, to add a foreword to the reprint and make an overview about the role and contribution of the book to the inheritance of architectural culture of quadrangle dwellings and the development of repair cause of quadrangle dwellings in the past 20 years.

Third, to make necessary additions and adjustments to the related contents of gates, screen walls opposite gate, walls, construction techniques, oil decoration and color paintings related to the composition of quadrangle dwellings.

Forth, to rewrite Chapter 9. The original Chapter 9 *Prospects of THE ARCHITECTURE OF THE QUADRANGLE IN BEIJING* was deleted and replaced with *Protection of old town of*

Beijing and the restorative construction of quadrangle dwellings, focusing on the renovation and improvement of Hutong and quadrangle dwellings in the old town of Beijing, and the strategy and methods of "restorative construction", under the guidance of *Beijing Urban Master Plan (2016–2035)*.

The revised book completely keeps the original basic content, and will explain in a more systematic and detailed way about how to implement "restorative construction", with the expectation of playing its due role in the current course of the protection and construction of quadrangle dwellings.

20 years, long and short.

Looking back on the case that the book was dedicated to the 20th World Architects Conference and numerous readers 20 years ago, it seems like yesterday. At that time I was only 52 years old, and now I've got in my seventies.

Looking back at the past 20 years, I can't help sighing deeply. And at the same time, I also feel gratified that the book *THE ARCHITECTURE OF THE QUADRANGLE IN BEIJING* can make a contribution to inheriting and carrying forward the traditional architectural culture of our country. Here, I would like to express my sincere gratitude to my colleagues and friends who have been fond of quadrangle dwellings for 20 years and have made a contribution to inheriting and developing the architectural culture of quadrangle dwellings. We shall go to great lengths to work hard and shoulder to shoulder, to make new and continuous contributions to the protection, inheritance and development of quadrangle dwellings of Beijing and the development of Chinese traditional architectural culture!

Ma Bingjian

In Yingchen Studio, Beijing, 1 May, 2019

前言

北京四合院是老北京人世代居住的主要建筑，是中国传统居住建筑的典范。

北京四合院有着十分悠久的历史，它的雏形产生于商周时期，元代时作为主要居住建筑大规模出现在北京，明清两朝得到长足发展。

北京四合院有优于其他任何住宅形式的居住环境，它有宽绰疏朗、起居方便的中心院落，有高度的私密性和亲合性，非常适合独家居住。

北京四合院以东、西、南、北四面房屋及其围合的院落为基本单元，可向纵深和两侧任意发展，适合于各种不同规模的家庭居住。

北京四合院的建筑格局和空间构成体现着以家长为中心的封建家庭秩序，是中国封建社会的基本组成细胞。

北京四合院有着地道的京韵京味儿，展示着老北京人传统的民风民俗，具有浓郁的民族风格和地方特色。

北京四合院的建筑构造和工艺技术，反映出我国民居建筑技术所达到的高超水平，是一部生动的中国居住建筑技术史。

北京四合院的建筑装饰具有极高的艺术价值和观赏价值，它反映出中国传统民居建筑装饰艺术的辉煌成就。

北京四合院有丰富的文化意蕴，它的每一块砖石构件、每一处雕饰彩绘、每一幅匾额楹联，都与社会的伦理道德、行为规范，人们的信仰追求、文化修养有直接的联系。

作为中国传统建筑重要组成部分的北京四合院，近年来受到各方面越来越多的关注。政府和规划部门关注着它——它是体现北京六朝古都传统风貌的重要内容，要保护古都风貌，必须认真保护北京传统四合院；文物部门、旅游部门关注着它——它是北京重要的文物遗存和不可缺少的旅游资源，受到中外人士的青睐；旅居海外的华人关注着它——在祖国日益强大的今天，他们渴望落叶归根，并想重新住进心目中留有深刻印象的四合院；改革开放中先富起来的酷爱中国传统文化的人们关注着它——他们不惜重金，购买旧院，建造新宅，过起独门独院的安逸生活；对中华文化情有独钟的外国人关注着它——他们以能住进中国传统四合院为时髦和荣耀，纷纷在京城买地建宅，心安理得地当起了“北京人”；房地产商也关注着它——他们看准了北京四合院这具有中国民族风格的高档住宅的潜在市场，纷纷投资，改造旧街区，开发新街区，把资本转向四合院的开发建设……凡此种种都表明，在高楼大厦林立的都市中，正涌动着一股传统民居四合院复兴的热流。

这种情况要求我们对传统民居四合院进行全面的、系统的、深入的研究；对它的价值应当重新认识；对它的营造、保护技术以及深刻的文化内涵应当加以继承；对四合院在新的历史条件下的发展趋势要作出科学的分析和展望。

自1980年以来，我们在从事古建筑研究的过程中，对北京四合院进行了大量的实物调查和资料收集工作，在从事传统建筑设计的工作中，又进行了不少四合院的修缮、复建设计，并从中学到了许多知识，悟出了不少道理，积累了一定经验，同时也感受到人们对四合院的深深眷恋和对四合院文化断层的无比痛惜。于是萌生了要写一本全面介绍四合院的书籍的想法，以使这一优秀的传统建筑文化得到继承和弘扬。这个想法得到领导和同行们的普遍赞同，于是，将这个课题列入计划之中，并拟于2001年完成付梓。

1998年年初，天津大学出版社的同志找到我，向我了解出书计划。当听到有关北京四合院的选题时，他们尤感兴趣，并向我约稿。由于当时承担着繁重的仿古建筑设计任务，未敢贸然答应。5月，天大出版社又找到我，并表达了要将这个选题作为向第二十届世界建筑师大会献礼项目的意向。将中国优秀传统建筑文化介绍给世界，是我们多年来奋斗不息的事业。第二十届世界建筑师大会在北京召开，是一个千载难逢的机会。此时，我们承担的设计工作也基本完成，时间安排上有了一些松动。我考虑再三，决定接受出版社的稿约，提前两年实施《北京四合院建筑》的出版计划。

《北京四合院建筑》一书内容涉及四合院的历史、文化、风水、格局、空间、构造、装修、装饰、设计、施工、保护、修缮等全部内容，是迄今为止包含内容最全，技术、艺术信息含量最大的一部专业学术技术专著。它不仅能帮助一般人士了解和认识四合院，而且能帮助专业技术人员、大中专学生学习和了解四合院，并对四合院的设计、营造、保护、维修也有着重要参考价值。

本书共分九章，第一章介绍传统四合院住宅产生、形成、发展、衰败的历史以及在新形势下复苏的状况。第二章介绍四合院的基本格局，其中既有理想的标准院落，又有不甚规则的实物例证，可使读者对四合院的类型、布局有比较全面的认识。第三章讲述有关风水方面的一些常识，对其中比较实用的内容作了较为详尽的介绍。这三章，对于介绍四合院的书籍来说，是必不可少的内容，但它们不是本书的重点。这些内容，许多先行者都已做过深入研究，并发表过不少论著，本书只是根据内容需要将前人的成果进行了摘引和重新组织。第四章分析了四合院的建筑空间组合模式，以及这种模式与封建制度之间的内在联系，将读者的视线由表象引向深层。第五章和第六章是本书的重点章节。第五章讲述四合院建筑及其构造，其中用了三节篇幅介绍四合院的主要建筑及附属建筑，如正房、厢房、

耳房、倒座房、后罩房、宅门、垂花门、屏门、游廊、影壁以及上马石、拴马桩等的构造、尺度、色彩、建筑造型和艺术特点；用两节篇幅对室内外木装修作了详细介绍。第六章，四合院建筑的装饰和室内陈设，主要谈四合院的建筑装饰以及艺术成就，重点写四合院的砖雕、石雕、木雕、油饰彩画以及室内家具与陈设。通过对这些艺术形式和内容的介绍，揭示其中的文化内涵，给人们以有益的启示和借鉴。其中家具和彩画两节，分别聘请对家具陈设及油饰彩绘有精深造诣的王希富、蒋广全二位先生执笔，使本书大为增色。第七章讲述私家园林和庭院绿化，专门介绍从属于四合院民居的私家园林的规模、功用、造园手法、艺术特色及文化内涵；对庭院的绿化种植及其讲究也作了必要的阐述。五、六、七三章内容比较详尽，配合文字选用了大量实物照片和墨线图，给人以具体形象、深入确切的感受和印象。第八章，四合院的营造与保护，也是本书的重点，它的内容对四合院的设计、施工、保护、修缮将提供具体的帮助。最后一章是对四合院现状分析和对未来的展望。在这一章中，我们结合近年的工作体会和对四合院的社会需求状况，分析了四合院的发展前景，作出了新的历史条件下四合院不仅可以保留，而且还会继续发展的客观估计。

《北京四合院建筑》一书的编著工作是异常紧张的。从与出版社签约到完稿仅有近半年的时间。这期间，除著书之外，还要兼顾工程设计、课题研究、《古建园林技术》杂志的编辑和其他日常工作，可谓杂事繁冗，重负难堪，加之本人学识所限，难免论证失当、谬误失察、挂一漏万。尤其对四合院文化内涵的发掘更显肤浅单薄。欠缺之处，敬希广大读者和专家、同仁赐教斧正。

我们是怀着对祖国优秀传统建筑文化的挚爱之情撰写本书的。我们希望通过它能使更多的中外人士了解北京四合院，不仅了解它的形式和艺术，更了解它的内容和价值，珍惜它的存在和发展，使这一优秀的传统居住形式更好地服务于今天的社会，使中华传统建筑文化在中国、在世界得以更加广泛地传播和弘扬。这就是我们奉献此书的主要目的。

马炳坚
1999 年 3 月
于北京营宸斋

Preface

The siheyuan (quadrangle) is the main architectural form of dwelling houses where elder Beijing inhabitants live from generation to generation and the paradigm of Chinese residential buildings.

The quadrangle has a very long history. It took shape in the Shang-Zhou Periods(c.16th century-256 B.C.). It appeared on a large scale as the principal dwelling in Beijing in the Yuan Dynasty (1271-1368) and developed rigorously in the Ming and Qing Dynasties (1368-1911).

The quadrangle has a living environment superior to any other forms of dwellings for its central courtyard, spacious and convenient to live, high privacy and affinity fit for a family to reside by itself.

The basic units of the quadrangle are the rooms in the east, west, south and north of the quadrangle and the courtyard enclosed by the above-mentioned rooms. It can develop in depth and on both sides, suitable for families of different sizes.

The building pattern and spatial composition of the quadrangle embody the order of a feudal family with the head of the family as the core, and the quadrangle is the basic cell of the Chinese feudal society.

The quadrangle exhibits the traditional folk custom, the typical taste of Beijing inhabitants and national style and local features.

The structure and technology of the quadrangle reflect the high level which the building technology of folk houses has ever reached in China. The quadrangle is the vivid history of technology of Chinese architecture.

The decoration of the quadrangle has high artistic value to view and admire. They represent the brilliant achievements of the architectural decoration of the Chinese folk houses.

The quadrangle possesses rich cultural connotation. Every brick and stone member, every carving and colored drawing, every horizontal inscribed board and the couplets hung on the pillars of a hall are all directly associated with social ethics and morality, codes of conduct, people's belief and pursuit, and artistic appreciation.

As an important integral part of Chinese traditional architecture the Beijing quadrangle is attracting more and more attention from all walks of life. The government and planning departments are concerned with it — it is the important entity which embodies the traditional style and features of the ancient capital of six dynasties. To preserve them the quadrangle must be preserved in earnest; the departments of cultural relics and tourist administrations attach importance to it — it is the vital cultural monument and the indispensable tourist resource which is admired by Chinese and foreign tourists; the oversea Chinese have a tender feeling for it — when the motherland is becoming increasingly strong and rich they are longing for returning to their birthplace and living once again in the longcherished quadrangles; those who have become wealthy thanks to the Party reform and open policy and have a deep love for the traditional culture pay attention to the quadrangle, spare no money to buy old quadrangles to reconstruct new ones, living a comfortable life in a detached house which has its own entrance and courtyard; those foreigners who show affection for Chinese culture take interest in the quadrangle — they are proud of living in the traditional Chinese quadrangle. They are busy buying land to build their quadrangle and feel at ease and justified to become Beijing citizen. The real estate investors set store by the quadrangle — they find the potential market in the high-grade residence — the quadrangle of national style and invest their money in the transformation of old blocks and the development of new blocks, in the construction of quadrangles … all these show that a surge of reviving the traditional quadrangles is welling in Beijing — the metropolis which has already had throngs of high rise buildings and magnificant mansions.

Such a situation requires that we should carry out comprehensive, systematic and deep studies on the traditional folk houses — quadrangle; re-appraise the value; inherit its technology of construction and protection and its profound cultural connotation; make scientific analysis and prospect of its developmental trend under the new historical conditions. Since 1980 along with studying the ancient architecture we have made extensive on-the-spot surveys and information gathering work. In the course of designing, repairing and reconstructing historical buildings we have learned and realized a great deal and accumulated a certain amount of experience. Having been impressed by people deep affection for the quadrangle and strong pursuit, we decided to write a book on the quadrangle in detail to the end that the excellent traditional architectural culture can be inherited and expanded. Our idea was supported by the leadership and colleagues and the topic was included in our plan. It was scheduled to come out in 2001.

At the beginning of this year (1998) editors of Tianjin University Press came to me and asked me for our plan of publication. When they heard of the topic of the quadrangle in Beijing they were interested and asked us to give them the manuscripts. As we were engaged in the heavy task of designing buildings in imitation of ancient buildings we did not make hasty decision. In May they came to me again and told me that they would offer this book as a gift to the XX UIA Congress Beijing 1999. Introducing the excellent traditional architectural culture to the world is the cause we have incessantly struggled for. The convening of the XX UIA Congress in Beijing will be a golden opportunity for architects coming from all over the world to know Chinese architecture. By that time the designing work we are doing will come to a close and the time will not be pressing. I thought over again and again and finally accepted the offer. The publication of *THE ARCHITECTURE OF THE QUADRANGLE IN BEIJING* would be realized two years ahead of the schedule.

THE ARCHITECTURE OF THE QUADRANGLE IN BEIJING deals with the whole contents of the history, culture, fengshui (localization) theory, space, pattern, structure, fitting, decoration, design, construction, protection, and repair. It is an academic and technical work covering all fields of the quadrangle, and providing information available about arts and technology. It helps common readers know the quadrangle. It will also prove valuable to the professionals and students of technical schools in the design, construction, protection, and repair of the quadrangle.

This book consists of nine chapters. The first chapter introduces the history of the birth, growth, development, and decline of the quadrangle and the revival under the new conditions. The second chapter gives the basic patterns of the quadrangle, in which there are both ideal standard courtyard and practical examples of less regular real quadrangles. They provide readers with relatively all around knowledge of the type and layout of the quadrangle. The third chapter tells about some common sense about fengshui theory and a detailed description of some of the practical points. Though the first three chapters are fundamentals of those books on the quadrangle they are not the focal points of this book. These contents have been studied by many forerunners, and many papers have been published. This book is only the quotations and re-organization of their fruits according to the needs of content. The fourth chapter analyzes the combination mode of building space and the internal relation between this mode and the feudal system, leading readers from the superficial phenomena to internal connotation. The fifth and sixth chapters are the key chapters. The fifth chapter describes the building and structure of the quadrangle. Three sections are devoted to the structure, scale, color, modelling and artistic features of the principal and subordinate buildings, such as the principal room, wing rooms, side rooms, rear rooms, houzhaofang (the last row of rooms), gate, decorative door, screen door, veranda, screen wall as well as shangmashi (a stone block for ascending a horse) and horse post; two sections are devoted to the detailed description of the indoor and outdoor wooden fitting. The sixth chapter dwells on the ornament and interior design of the quadrangle, with stress placed on the ornaments and their artistic attainments, for example, brick carvings, stone carvings, wood carvings, colored drawings and the indoor furniture. Through the description of these artistic forms and their contents we reveal the cultural connotation and present people helpful inspiration and reference. The two sections about the furniture and colored drawings written by two experts of great attainment Mr. Wang Xifu, and Mr Jiang Guangquan, do credit to this book. The seventh chapter specializes in the study of the scale, function, techniques of gardening, artistic features and the cultural connotation of the private garden in a quadrangle; it also gives necessary guidance and particulars about the trees and flowers to be grown in the garden. Chapters 5, 6, and 7 are full of detailed statements and rich in illustrations of photos and drawings. They will impress the readers with concrete images, exact experience and understanding. The eighth chapter discusses the design, construction, protection, and maintenance of the quadrangle in three sections. This chapter is also the focal point of this book. What is included in this chapter can be useful to the design, construction, protection and repair of the quadrangle. The last chapter of this book ends with the analysis of the present situation and the future prospect of the quadrangle. In this chapter basing on what we have realized in our work and what we know about the social demand for the quadrangle in these years, we analyze the developmental prospect of the quadrangle and make objective estimates that the quadrangle should be not only retained but will be developed under the new historical conditions.

The preparation of *THE ARCHITECTURE OF THE QUADRANGLE IN BEIJING* is a strenuous work. Tight was the time — half a year from the signing of the contract with Tianjin University Press to the completion of the manuscripts. During this period in addition to writing we were required to give consideration to the project design, topic study, the editing of the journal *TRADITIONAL CHINESE ARCHITECTURE AND GARDENS* and other routines. We were swamped with work. Fortunately we have successfully accomplished our task.

It was with the sincere love for the excellent architectural culture of our motherland that we prepared this book. We hope that this book will help more Chinese and foreign readers to know the quadrangle, about its form and art, its content and value, treasure its existence and development so that the excellent traditional form of residence can better serve the society and the traditional architectural culture of China can be disseminated more widely. This is the main aim of our offer of this book.

Ma Bingjian

Beijing, March 1999

北京四合院雪景
李有杰　摄影

第一章
北京四合院的历史沿革

四合院是北京人世代居住的主要建筑形式。它作为中国传统居住建筑的典范，驰名中外，世人皆知。

所谓四合院，是指由东、西、南、北四面房子围合起来形成的内院式住宅。老北京人称它为四合房。

谈及北京四合院，人们往往会想到元代建都北京及其街坊胡同形成的历史。其实，在中国这种四合院式居住建筑形式的形成有十分悠久的历史。从目前能够见到的史料看，早在商周时代，我们祖先的居住建筑就已采取了四合院的形式。岐山凤雏的西周建筑遗址平面呈矩形，中轴线上由南至北分别为门道、前堂、后室。前堂与后室之间有廊相通，院两侧为前后相连的厢房，中间形成两组院落，是一座相当工整的四合院（图1－1）。这是迄今为止发现的最早的一座四合院。到了汉代，这种四合式院落的发展已很普遍。成都出土的东汉画像砖上的庭院，很明显地看出四合院的格局（图1－2）。汉代有钱人的宅第常有前堂、后寝、大门、中门以及楼、阁、室、井、灶、庑、囷等内容，由一个或多个四合式院落构成。在汉代明器中还可看到坞堡式四合院落（图1－3）。隋、唐时期四合院式住宅的史料更加丰富，无论从绘画、明器，还是壁画、绢画中，均可看到这种四合院式宅第（图1－4）。至于宋代留下的有关四合院式住宅的资料就更多了。无论是宋画《文姬归汉图》中的大型住宅，还是王希孟《千里江山图》中的小型住宅，从中都可以看出四合院式的格局（图1－5）。可见，四合院这种居住建筑形式的形成和发展，在我国已有两千多年的历史。

有人把北京四合院的历史追溯到元代在北京建都，这是有一定道理的。元代建都北京，开展了大规模的城市建设。公元1276年，在金中都原址东北部建大都城，元世祖忽必烈“昭旧城居民之迁京城者，以赀高（有钱人）及居职（在朝廷做

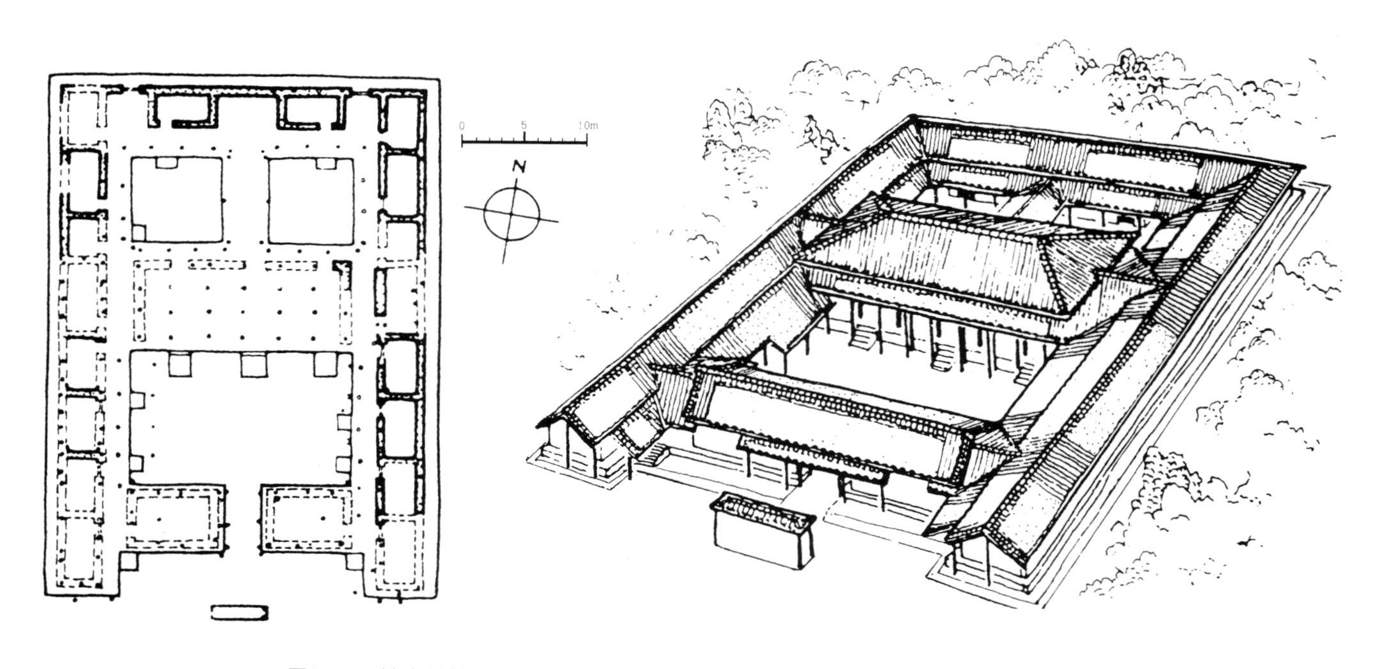

图1—1　岐山凤雏西周建筑遗址及复原图（本图转引自《中国古代建筑史》）

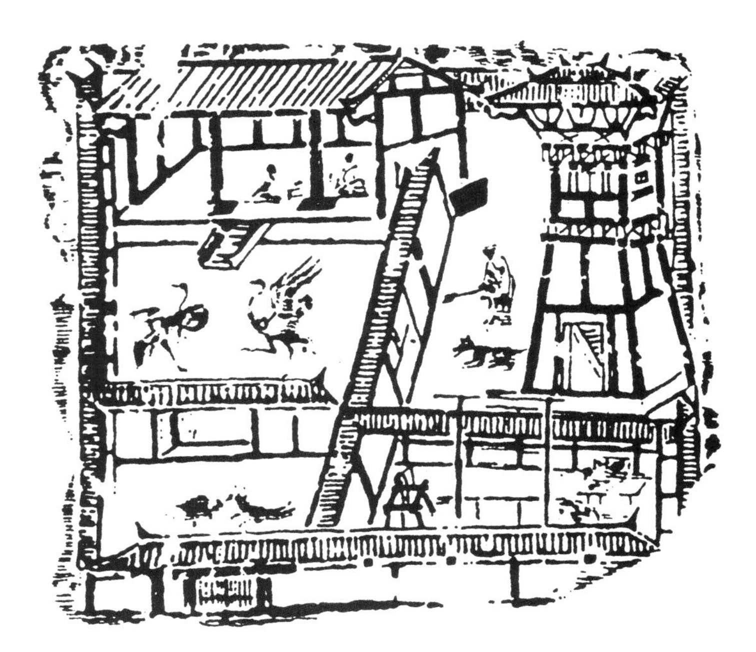

图1—2 成都出土东汉画像砖上的庭院形式（本图引自《中国古代建筑史》）

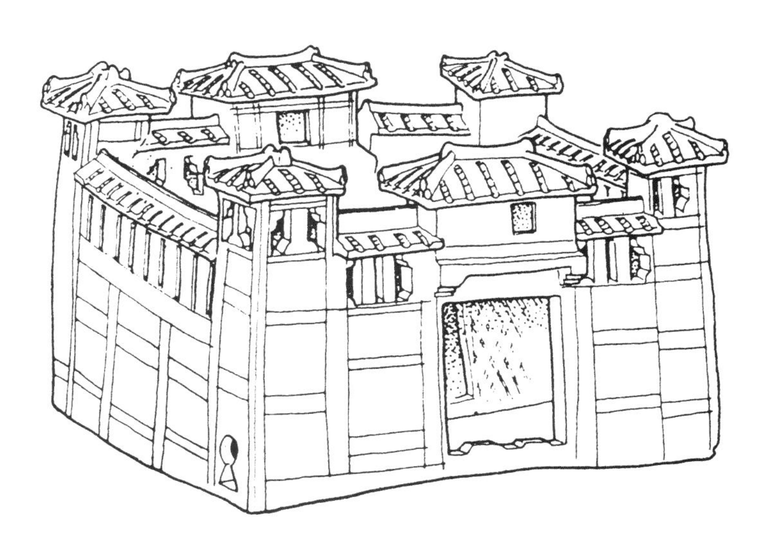

图1—3 广州麻鹰岗东汉建初元年出土的陶坞堡（本图引自《中国古代建筑史》）

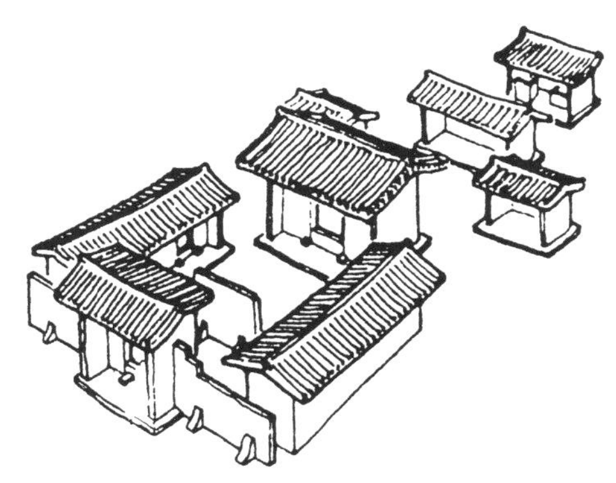

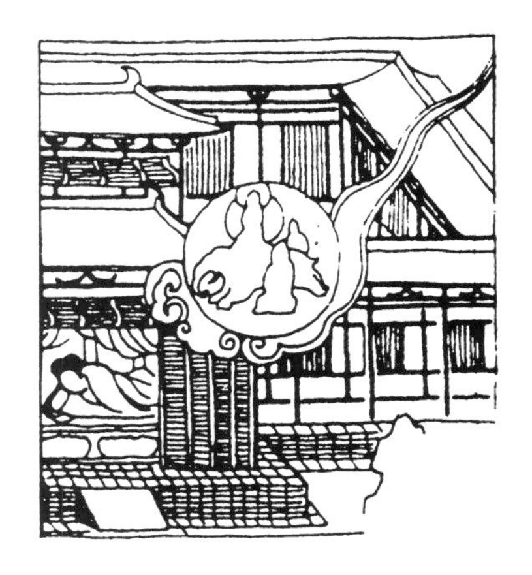

图1—4 展子虔《游春图》中的四合院住宅、唐大历六年（公元771年）王休泰墓出土明器上画的住宅、敦煌出土绢画佛降生图中的住宅（本图引自刘致平、王其明《中国居住建筑简史》）

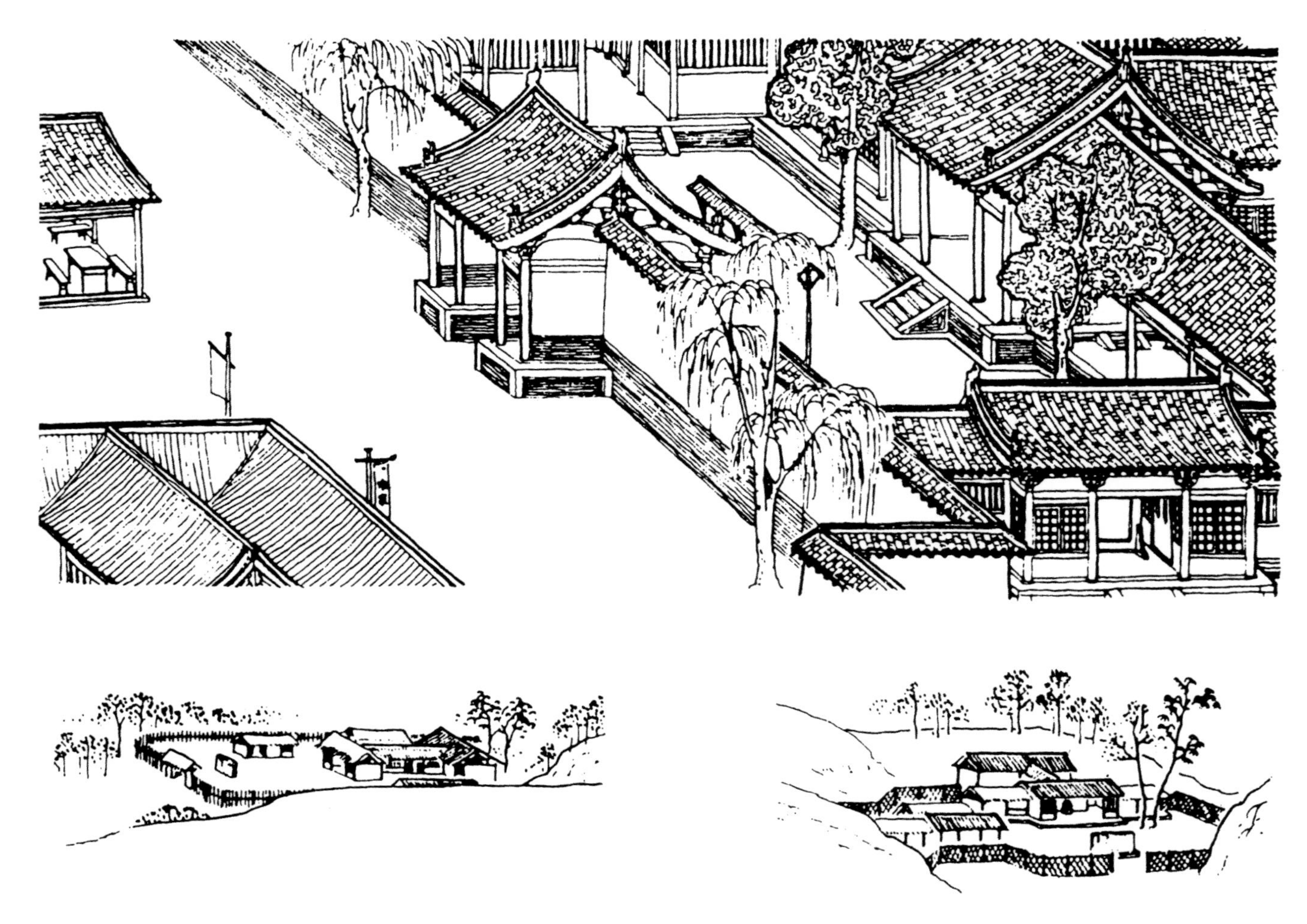

图1—5　宋画《文姬归汉图》中的大型住宅、王希孟《千里江山图》中的小型住宅　（本图引自刘致平、王其明《中国居住建筑简史》）

官）者为先，乃定制以地八亩为一分……听民作室”。于是，元朝的贵族、官僚就按此规定在大都城盖起一座座院落。院落与院落之间是供人行走的通衢街道，称为“衖通”。按元末熊梦祥所著《析津志》载：“大都街制，自南以至于北谓之经，自东至西谓之纬。大街二十四步阔，三百八十四火巷，二十九衖通。”这里所谓“衖通”即是我们今天所称的胡同。胡同与胡同之间，就是大片的四合院住宅。

元代四合院目前在北京已无实物，唯一能供参考的就是在元大都旧址上发掘出来的后英房元代住宅遗址。这座遗址所反映的院落布局、开间尺寸、工字厅、旁门等内容，与历代的四合院十分近似，说明元代四合院与历代居住建筑间密切的承袭关系（图 1—6）。除去这个例子之外，从著名的山西芮城县永乐宫纯阳殿元代壁画中也可看到元代四合院住宅建筑的格局和形式（图 1—7）。

明王朝建立后，社会经济得到较快发展。明都城从南京迁到北京，并分别从浙江、山西等处迁进数以万计的富户，从而有力地推动了北京经济的发展。在明代，制砖技术空前发达，这也促进了建筑业和住宅建设的发展。这个时期出现的《鲁班经》《三才图会》等书籍，说明明代不仅宅第建设的实践活动十分活跃，而且有理论方面的指导。

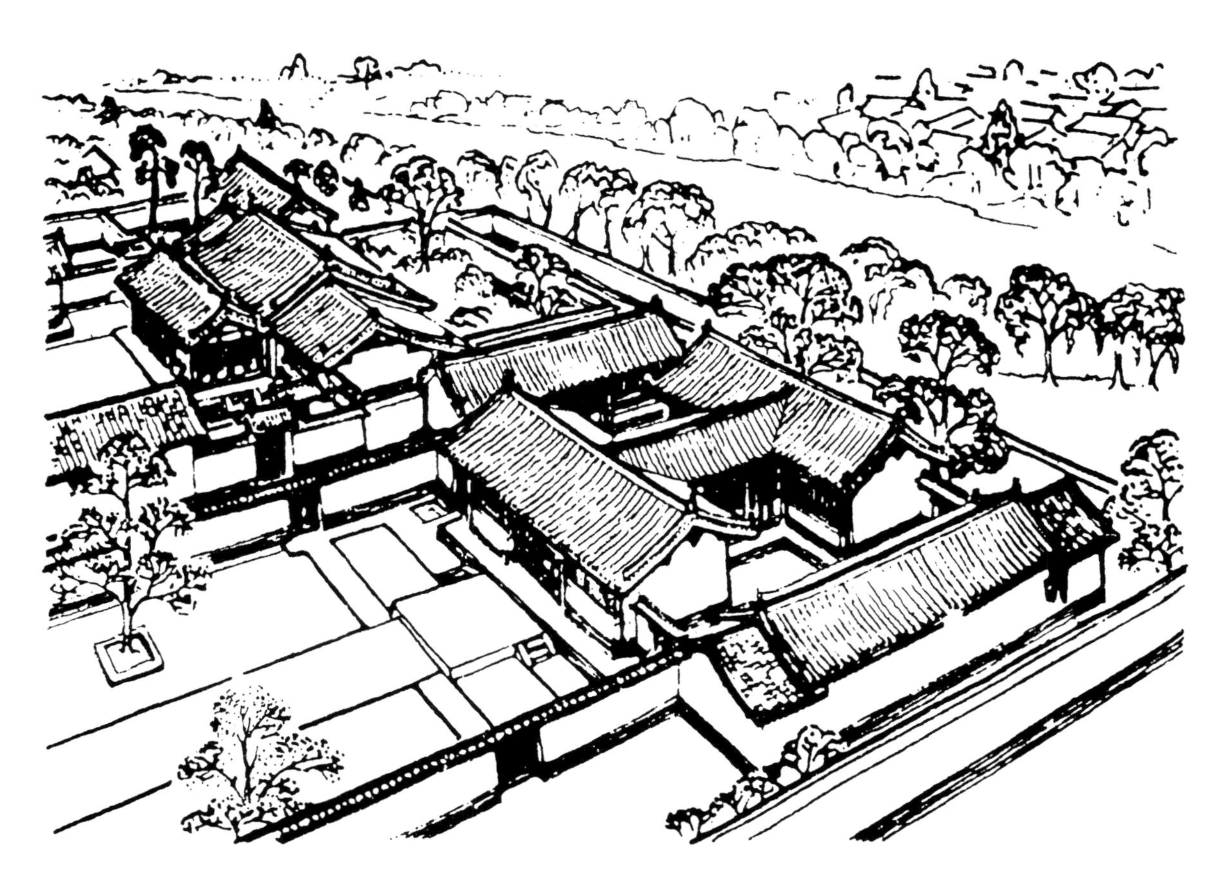

图1—6 北京后英房元代住宅遗址复原图（本图引自《中国古代建筑史》）

图1—7 山西芮城永乐宫元代壁画中的住宅（本图引自刘致平、王其明《中国居住建筑简史》）

为维持封建秩序，明代对各阶层人士的居住建筑从制度、规模、色彩各方面均作了严格规定。如洪武二十六年定制，官员营造房屋不许歇山转角、重檐、重栱及绘藻井……庶民庐舍不过三间五架，不许用斗拱、饰彩色……这些规定，为当时不同阶层人士营造宅第的有序发展，起到了积极作用。

明代住宅保留至今的，除安徽等地有少量遗存以外，在北京已极难见到实物。从《三才图会》《鲁班经》等书中的插图可以看到，明代住宅仍沿袭了元代四合院的形式。这从北方的山西、南方的许多地区保留至今的明代住宅可窥见一二（图1－8）。

清代定都北京后，大量吸收汉文化，完全承袭了明代北京城的建筑风格，对北京的居住建筑四合院也予以全面继承。清王朝早期在北京实行了旗民分城居住制度，令城内的汉人全部迁到外城，内城只留满人居住。这一措施客观上促进了外城的发展，也使内城的宅第得到进一步调整和充实。

清代最有代表性的居住建筑是宫室式宅第，这就是官僚、地主、富商们居住的大中型四合院。称之为宫室式宅第，主要是因为它在规制、格局方面承袭了古代宫室建筑的特点。这种大中型四合院均设有客厅、饭厅、主人房、佣人房、车轿房等建筑，院落二三重乃至多重（图1－9），气派而豪华。

明清北京四合院与元代四合院相比有较明显的变异，这主要表现在院落布局的变化、工字形平面的取消以及占地面积的减少。元代北京后英房等四合院遗址中，前院面积较大，明清四合院前院（外宅）面积较小，后院（内宅）面积增大，使院落面积的分配更趋合理。明清四合院还取消了前堂、穿廊、后寝连在一起的工字形平面布局（笔者发现，在北京现存的府第中还有极少数例外），代之以正房、厢房、抄手游廊组成的四合院式布局。这种变化同明清两朝北京城居民成分的变化及由此带来的东西南北文化交流是分不开的。此外，由于明清时期北京人口增长较快，元代每户八亩地的大院落已不敷分配，明清四合院占地普遍较小，小者一亩，大者也不过三五亩（王府等大型府第除外）。这些是明清四合院与元代四合院的主要区别。

图1—8 襄汾丁村丁宅、东阳卢宅（本图转引自刘致平、王其明《中国居住建筑简史》）

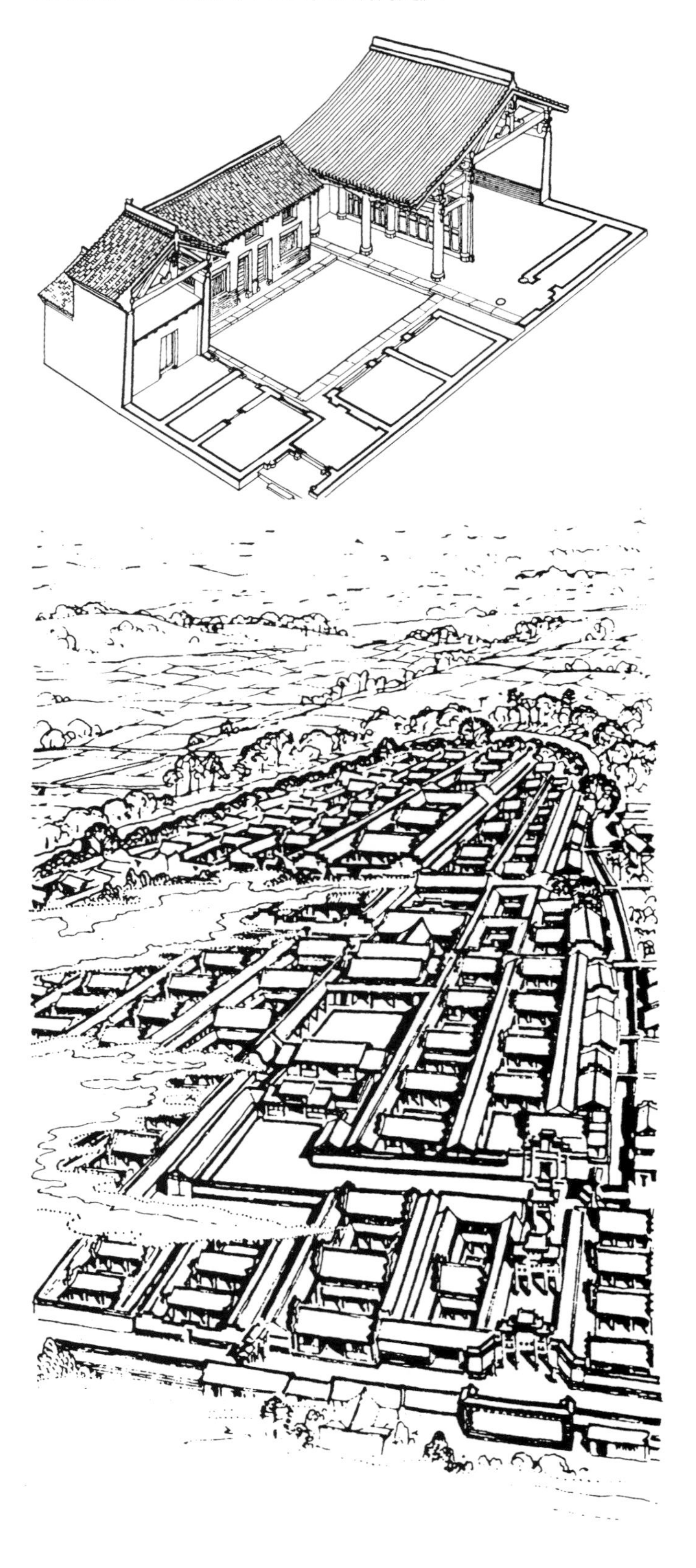

清代四合院在北京的遗存很多，至今仍在沿袭使用，成为当今北京古都文明风景线的重要内容。

北京四合院是我国古代诸多传统民居形式中颇具代表性的一种。它集各种民居形式之长，在中华诸种民居建筑中堪称典范。它的这些特点的形成，与北京作为六朝古都的特殊政治历史地位是分不开的。长期居住在北京这

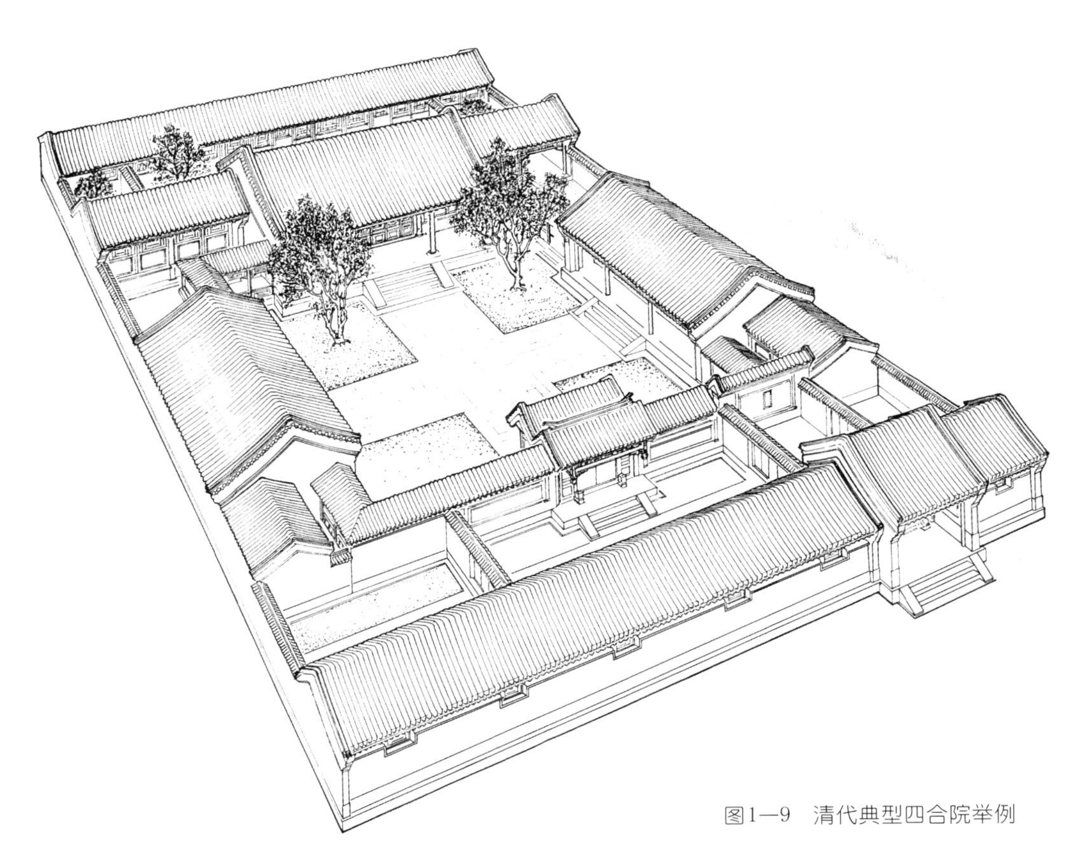

图1—9 清代典型四合院举例

块土地上的各朝代贵族、士大夫阶层对家居环境有着相当高的要求，这就从各个方面促进了北京四合院的发展与完善；加上北京地区的地理位置、气候特点和传统民俗，共同构成了北京四合院独具特色的传统居住建筑文化。

清代是北京四合院发展的巅峰时期。自清代后期起，中国逐渐沦为半殖民地半封建社会，北京四合院的发展也开始逐步走下坡路。

在外敌入侵和西方文化渗入的影响下，北京传统住宅建筑也受到一定影响。这个时期建造的四合院，有的或多或少加进了一些西洋建筑的装饰成分，最典型的就是圆明园式随墙门的出现（图1—10）。受“西学东渐”之风影响较深的人为标榜自己为“新派”代表，也有一些在宅内兴建“洋楼”的例子，但为数不多。总的来说，这个时期，北京的传统民居基本保持了明清型制。

日本帝国主义侵华，使中国社会发生了很大变化。由于通货膨胀，物价上涨，市民经济状况每况愈下，很多原来住独门独院的居民已没有能力养更多的房子，只好将多余的房子出租，以租金来补贴生活。居民的住房越来越少，院里的房客越来越多。独门独户的四合院开始变成多户杂居的大杂院，四合院的居住性质发生了变化。

1949年以后，北京传统四合院在使用上出现了根本性变化。由于所有制的变更，很多清代遗留下来的王府、宅院由私产变为公产。它们不再为昔日的贵族所占有，转而成为国家机关、学校、医院、工厂、幼儿园、俱乐部等公用住房。使用功能的改变，使得建筑本身与使用者的需求之间产生了难以解决的矛盾，最终的结果，不是人服从建筑，而是建筑被人所改造。那些仍作为住宅用的院落，已不再为独家占有，变为多户居住的“大杂院”。这些用途上的变化，使四合院再难保持昔日的深邃、安谧、幽雅和温馨，四合院被分割、改造、瓜分成了普遍现象。

“文化大革命”是北京四合院罹难最为严重的时期。20世纪60年代末的红卫兵“扫四旧”，将四合院中精美的砖雕、木雕、石刻、彩绘尽行扫荡，无数价值极高的艺术品，或被砸成碎块，或被抹上泥灰，能得以幸存者为数寥寥。紧接着发生的为“备战”搞的全民挖洞运动，进一步破坏了四合院的原有格局和排水系统，造成严重后果。1976年唐山大地震更是雪上加霜。为避震灾，在已经很拥挤的院子里塞满了“抗震棚”，随着人口增长，这些抗震棚后来都成了永久性建筑，把四合院搞得面目全非。

图1—10 清末民初建造的带有西洋建筑形式的宅门

"文革"以后，北京城市发展总体规划中确定了四合院保护方针，使上述恶化趋势有所控制，但紧跟而来的大规模的旧城改造与四合院的保护之间又出现了尖锐的矛盾，建设性破坏时有发生，如何解决好旧城改造和四合院保护的矛盾，仍是一个需要认真研究的课题。

在80年代初至今的十多年中，北京的四合院聚集区又出现了一种新的景象：一些移居北京的侨民和在改革开放中先富起来的人们，抱着怀旧心理和对传统文化的强烈追求，在四合院集中的地方买下旧院，重新翻建新四合院。有些长期居住在中国的外国人也征地建房，住进了传统的中国民居。这种新四合院大多采用传统建筑的外形和色彩，室内则是暖气、上下水、卫生间、空调等现代化设施和高档装修。经过翻建的四合院，由多家居住的大杂院重新变成一家一户的私宅，院落宽敞，花草丰茂，景致幽雅，安恬静谧，四合院似又回到了它的鼎盛时期。这种新的景象近年来愈演愈烈，大有蓬勃发展之势。这种现象，是北京四合院的复苏，还是传统住宅建筑在新的历史条件下的发展，目前尚未有人进行研究和界定，但它作为一种不可忽视的历史现象，已在北京四合院的发展史上书写了新的一笔（图1－11）。

北京四合院是我们祖先经过几千年的实践、改革、遴选、优化而创造出来的一种优秀的住宅建筑形式。它是中国人民祖祖辈辈辛劳智慧的产物，是一代代中华儿女艺术才华的结晶，它记载着中华民族悠久的历史，寄托着中华儿女浓厚的民族感情，是中华民族一笔丰厚的历史文化遗产。

在改革开放不断深入的今天，如何在继承中华优秀文化传统的前提下，有分析地吸收外来文化为我所用，是摆在我们面前的重要课题。我们要认真学习传统、弘扬传统，为创造具有中国民族风格的现代建筑而努力。

图1—11 新四合院举例——北京国子监街某宅

图1—11．1 外院

图1—11．2 从正房南望

图1—11．3 正房及庭院

第二章
北京四合院的基本格局

要了解北京四合院，首先应了解它的基本格局，而这种格局又与它的方位有直接关系。

一、四合院的基本方位

四合院是排列分布在胡同两侧的，胡同的走向与四合院的方位有直接关系。

北京的胡同是以东西走向为主，这在北京内城尤为明显。如现在仍保留比较完好的西四北一条至八条地区、东四一条至十条地区、东城区南锣鼓巷一带就是典型例证。在这样的胡同内，四合院住宅分列在胡同南北两侧，形成坐北朝南的街北院落和坐南朝北的街南院落（上述坐向指宅门而言）。除去这些东西走向的胡同外，还有一些沟通两相邻胡同的直胡同，即南北走向的胡同。而分布在南北走向胡同里的宅院，就成为坐西朝东的街西院落和坐东朝西的街东院落了。这样，北京四合院住宅就出现了街北、街南（这两类为主）和街西、街东（这两类为辅）这样四个基本方位。

地处外城的一些居住区，情况较为复杂。如宣武、崇文的一些地区，大多在历史上未经认真规划，是自由发展形成的，街道杂乱无序，出现很多斜胡同、窄胡同、直胡同、短胡同，因此也就造成这些地区出现大量的“非标准”住宅，院落的方位也呈现各种复杂情况（图2－1）。

图2—1 东城区南锣鼓巷一带、西四北一条至八条地区、前门大街琉璃厂一带、前门大街至崇文门一带的街道胡同排列状况比较（本图为1949年以前北京城区状况）

图2—1．1 东城区南锣鼓巷一带街道胡同排列状况

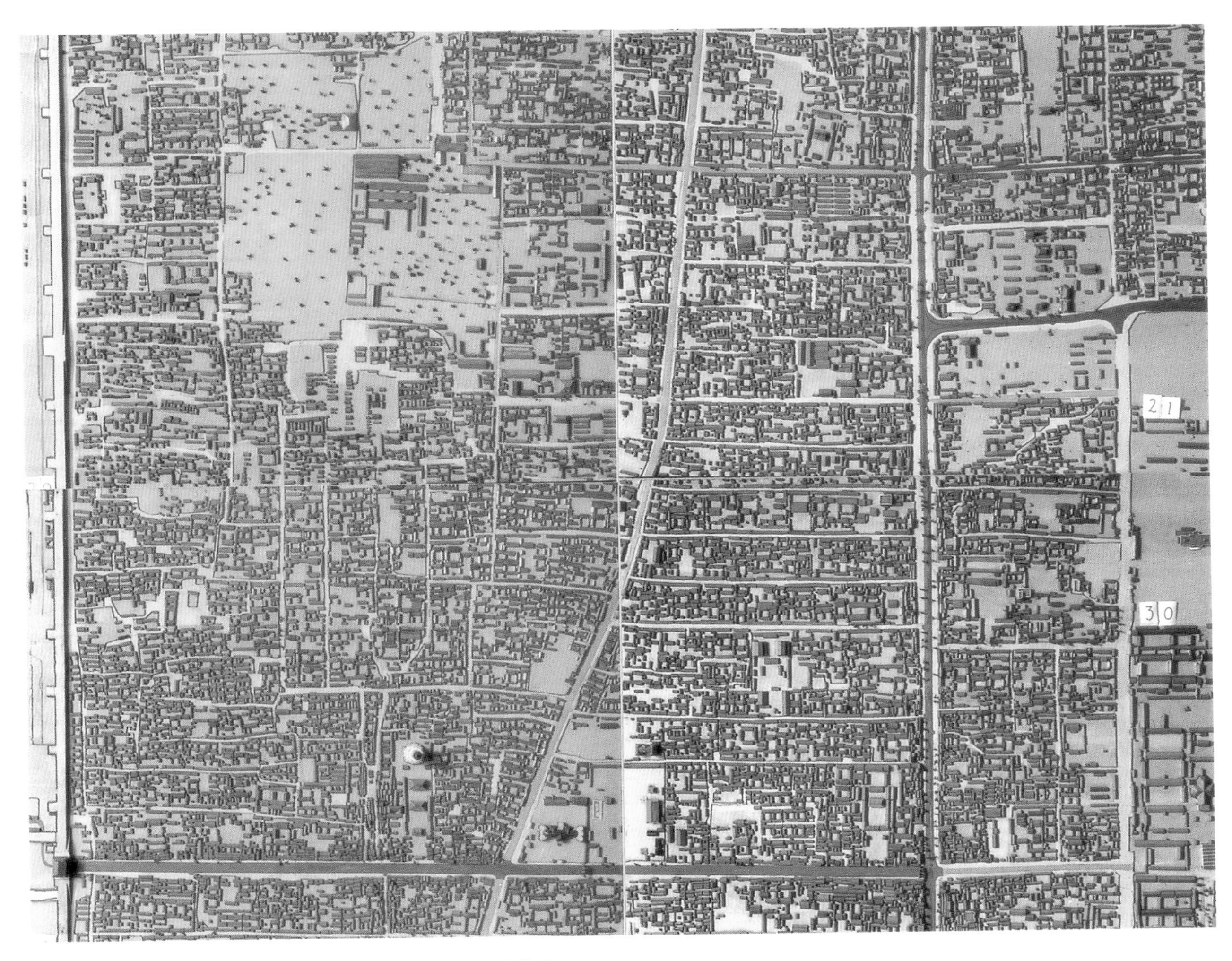

图2—1．2　西四北一条至八条地区街道胡同排列状况

图2—1．3　前门大街琉璃厂一带、前门大街至崇文门一带的街道胡同排列状况

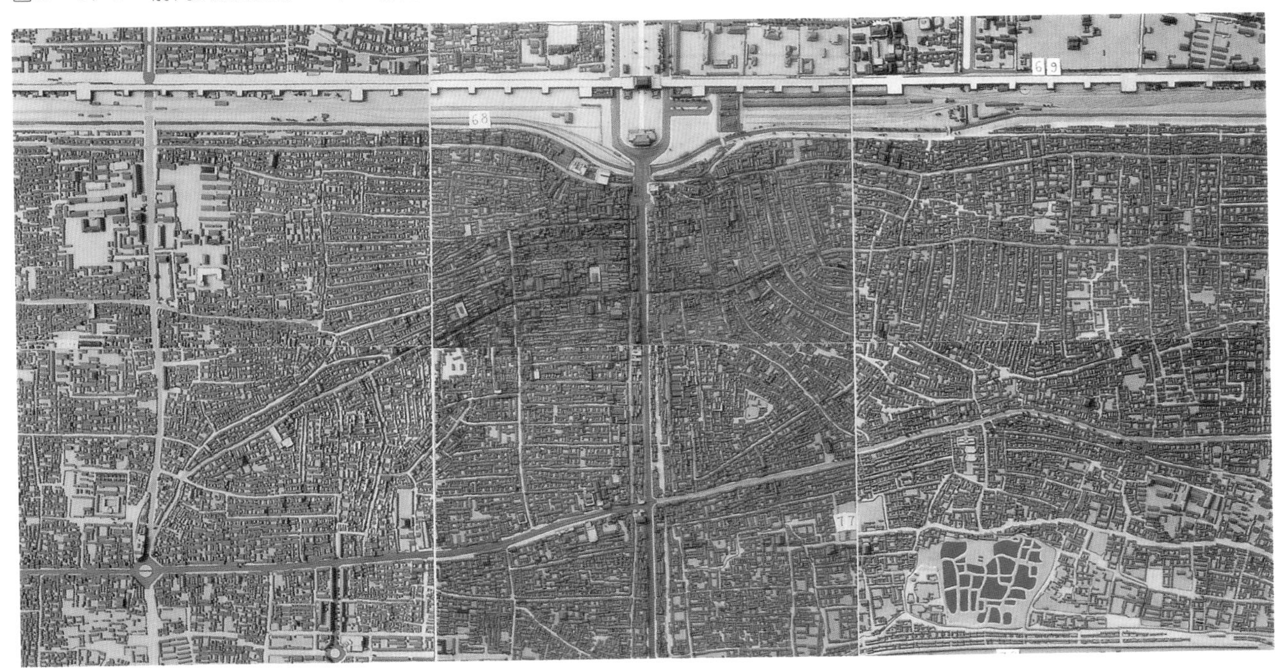

由北京地区的地理位置和气候条件所决定，北京的房子以坐北朝南的北房为最好，其次为坐西朝东的西房，东房和南房的朝向较差，不是理想的居住方位。北京人“有钱不住东南房，冬不暖来夏不凉”的民谚，说的就是这种情况。所以，只要条件允许，人们建宅时，一般都要将主房定在坐北朝南的位置，然后再按次序安排厢房和倒座房。

在坐北朝南的院落中，正房很自然地处在坐北朝南的位置，不用进行人为调整；而坐南朝北的院落为取得朝南的效果，则要采取一些调整措施。图2－2是四合院四种基本方位的示意图（图2－2）。图中的街南四合院是一座比较标准的二进院落，院子的最南端为倒座房，向北依次为外院、垂花门、内院、正房（北房）。由于这种四合院的宅门开在西北角，所以，进入宅门后，要经过院子西侧一条狭长的通道才能走到院子西南角，再由西南角的侧门进入外院，尔后，再进入垂花门、内院。这种调整办法，无疑要浪费一定面积的宅基地，且使用起来亦不十分方便。当然，同类情况下也有将宅门置于东北角的例子 （图2－3）。前公用库某宅宅门即位于院东北角。进入宅门后，经过院东侧一条狭长通道，达于东南角的侧门，进侧门后再向西进前院，向北进垂花门，进内院。可见，同样方位的院落，调整的方法可以不同，要根据宅院位置、环境等情况而定，并无僵死模式。

如果这种街南院落是一进院落的话，就不需要进行这样的调整了。进入位于西北角的宅门后，正对的是块镶砌在西厢房北山墙上的照壁，照壁东侧为屏门，入屏门进内院。内院坐北朝南为正房，两侧为厢房，南面为倒座房（图2－4）。

在北京四合院中，坐南朝北的院落处在非主要位置，一般规模不大，以一进院落者居多，因此，通过狭长通道来调整宅院朝向的例子不是很多。

处在南北走向胡同中的宅院，院门只能开在东西两侧。这种非“正规”的院落，居住者一般都是平民百姓。这种宅院，仍可在坐北朝南的位置安排正房，以取得最好的朝向。位于宣武区山西街的荀慧生故居就是这种格局。它位于山西街西侧，宅门坐西朝东，进门迎面为南房（倒座）的东山墙，山墙迤北与东厢房南山墙相邻处设一旁门，院内正房为北房五间，东西厢房各三间，是一座一进四合院。位于煤市街培英胡同 （原大马神庙胡同）的王瑶卿故居北院，也是一座坐西朝东的四合院，宅门开在东厢房南侧，院内以北房为正房，南房为倒座（图2－5），取得了最好的朝向。

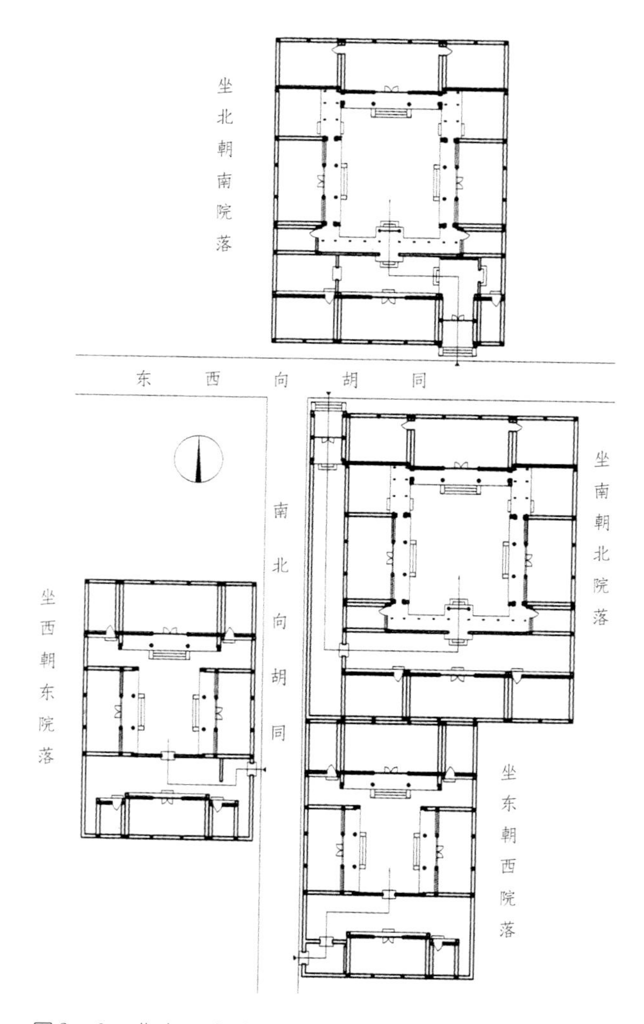

图2—2　北京四合院的四种基本方位示意图

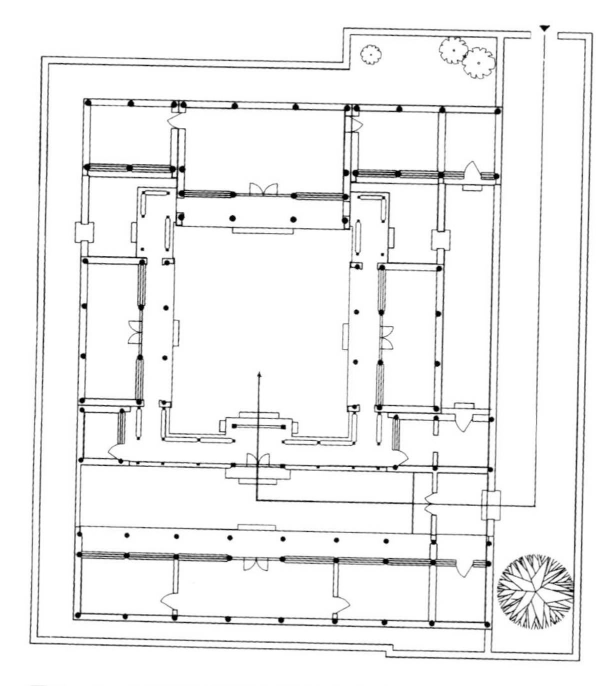
图2—3　前公用库某宅路南四合院方位调整示意

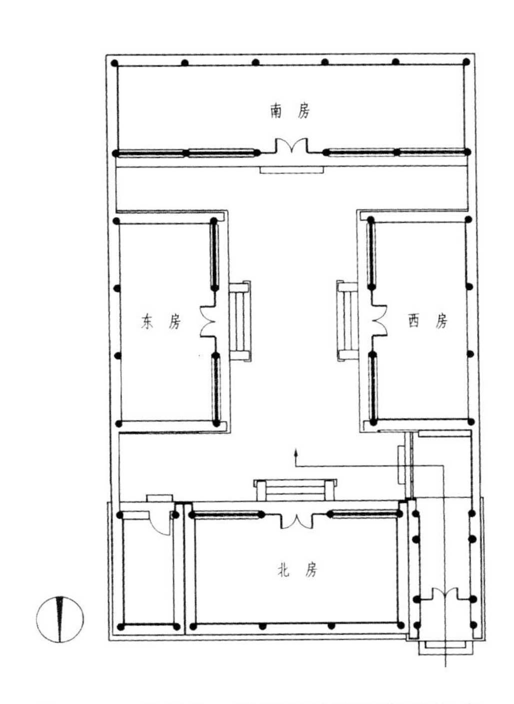

图2—4 常见的一进院落的路南院落构成

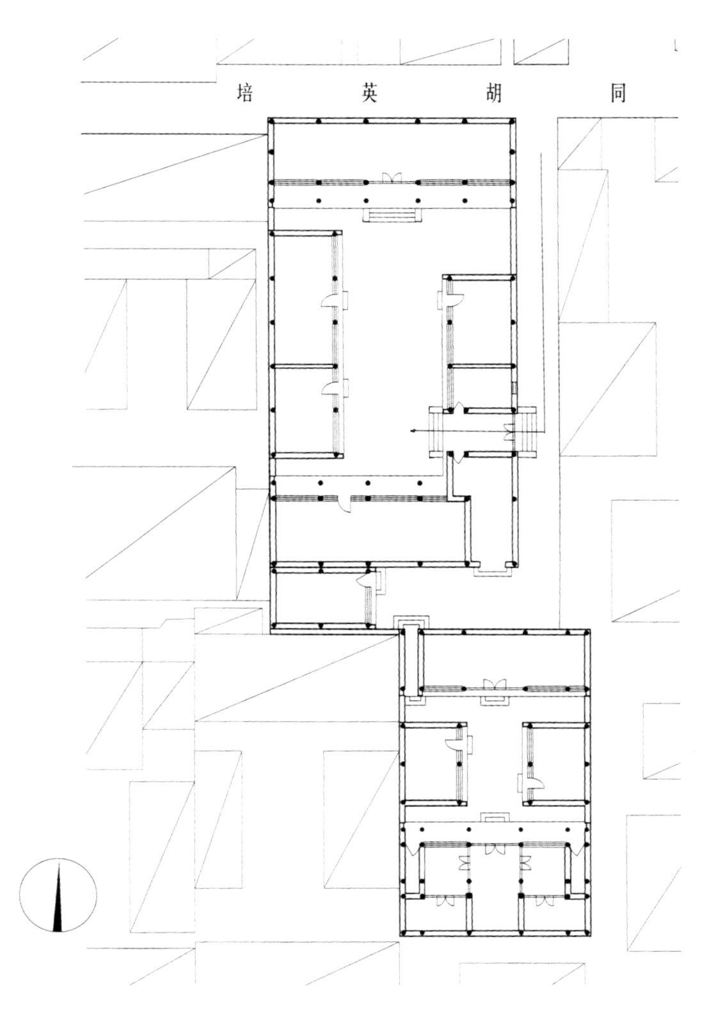

图2—5 路西院落三例

图2—5．2 王瑶卿故居北院

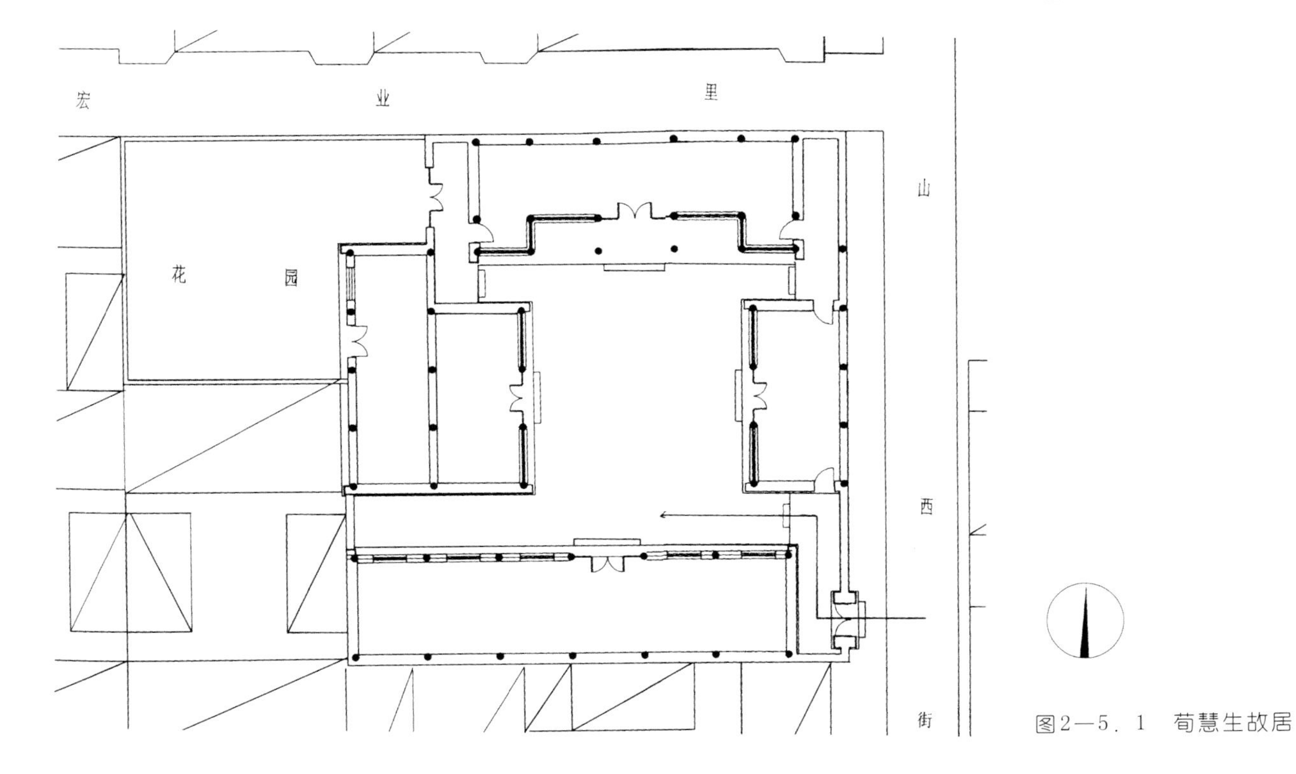

图2—5．1 荀慧生故居

但情况也不尽然。位于宣武区南半截胡同的绍兴会馆（鲁迅故居）就存在另一种情形。该会馆坐西朝东，由南、北、中三组院落组成。北院坐北朝南，取得了最佳朝向；而中院和南院均以西房为正房，处在坐西朝东的方位。这大概是由于这两座院子东西方向尺寸大而南北方向尺寸小，无法建正院的缘故。可见，处在东西朝向的院落，其方位的确定也要因地制宜，不能一概而论（图2－6）。

宅院的方位问题是首要的，只有确定了方位，才能确定建筑的格局。

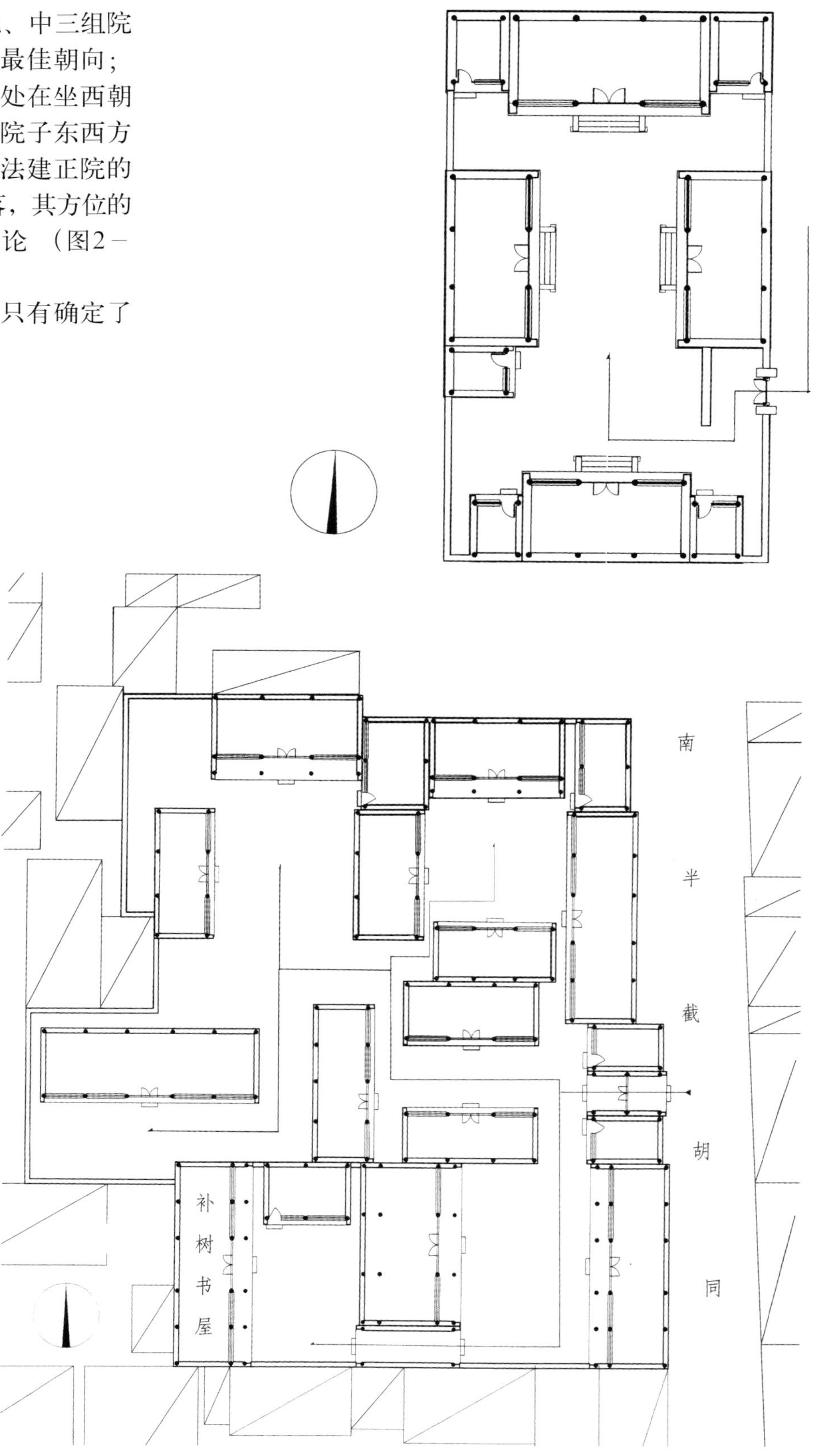

图2—5.3　西城区大金丝胡同某宅

图2—6　宣武区南半截胡同7号绍兴会馆（鲁迅故居）的朝向和方位

二、四合院的基本格局

除上述方位的不同外，北京四合院还因有规模大小、等级高低的差别而形成了多种类型。常见有以下几种：一进院落（又称基本型）、二进院落、三进院落（又称标准四合院）、四进及四进以上院落（可称为纵向复合型院落）、一主一次并列式院落、两组或多组并列式院落、主院带花园院落等。

现以坐北朝南的院落为例，分述如下。

1. 一进院落

一进院落又称基本型院落，这是一种由四面或三面房子围合组成的四合院或三合院。这种院落的特点是有正房（北房），一般为三间，正房两侧各有一间耳房，成为三正两耳，共五间。如果院落窄小，仅有四间房的宽度时，三间正房的两侧可以各置半间耳房，呈“四破五”的格局。正房南面两侧为东西厢房，各三间，与正房成“品”字形排列。正房对面是南房，又称倒座房，间数与正房相同。这样由四面房子围合起来形成的院落叫四合院。如果没有南房，则称三合院。

这种一进院落的小型住宅，宅门开在东南方位。若是四合院时，宅门一般采取门庑式，占据倒座房东头的一间或半间。进门后迎面是镶砌在东厢房南山墙上的坐山影壁（照壁）。向西通过屏门便可进入内院。如果南面没有倒座房而仅有院墙时，则在东南方位做墙垣式门（又称随墙门、小门楼）（图2－7）。这种典型的一进院落，是北京四合院的基本单元。

2. 二进院落

二进院落是在一进院落的基础上，沿纵向扩展而形成的。四合院由一进院扩展为二进院时，通常是在东西厢房的南山墙之间加障墙（又称隔墙），将院落划分为内外两重。障墙合拢处设二门，以供出入。二进院落属小型四合院，占地面积一般较小，东西宽度不过十五六米，南北深不过二三十米。这样的小院没有抄手游廊，二门多采用屏门的形式，既很美观，也很经济。北京宣武区椿树上头条余叔岩故居的中路建筑就是这样一座比较典型的二进院落。但由于该院是一座一主一次并列式格局的宅院，宅门没有开在中路建筑的东南角，而是设在了

图2—7 典型的一进院落——四合院和三合院举例

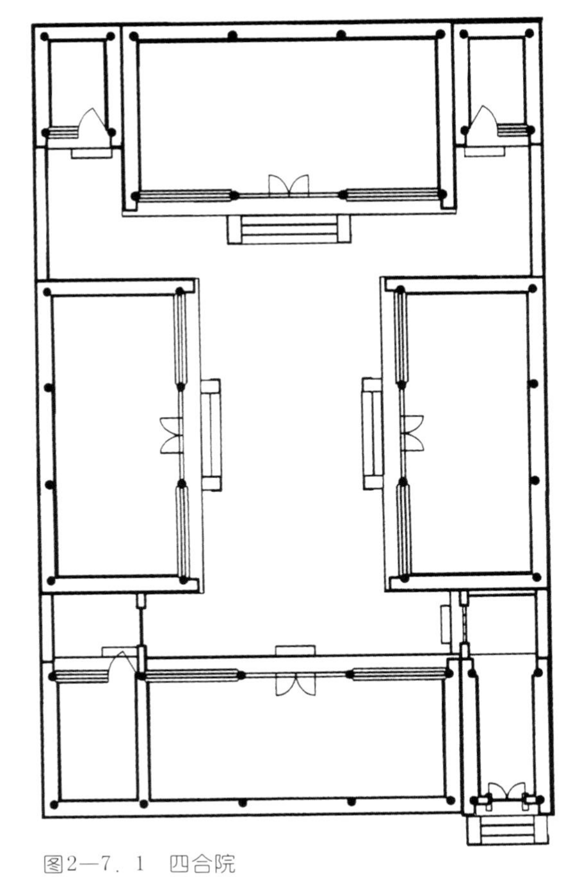

图2—7．1 四合院

图2—7．2 三合院

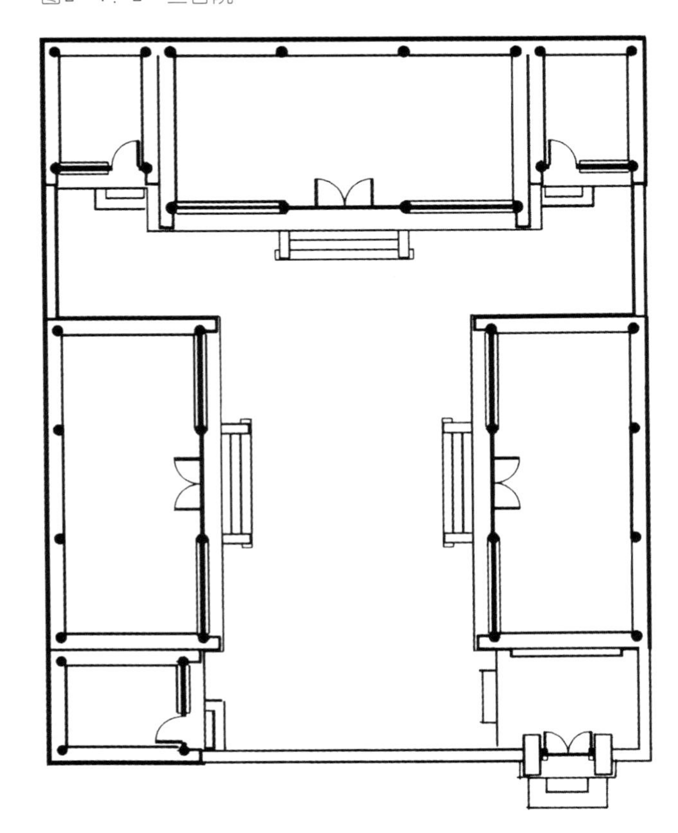

东路的南端。该院的二门采用了独立柱垂花门形式，功能与屏门相同，形式却讲究许多（图2－8.1）。

在二进院落的四合院中，也有规模较大、格局比较讲究的。它与上述小型的二进院落相比，主要差别在占地宽窄的不同。占地较宽的二进院落（宽22米左右，进深30米左右），北房可以排出七间，即正房三间，两侧耳房各两间，成为三正四耳。这种四合院正房、厢房都可设外廊，外廊之间由抄手游廊连接。如果院落的纵深方向有余量，还可以在东西厢房南侧各设一间厢耳房，分隔内外院的障墙设在厢耳房南山墙一线。抄手游廊由厢房南侧接转，沿障墙内侧延伸并交于二门。二门采用四柱垂花门形式，与两侧游廊相接。这样，由正房、厢房的外廊、抄手游廊和垂花门共同构成内院的环形通道，这是一条可以避雨雪的交通系统（图2－8.2）。

这种有抄手游廊和垂花门的四合院，已不是一般平民百姓居住的小型住宅，而是具有一定规模、相当讲究的宅院了。

图2—8　二进院落四合院举例

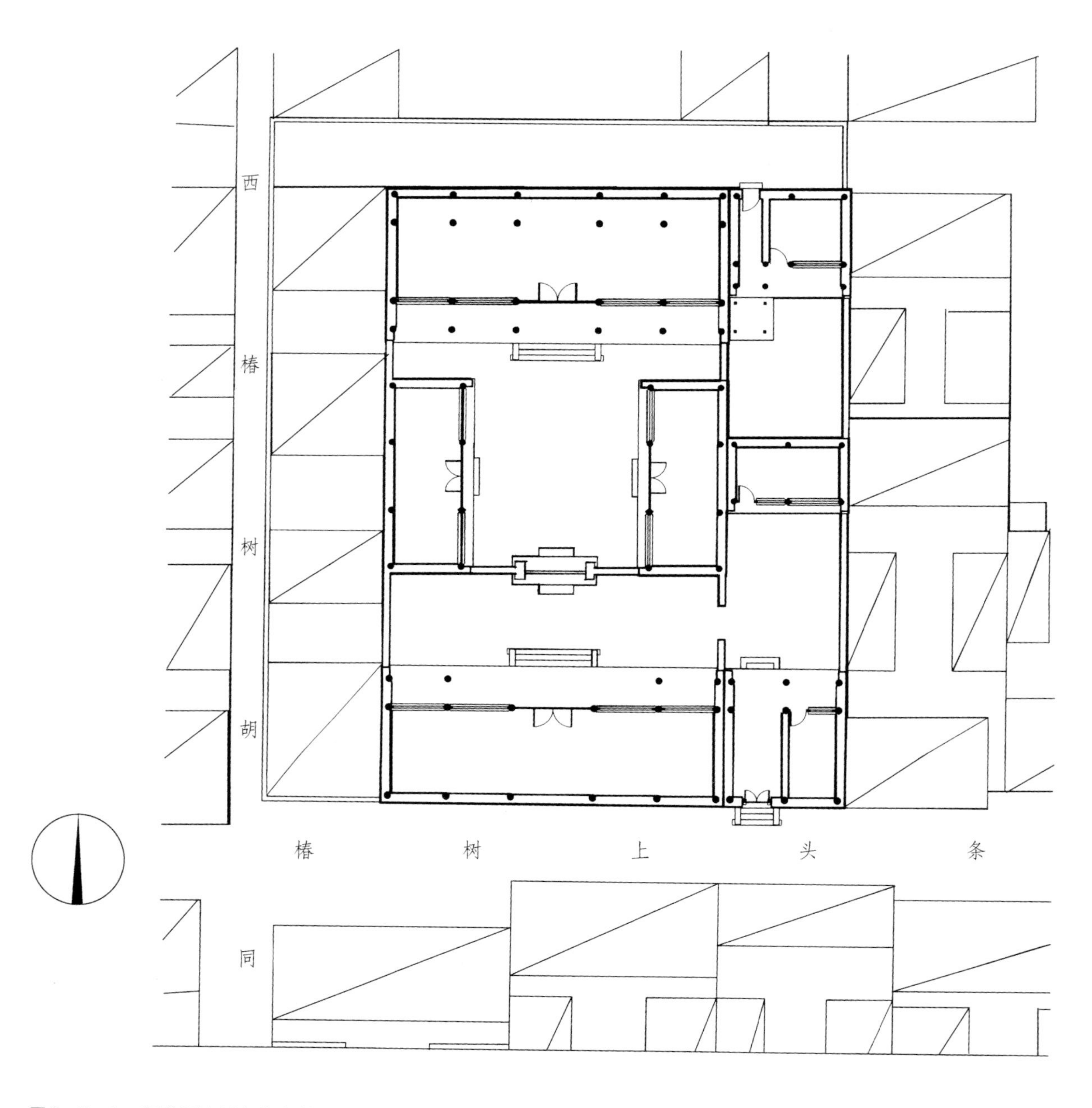

图2—8．1　宣武区椿树上头条某号余叔岩故居

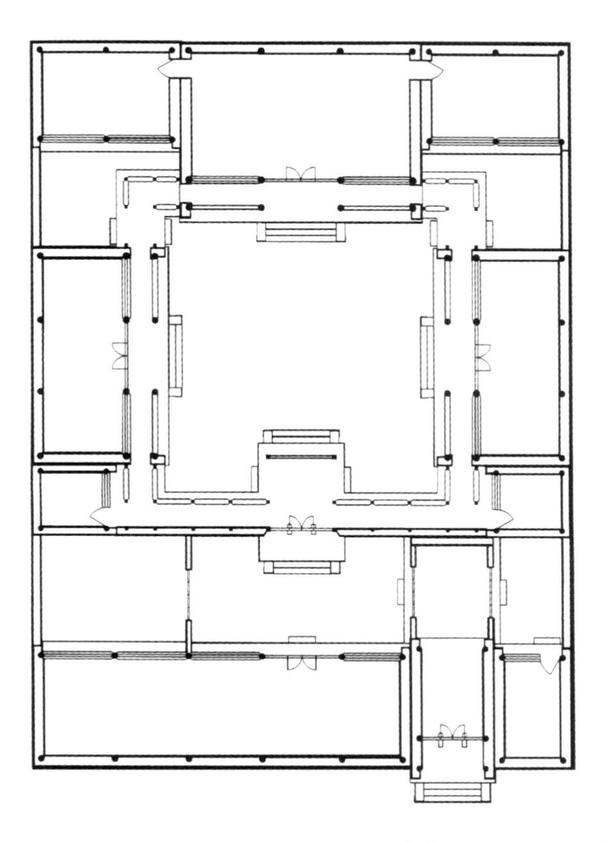

图2—8．2 带抄手游廊的二进院落

图2—9 三进院落举例

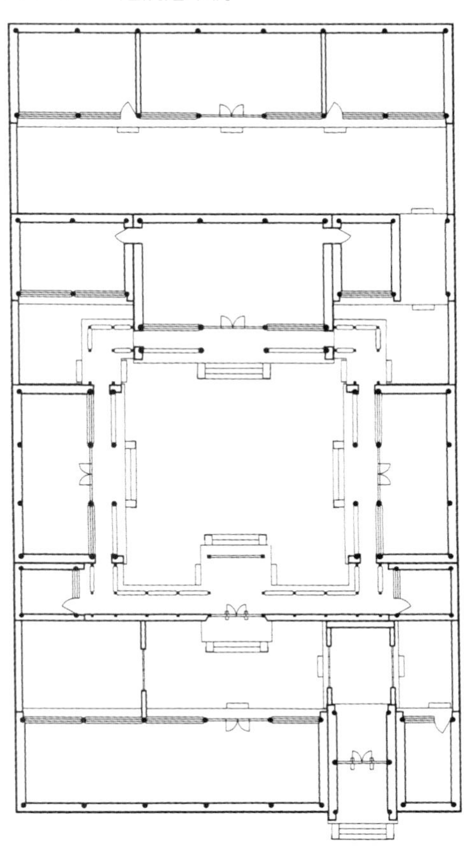

图2—9．1 标准的三进院落

图2—9．2 东四四条某宅

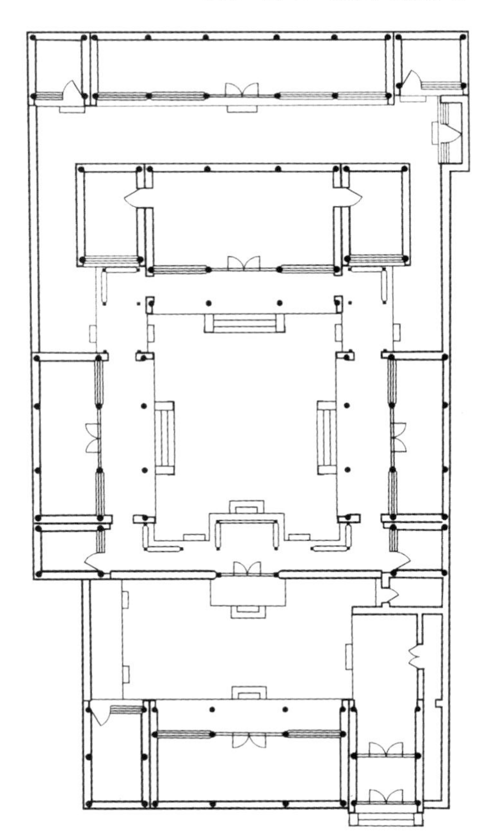

3. 三进院落

三进院落是在二进院落的基础上再向纵深发展而形成的。一般是在正房的后面加一排后罩房，后罩房与正房之间形成狭长的后院。后院与中院之间通过正房东耳房尽端的通道来沟通，宅人可以经过这个通道进入后院。这种在正房后面加一排后罩房的布局，被认为是比较理想的三进院布局，被人们称为“典型”的或“标准”的四合院。实际上，这种三进院的例子很多，如东四四条某宅、大草场某宅都是这样的院子（图2－9）。

三进院落还有另一种格局，也是比较常见的。它是按照第二进院落的模式，在正房后面再加一重院落。第三进院也同中院一样有正房、耳房、东西厢房、抄手游廊等。这种布局中，二、三院之间的沟通有两种方法，一种是将二进院正房明间做成过厅，宅人从中路进入后院；也可以在耳房一侧开设通道以供通行。宣武区前孙公园胡同35号四合院就是这种格局（图2－9.4）。

三重院落的四合院属中型住宅，已经具有相当规模了。

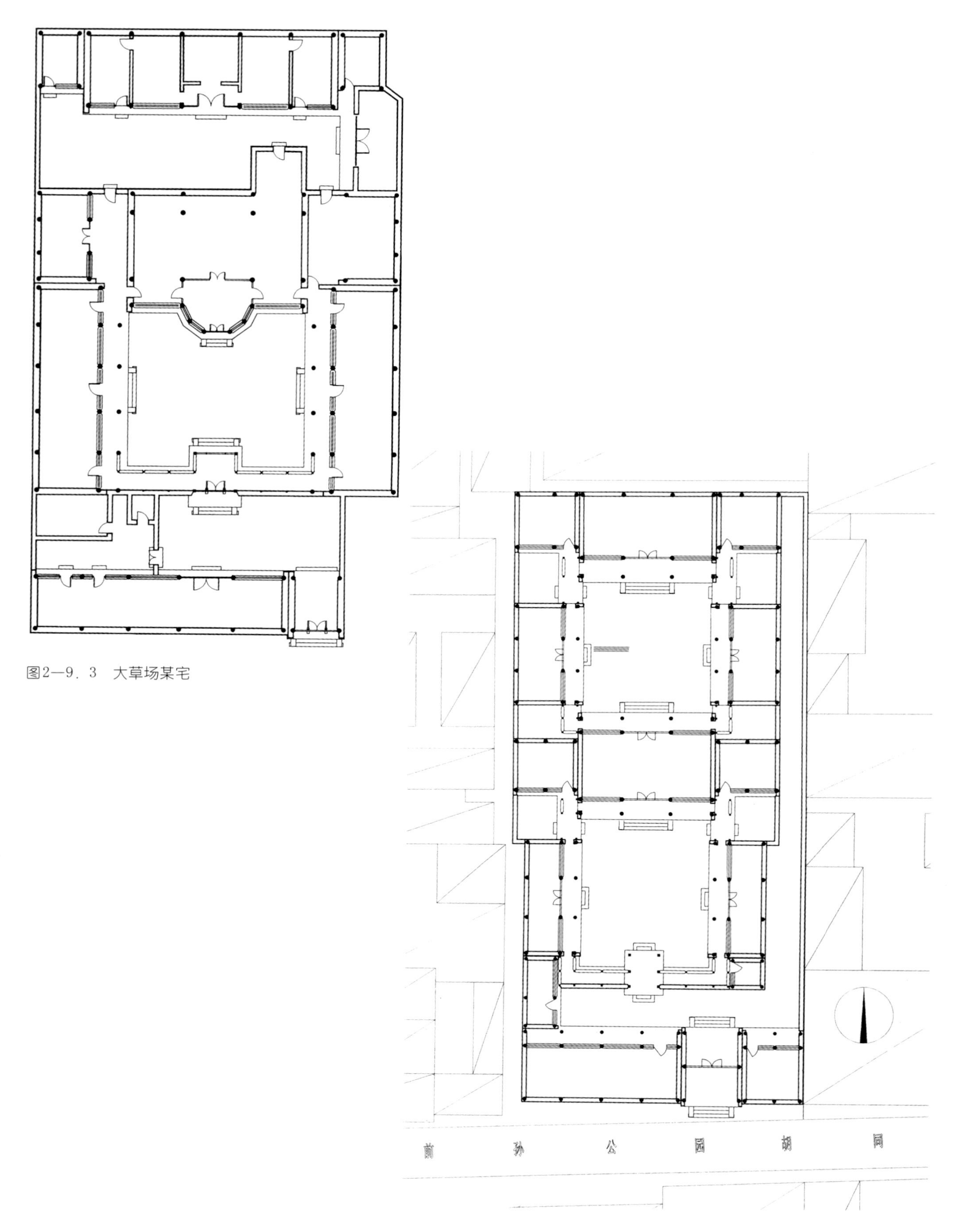

图2—9．3 大草场某宅

图2—9．4 宣武区前孙公园胡同 35 号三进院落四合院

图2—10 四进院落举例

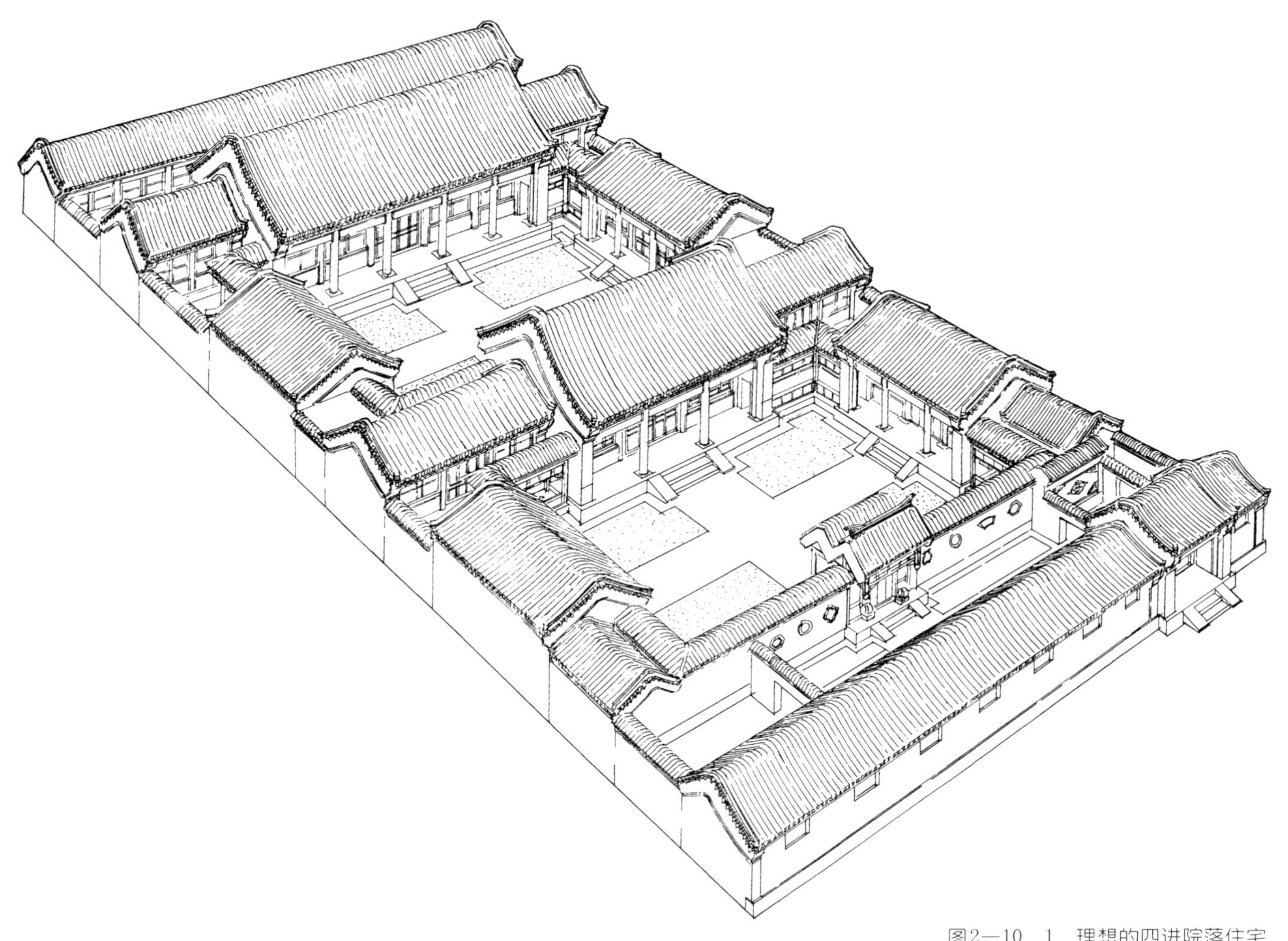

图2—10．1 理想的四进院落住宅

4. 四进院落

四进院落是三进院落沿纵深方向的进一步扩展。一般做法是，在三进院后面加一排后罩房。但这种三进院落不是带后罩房的那一种，而是第二、三进院格局相同或相似的那种（如上文提到的宣武区前孙公园胡同35号）。四进院落中轴线上的建筑，由南至北依次为：倒座→第一进院→垂花门→第二进院→正房或过厅→第三进院→正房→第四进院→后罩房（图2－10.1）。

现实当中的四进院落并不像上述例子那样规整，因受地形、尺度、功能各方面条件的制约，往往会有一些变化。如东城区南锣鼓巷帽儿胡同某宅，就是一座比较典型的四进四合院（图2－10.2）。它的宅门开在东南方向，进门左转向北为垂花门，进门后是内院。这座内院的格局与一般院落不同，它的正房是坐落在院子当中的三间客厅，客厅两侧各有两幢连列的厢房，南边一幢三开间，北边一幢五开间，东西两面对称布局，客厅与厢房之间为院当。客厅北面又一座垂花门，这两座垂花门之间的部分，是宅子的第二进院。从第二进院房间的格局和分布看，这不是供眷属居住的内宅，而是兼有议事、接待、娱乐、留宿亲朋好友等多种功能的院落。进入第二道垂花门是第三进院落，这是供家人居住的内宅。第四进院落的院当比较开阔，有10米左右，最北面有一幢后罩房，房前面单独圈了一道院墙，使后罩房自成体系，这显然也是按功能要求而设计的。

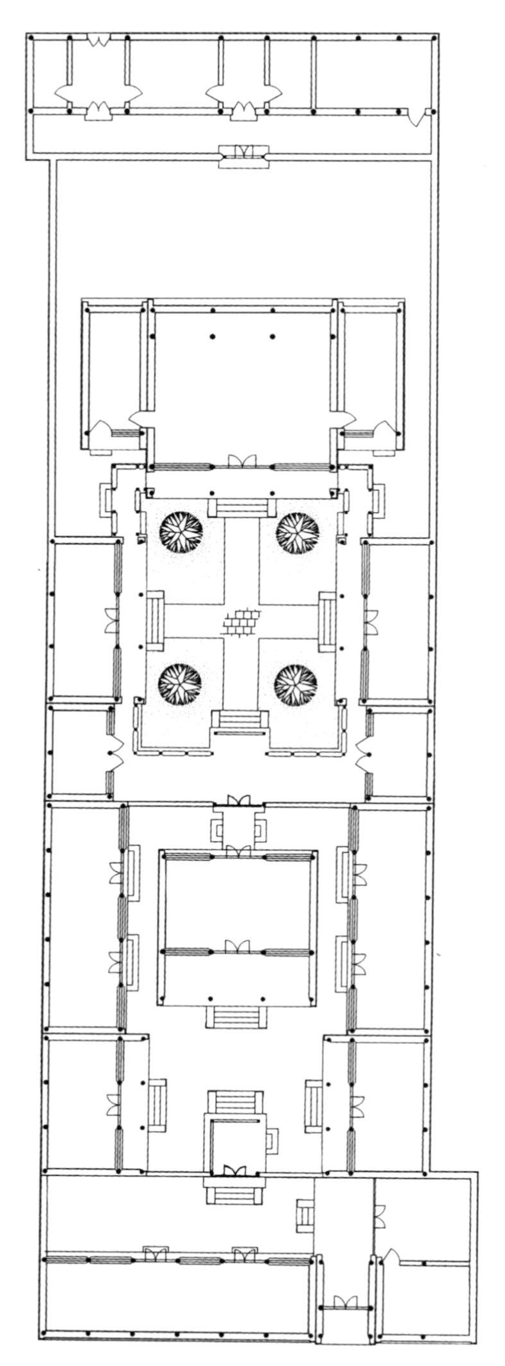

图2—10．2　东城区帽儿胡同某宅

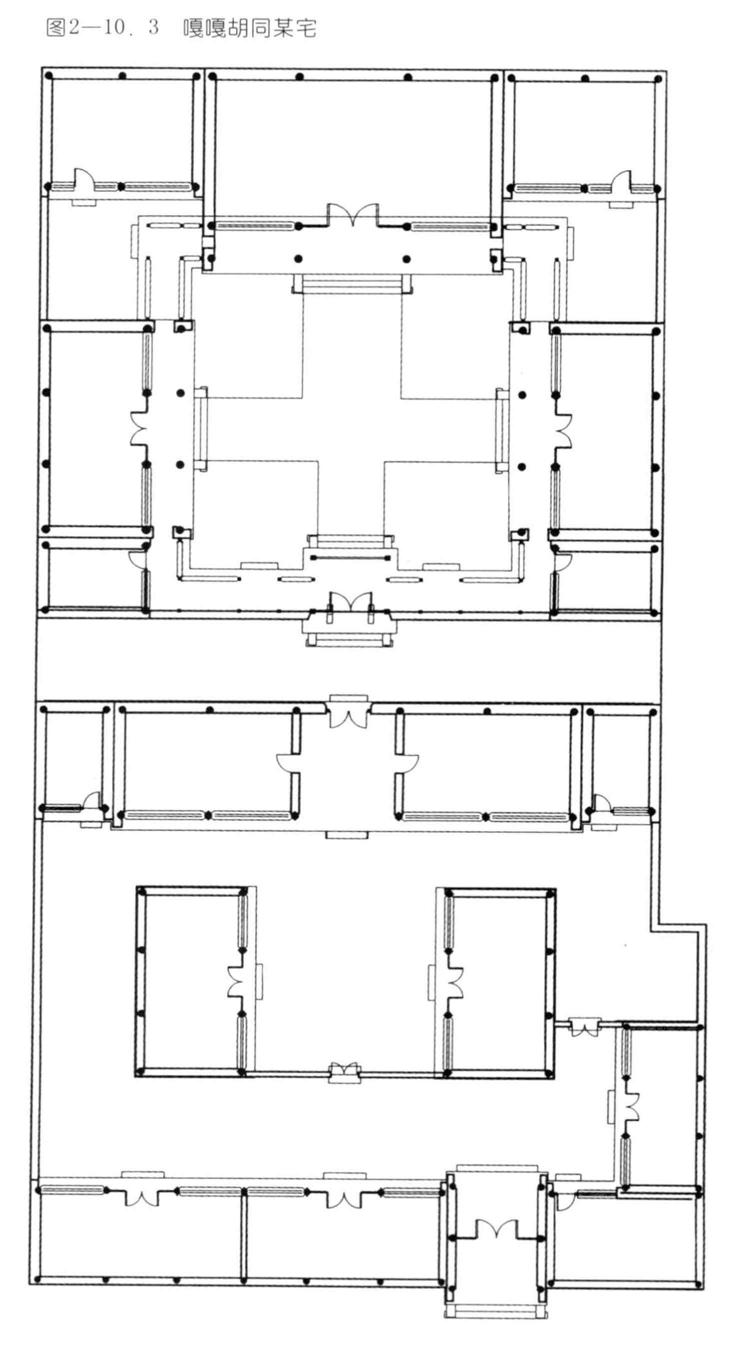

图2—10．3　嘎嘎胡同某宅

嘎嘎胡同某宅也是一座四进院落的宅院（图2－10.3）。它的格局与帽儿胡同某宅又有不同。这座院落中轴线上的二门是一座屏门，第二进院正房五间，两侧各一间耳房，共七间，东西厢房各三间。正房明间为过厅，通过过厅进入第三进院。第三进院是一个东西狭长、南北仅四五米的院落，两侧无建筑，北面正中是进入内宅的垂花门。这个扁长的院子很像倒座房北面的第一进院落。进入垂花门后是第四进院，这是内宅，正房三正四耳，两侧东西厢房并厢耳房各四间，各房之间有抄手游廊相连接。从这座院落的格局看，第二进院显然也是用来接待客人、交际朋友的场所，与前例功能相似，但格局却大不相同。这说明四合院在布局上并无僵死模式，需根据功能要求和主人的好恶而定。

四进院落的宅院，已属大型住宅，是典型的深宅大院。

5. 一主一次并列式院落

如果在一座独立院落的东侧或西侧再加一排房子和一个院子，就成了一主一次并列式院落（图2－11.1）。一主一次并列式院落的出现，盖因房基地的宽度大于一个而又小于两个标准院落的宽度，为充分利用宅基地，便在确定了主院尺度格局之后，将剩下的部分作为一个附属院落。关东店某宅就是这样一座一主一次并列式宅院（图2－11.2）。这是一座路西院落，宅门位于宅院的东南角，坐西朝东，总宽度约24米，主院约占总宽度的五分之四，附院约占五分之一。在这一狭长的地段内，加盖了一幢东房，共九间，其中一间为宅门。东房与主院之间留有南窄北宽的夹道。附院单独有门与主院相通，形成一主一次并列的格局。

位于宣武区椿树上头条的余叔岩故居，也采取了主次并列的格局。原宅为一主二次并列式，主院居中，是一座二进院落，东西两侧各一座附院。东侧的附院也是二进，分别与主院相通。这样安排，大概是为使用起来方便。西侧附院经改建已非原貌，现仅剩中院和东附院，变成一主一次的形式了（参见图2－8.1）。

图2—11 一主一次并列式院落

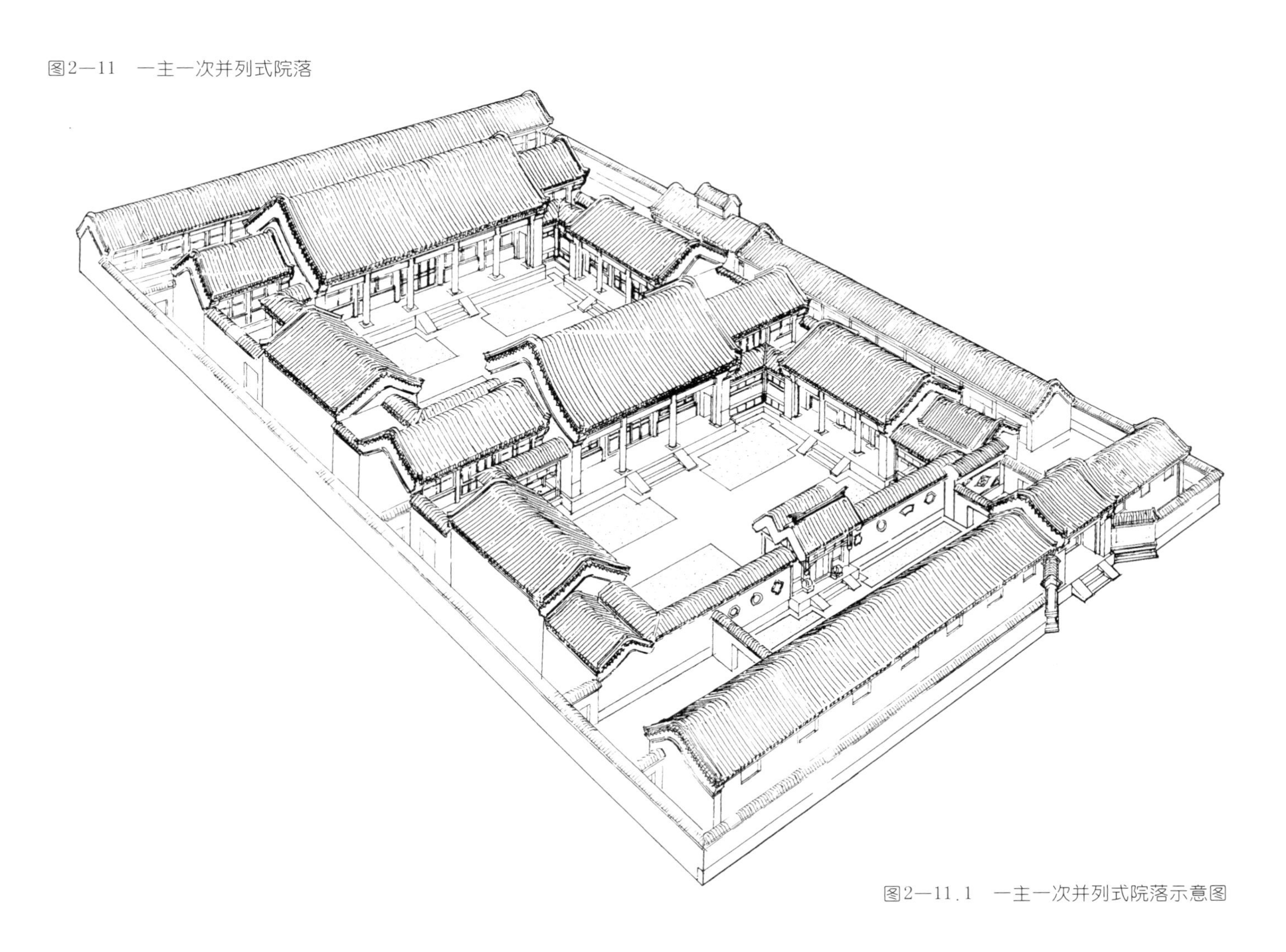

图2—11.1 一主一次并列式院落示意图

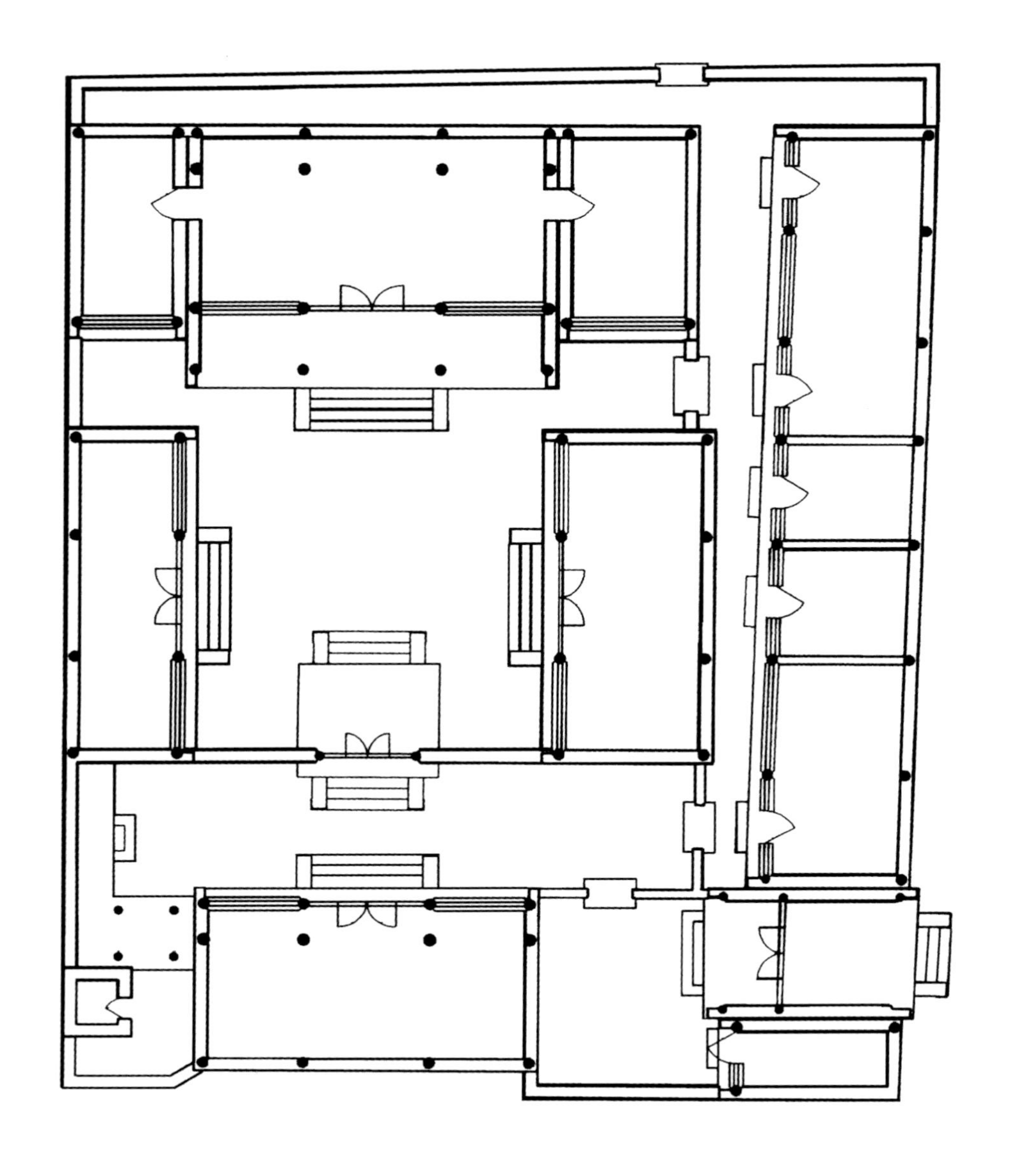

图2—11．2　关东店某宅

6. 两组或多组并列式院落

两组并列式院落，是由两个大小相等或相近的院子并列在一起形成的。这是豪门大户的宅院采取的一种形式。

封建社会的大户人家，有兄弟二人同居一处的。这样，在建宅时就往往建两座，兄弟二人各一座。两座宅院大小、格局相同或相近，既各自独立又相互沟通，这是两组并列式住宅出现的一个原因。当然还有许多其他原因。北京东城区秦老胡同某宅就是一个典型例证（图2－12.1）。这两座宅子都是纵深五进院落的大宅院，西边一座略宽，正房九间；东边一座略窄，正房七间。两院均为东南向开门，两院的第一进院落并无截然分隔，仅在中间卡了两道屏门，既分隔又沟通。第一进院北面各建有一排房，但两排房的朝向不同。西院房坐北朝南为正向，中间一间为过厅。从功能上看，这个院子是供会客、接待和下人用的。东院的房子坐南朝北，呈倒座形式，中间亦为过厅。这幢房的北面又是一层院落。这是把第一重倒座（为杂役仆人使用）和第二重倒座（接待会客之用）划分成了两个区域，可以互不干扰。自垂花门以北，两院格局大体一致，分别为三进院→过厅→四进院→正房→五进院→后罩房。

较小型的四合院也有两组并列的做法，宣武区椿树下二条的尚小云故居即是一例。该院分东西两宅，东侧为一座三进院落，宽约16米，深约42米，西侧为一座二进院落，宽约16米，深约30米，但二门已无存（图2－12.2）。

两组并列式院落是宅院向横向发展形成的格局，一般只有豪门大户才建得起这样的宅子。

图2—12 两组并列式宅院举例

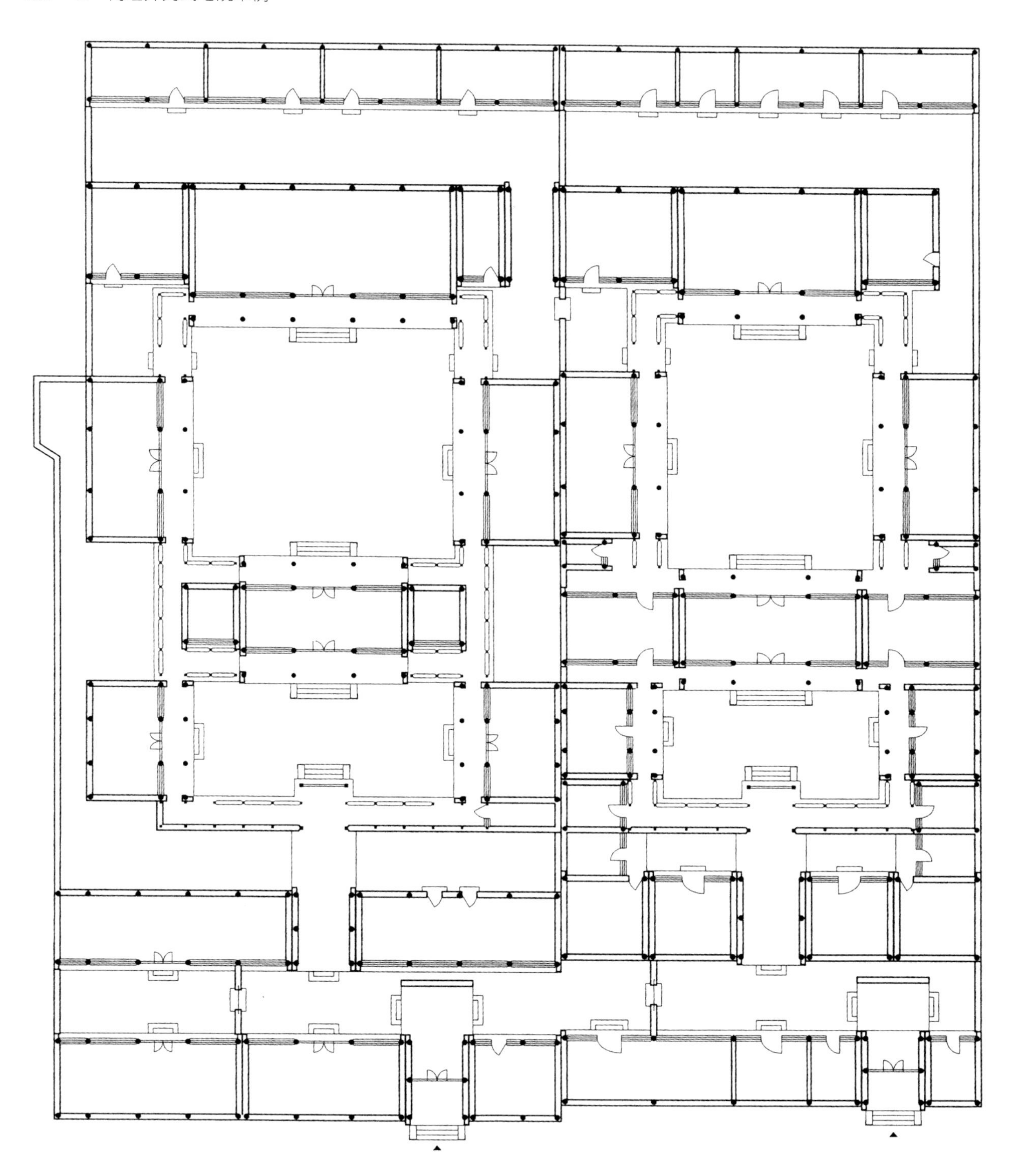

图2—12．1 秦老胡同某宅

图2—12. 2 宣武区椿树下二条尚小云故居

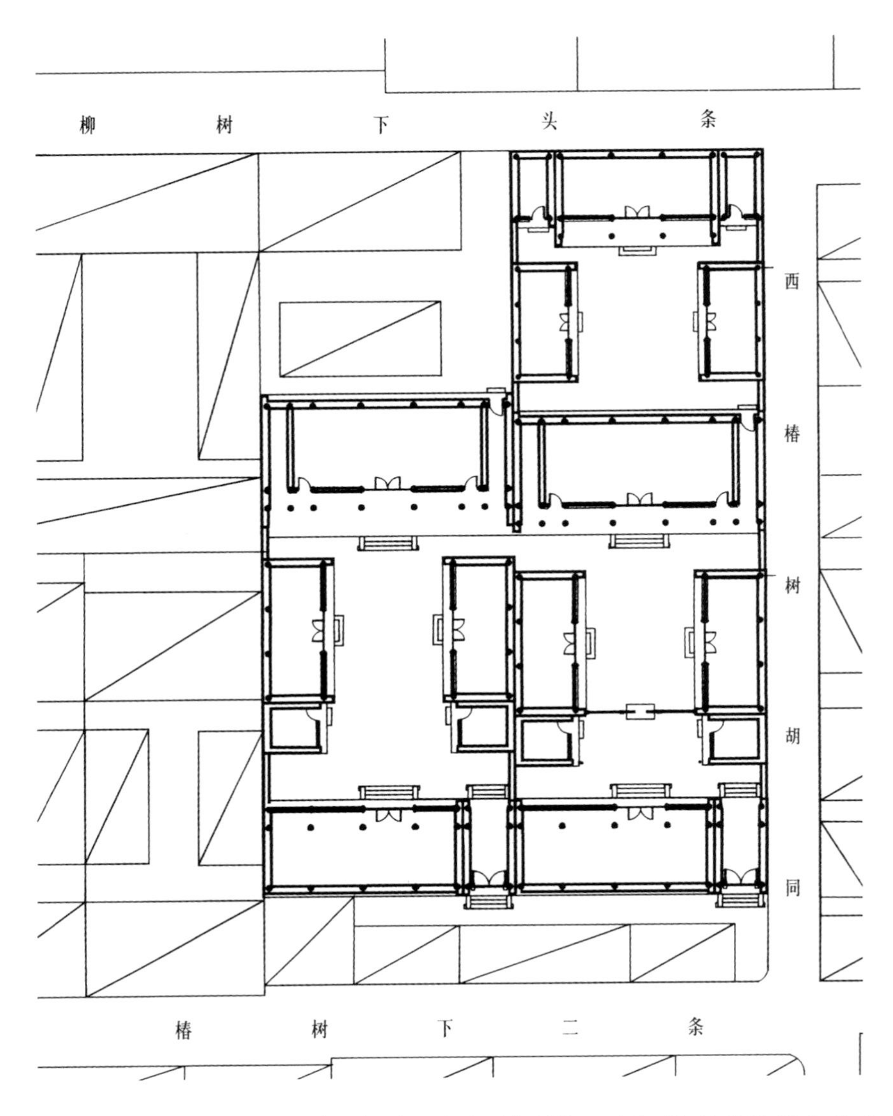

在横向发展的宅院中，还有多组（一般为三组）并列的。多组并列式院落在北京一般宅院中所见不多，且不典型，目前保留比较完好的只有沙井胡同15号、17号、19号，据说原为光绪时某中堂府。典型的三组或三组以上院落并列的，在王府中比较多见。著名的恭王府就是三组院落并列而构成的大建筑群。位于后海北岸的醇亲王府（又名摄政王府），则是由四座大型院落并列组合而成，其规模之大，建筑之宏伟，都是无与伦比的。但王府不是私产，属皇家所有，归内务府管理。从功能上看，它是集办公、居住为一体的建筑，将它归到住宅一类并不完全恰当，故此不复赘述。

7. 带花园住宅

北京的大中型四合院，不乏带花园者。在清代，皇室的造园活动盛行。在皇室建园之风的影响下，上层贵族、官僚及富商巨贾也纷纷仿效，划出地段营造私家园林。

北京的私家园林以王府花园最为典型。著名的清恭王府花园——萃锦园和摄政王府花园——鉴园（现为宋庆龄故居），是王府花园的代表。这种花园占地面积大，而且有较大的水面，亭台楼阁分布其中，异兽珍禽豢养其内，奇花名木扶疏繁茂，夏有鸟啾，冬有鹿鸣。这种规模的园林，在私家花园是不可能见到的。中下层官僚和殷实富户的花园一般占地面积很小，园中景致多表现一个主题。如附属于秦老胡同某宅的花园，布局简洁灵巧，园门上“西园翰墨”的题额，标志这是一座以书房为中心的花园；而牛排子胡同某宅的半亩园，则为清初戏曲理论家、作家李笠翁所建，以“富贵而有书卷气”闻名。

图2—13 带花园住宅举例

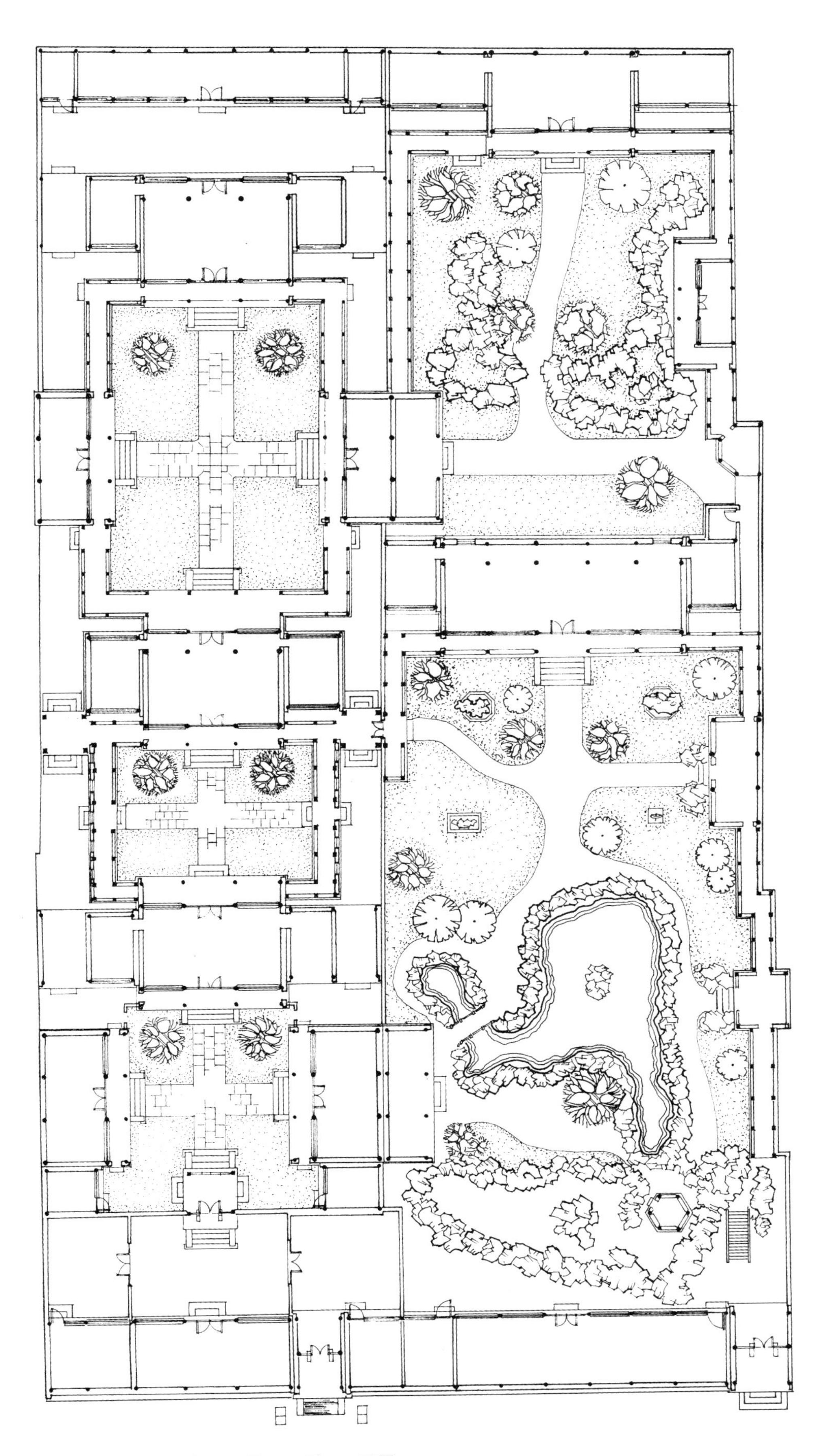

图2—13.1 帽儿胡同9号、11号——可园

私宅中保留最完好的花园，要数帽儿胡同9号、11号的可园。据说这是清末慈禧宠臣荣禄之弟荣源的宅园。这座宅园西侧是一座五进院落的豪宅，院落重重，幽雅深邃。东侧是花园部分。该花园以大花厅为界，将园子分为两部分。前园开阔，有假山、水面、小桥、碑记、花草树木以及小亭、游廊等园林建筑，并通过西侧敞亭和廊门与主院相通。后园则以山石花木为主，东侧有爬山廊、敞轩，西侧有通向西院的花厅。北面是供主人休息的地方。三间正厅，两侧各两间耳房。所有建筑都凭借游廊串连相接，直达前面的大花厅。全园共占地约4000m^2,是私家花园中较大的一座（图2－13.1）。

帽儿胡同的另外一处带花园住宅也颇具代表性。现为35号、37号的宅院原为宣统皇帝之后婉蓉出嫁前的住处，西侧为一座四进院落的宅院，东侧第二进院落是一座以山石为主题的花园。这种将宅院的一部分划出做花园的例子是比较多见的（图2－13.2）。

综上所述，由四面房子围合而形成的一进式院落，是北京四合院的基本单元。平民百姓的房子大部分是一、二进院落，属小型住宅。三进以上的住宅，都是由有官职地位的豪门大户所居住，因此也格外讲究。院落向大、中型发展，首先是向纵向发展，出现了三进院、四进院乃至五进院。纵向发展到一定程度时，就开始向横向发展，从而出现了一主一次、一主二次、两组并列、多组并列等多种宅院形式。宅院带花园，在清代曾为时尚，但民居花园都占地不大，且精致小巧，仅仅是宅院的陪衬和点缀，不似苏州园林那样占主要地位。

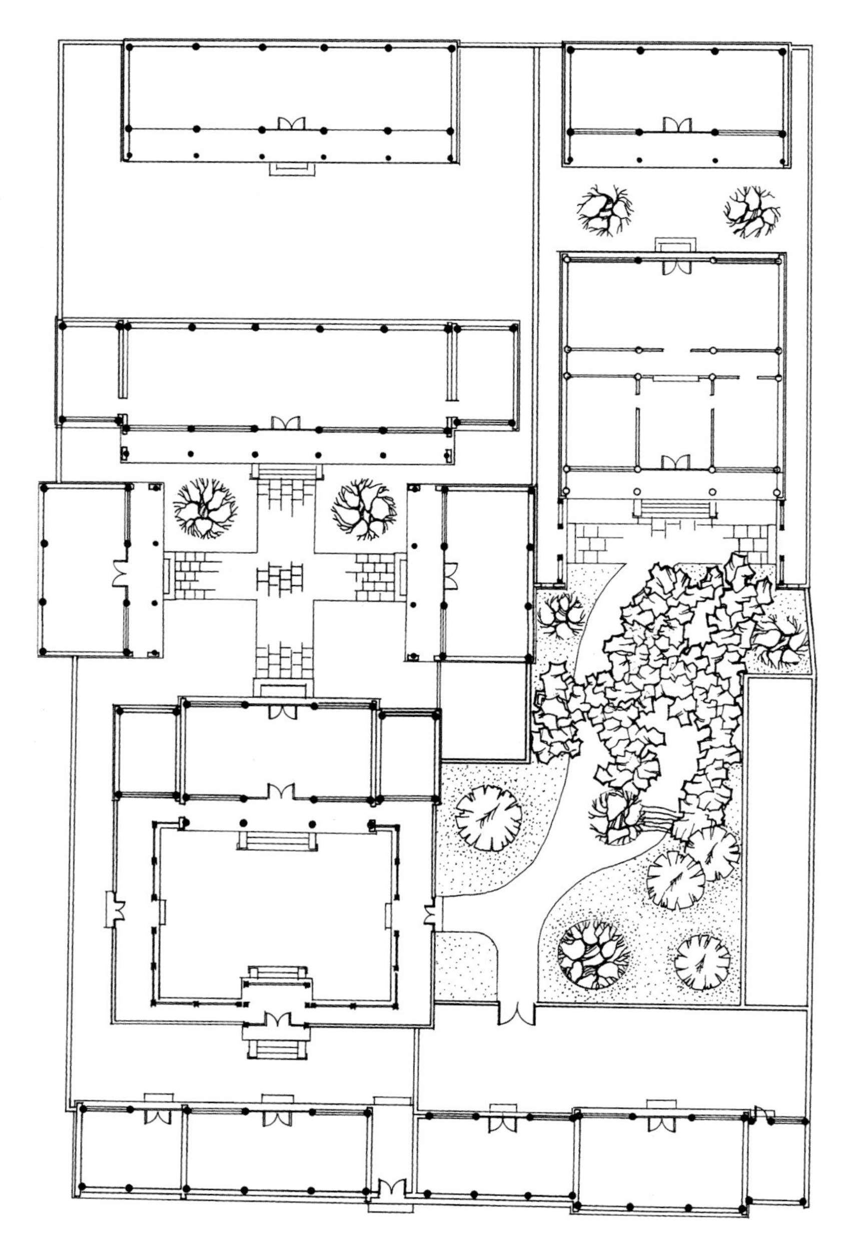

图2—13．2　帽儿胡同35号、37号

第三章
北京四合院的风水讲究

有关四合院的风水讲究，在中国传统风水术中称为“阳宅相法”。它经历千百载，久经传承，历代推衍，形成了庞杂的体系，对中国传统民居建筑有着广泛、深刻的影响，与人们的生活有着极为密切的关系。我们研究北京四合院，必然要接触到风水问题，而要了解这些内容，还要从基本的问题谈起。

一、风水学的来源与基本原则

风水学，是我国古代的建筑（地理）环境学。它是我国古代先民为建城镇、村落、住宅、寺院、墓地等，寻找吉祥地点的景观评价系统和选址布局理论。

根据国内外学者的考察分析及考古发现的实物例证，我国古代的风水学发源于黄土高原连绵起伏的山丘地区，是窑洞的居住者为寻找理想的洞穴而发展起来的。它虽然包含有较多的迷信成分，但它的基本原则却反映了早期中国人为寻找理想的居住地点，对环境条件的科学认识。

用于选择基址的风水理论的主要原则有以下几条。

（1）对周围地形的要求：它需要呈马蹄形的隐蔽地形，以马蹄形的山丘为靠背，前面有临水的开阔地带；位于山脊当中主峰山脚下的地方为最吉祥的地点——风水穴。

（2）对水的要求：吉祥地本身必须是干燥的，但在它前方不远的地方应该有水。

（3）对方位的要求：具备上述条件的地方应该面向吉祥的方向，也就是日照充足的方向。根据我国的地理位置，这个方向最好是南方。

以上三条，是适用于寻找城市、村镇、住宅、墓地、寺庙等建筑群基址的基本准则。这些基本准则可以概括为“负阴抱阳，背山面水”。

“负阴抱阳，背山面水”，具体地说就是，要求村落（或城市，或住宅、寺庙等）的后面有山脉及主峰，山脉称龙脉，主峰称来龙山；左右有次峰和冈阜，称左辅、右弼（又称青龙山、白虎山），山上有丰茂的草木植被。基址前有河流或水面，水的对面有作为对景的山岗，称案山，案山前面还可有山，名朝山。主峰与案山形成的轴线以南北方向为最佳，其他方向（如偏东、偏西或东西方向）只要具备这套格局也可以。基址正好处在山环水抱的中央地带，地势平坦而且有一定坡度。这就是一个背山面水、负阴抱阳的理想格局，是“藏风聚气”的优质环境，是人们所谓的“风水宝地”。

具备这种组合形式的自然环境，是一个较为封闭的空间。它有利于形成良好生态和局部小气候。根据我国具体的地理环境和气候特点，背山可以遮挡冬天北来的寒流，面南可以迎纳夏日南来的凉风，朝阳可以获得充足的日照，近水可以取得方便的生活、灌溉用水，还可便利水运交通，进行水中养殖。略有缓坡的平川可免受水患之灾，山上植被丰茂可以保持水土，改善小气候，并为人们提供燃料和其他多种资源。这种天造地就的优良环境，极适合人们休养生息，无疑会给人们带来世外桃源般的美好生活（图3－1）。

北京地区就是这样一块风水宝地，这也是它能够作为六朝古都的原因所在。唐代著名风水师杨益是这样来评价北京的风水格局的：“燕山最高，像天市，盖北干之正结，其龙发昆仑之中脉，绵亘数千里……以入中国为燕云，复东行数百里起天寿山，乃落平洋，方广千余里。辽东辽西两枝，黄河前绕，鸭绿后缠，而阴山、恒山、太行山诸山与海中诸岛相应，近则滦河、潮河、桑干河、易水并诸无名小水，夹身数源，界限分明。以地理之法论之，其龙势之长，垣局之美，干龙大尽，山水大会，带黄河、天寿，鸭绿缠其后，碣石钥其门，最是合风水法度。以形胜论，燕蓟内跨中原，外控朔漠，真天下都会。形胜甲天下，山带海，有金汤之固。”宋代朱熹也对北京地区的风水赞叹不已：“冀都是正天地中间，好个大风水。山脉从云中发来；

图3—1　理想的风水格局示意图

云中正高脊处，自脊以西之水则西流入于龙门西河，自脊以东之水则东流入于海。前面黄河环绕，右畔是华山耸立为虎。自华来至中原为嵩山，是为前案。遂过去为泰山，耸于左，是为龙。淮南诸山是第二重案，江南诸山及五岭又为第三、四重案，正谓此也。”清人吴长元也说：“京师前挹九河，后拱万山，正中表宅，水随龙下，自辛而庚，环注皇城，绕巽而出，天造地设。”

正如前人所分析的那样，北京正是处在这样一个优质大环境中：北有层峦起伏的燕山山脉，西有蜿蜒逶迤的太行山脉，两股山脉交汇聚结，奔腾起伏，形成巨大的“龙脉”。山峦之上，森林覆盖，云气聚积。来自黄土高原的桑干河和来自蒙古高原的洋河汇合成为永定河，从深山老林之中奔腾而出，在北京平原蜿蜒流淌，为北京带来了丰沛的水资源。河水溪流形同蛛网，淀泊湖海星罗棋布，形成北京地区温润丰饶、土肥水美的天府之地，构成了北京绝佳的风水外局。历代帝王正是由于看中了这块风水宝地特殊的地理和环境优势，才决定建都于京，使它成为六朝之都。

二、风水学中的“形法”和“理法”

以上仅是对风水的来源、基本原则以及风水师评价北京风水的简单描述，但风水学的内容绝不仅限于此。

中国传统的风水学说内容庞杂，流派纷呈，传世的著述很多，其中在世俗中影响较大、起主导作用的有两派，一派为“形势宗”，一派为“理气宗”。

形势宗的相宅之法为形法。形法注重自然、地理、环境、生态、景观诸因素的审辨和选择以及对应的处理方法。上文所谈及的风水原则，对北京风水环境的评价，都属形法范畴。理气宗的相宅之法为理法，它的基本依据是我国古代的《易经》和“天人感应”学说。这种方法是根据河图、洛书、阴阳五行、九宫八卦的宇宙图式，把天上的星宿和宅子的时空构成联系起来，通过分析相生相克的关系作出吉凶判断。这种方法同古代的占卜巫术有着千丝万缕的联系，故较之形势宗更多一些迷信色彩。

形法主要用于选址，其理论和方法是“辨形”和“察气”。辨形：无论是山川形法（评价山川环境的方法）还是阳宅形法（评价住宅周围环境的方法），都是以觅龙（寻找背山）、察砂（看周围山势）、观水、点穴为主要内容。这些，我们在上文已有简要叙述。风水说认为，察气的含义是“万物之生，以承天地之气”。这里所指的气不是空气，而类似气功学中所讲的“炁”。近年来，射电天文学家的研究成果提示，这种自然之气属于来自宇宙的微波辐射，也包括星体的电磁辐射。山环水抱的地形极像一口接受微波的天线，因而能大量吸收这种微波，形成特殊的气场。由于古代不能对这种现象给予科学的揭示，所以给“气”蒙上了一层面纱，成为风水学中最为神秘的部分。察气或望气是与辨形互为补充、互为印证的两个方面。

由于形法主要注重对宅外形和外部环境的审辨、选择与处理，对宅内部诸多具体、微妙问题的解决缺乏方法和理论。所以，有关宅（或村或寺院等）内形的评价与处理，很大程度上要依赖于理法。以“天人感应”“天人合一”为基本理论的理法，一向认为“夫宅者，乃是阴阳之枢纽，天伦之轨模……人因宅而立，宅因人得存，人宅相扶，感通天地”（《黄帝宅经》）。法天营居，参天作城，以求达到天人合一的至善境界，便成为承传千年的独特的“中国建筑的精神”。正如李约瑟博士所概括的那样：“再也没有别的地方表现得像中国人那样热心体现他们伟大的设想‘人不能离开自然’的原则……皇宫、庙宇等重大建筑当然不在话下，城乡中无论集中的或是散布在田园中的农舍，也都经常出现一种对‘宇宙图案’的感觉，以及作为方向、节令、风向和星宿的象征主义。”中国人世世代代都在依照这种“天人感应”理论去解决住宅方面各种具体细微的问题，尽管理论中有许多迷信成分和至今仍不能解释清楚的问题，但始终为人们所依赖。

三、阳宅相法的基本内容——形法和理法在相宅中的应用

形法用于阳宅的审辨，也以“觅龙、察砂、观水、点穴”为主要内容。如《阳宅爱众篇》说：“阳宅须教择地形，背山面水若有情，山有来龙昂秀发，水须转抱作环形。明堂宽大斯为福，水口收藏金满盈。关煞二方无障碍，光明正大旺门庭。”这段相宅歌诀，包含了“龙、砂、水、穴”全部内容。但对市井之宅——如北京城区的四合院而言，主要的环境影响并不是山川河流，而是宅院周围的墙垣、屋宇、道路、树木等环境因素。因此，在对市井之宅的审辨中，“龙、砂、水、穴”均被赋予了新的内涵。《阳宅集成》中就有这样的阐述：“一层街衢为一层水，一层墙屋为一层砂，门前街道即是明堂，对面屋宇即为案山。”这种在原意上的变通和引申，使形法的用途更加广泛。

察气之法用于相宅，主要有“一地，二门，三衢，四峤，五空缺”，称为五机。《金氏地学粹编·归厚录阳基章》讲道：“阴宅穴在地中，止穴内一气。阳宅穴在地上，不专以地气为用，兼取门气，盖清虚之上，气本横行，门户一启，气即从门而入，其力与地气相敌。……须得门、地两旺，然后可以招诸福。门地之处又看道路。道路局势朝归者，作来气断；横截者，作止气断。朝路比来龙，横路比界水，所谓三衢、桥梁同断。峤者，邻居高峻处。如艮方者有高屋，则气被障断，反从艮方还转气来，回向我宅，所谓回风反气。自高及下者，高屋多则气厚，高屋少则气浅。……空缺者，方隅孔窍，或在宅外，或在宅内，能引八风而入；关于祸福，不可不知。”这里所谓地气，指宅基地的大小、高下、土质、温度、湿度情况以及对于人们生理、心理的影响效果。地气过强或过弱都会使人在生理、心理上感到不适，这就需要用门气或其他气来进行调节。门，在阳宅相法中称为“气口”。门开启的方位、大小与宅内外景观的沟通、小气候的调节等有直接关系。如门气过强（门开过大或方位不对）则会冲淡地气，破坏宅内的亲和私密气氛。衢气，指宅外道路交通的走向及对宅内产生影响的各种因素。衢气适中，要求既交通便利，又避免受街外因素干扰。峤气，指以高屋为屏障而围合成空间，如围合适度（院的大小与周围建筑高度比例适度），人则会产生安适感，如比例失调，则会产生疏离感或压抑感。“回风反气”也包括高屋遮挡对宅内通风、排湿、降温、防寒等小气候调节的影响。

这里提到的气，不专指空气之气或宇宙之气。它既可理解为客观存在的物质，又可理解为人的心理感受。这种感受因时、因地、因人而异。

形法中，还有对宅外形优劣吉凶的判别内容。《阳宅十书》说："若大形不善，总内形得法，终不全吉，故论宅外形第一。"这些宅外形吉与凶的判别，是在辨形察气的基础之上归纳总结出来的，尽管有些虚幻迷信、荒诞无稽的成分，但也不乏科学合理的内容（图3－2）。例如，宅中北房高大、宅院南北深长有利通风纳凉；宅院西北高、东南低，有利于排水；宅外有路交通便利，但道路不宜多，不宜朝向宅门；宅院要远离城门、祠社、寺院、官衙、监狱、窑地或河川桥梁等人流集中、交通要冲地带，以及关于防火、防盗、防潮、避风、环境卫生、儿童安全等方面的考虑，都是关系人的生理需求的，是唯物的和科学的。又如，要求靠倚丘岗、环护左右，使人有安全和稳定感；宽敞的明堂及优美的对景能给人以美好的心情；避开"破、败、坏、断、残"的不良景物和容易引发不良想象（如门前有双池，形似"哭"字）的设施，使人保持良好的心态；宅形完整，左右对称，满足人的完美心理追求等等，这些都能给人以心理上的慰藉，也是科学、可取的。

图3—2 关于宅外形的判别标准举例（本图摘引自《中国古代风水与建筑选址》）

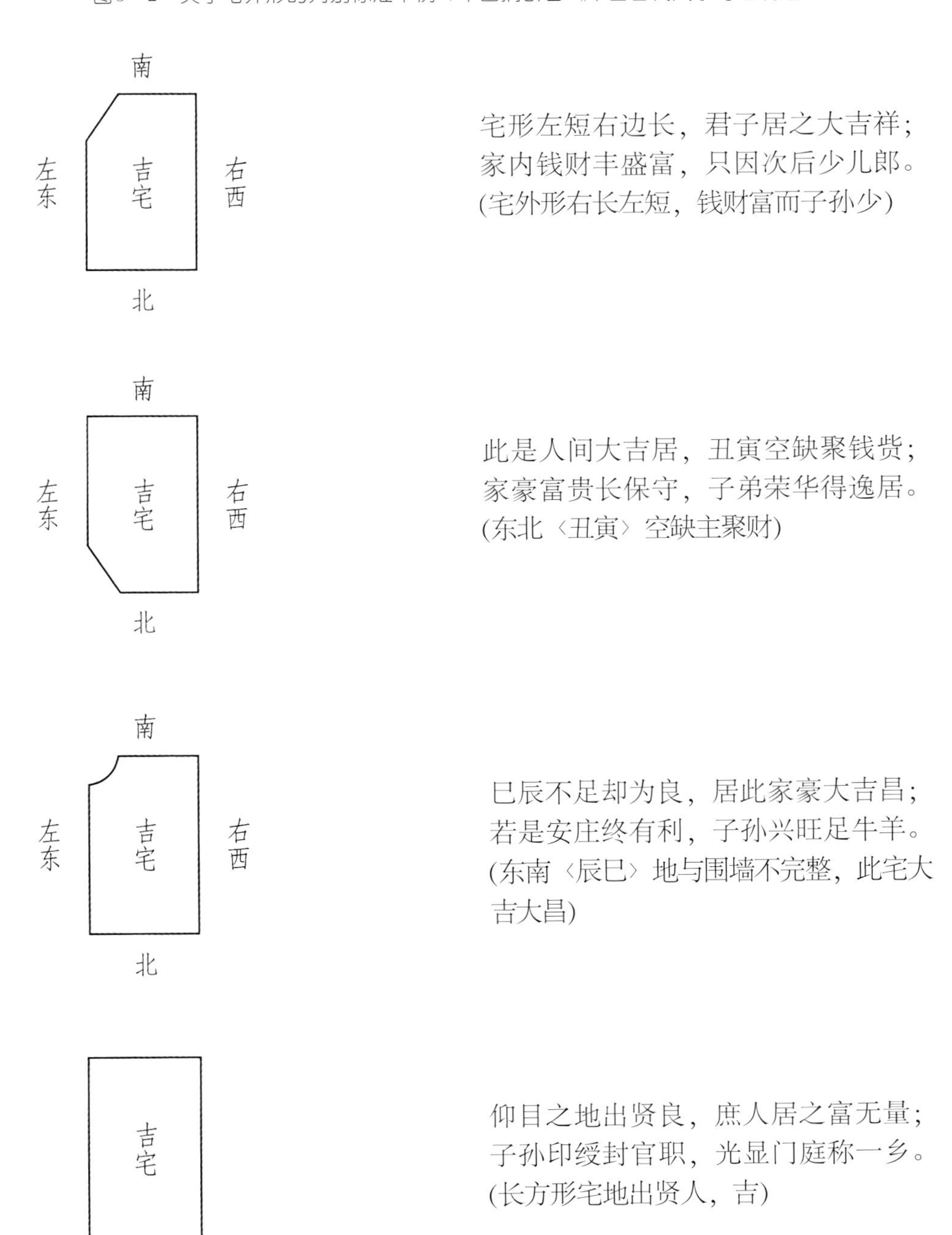

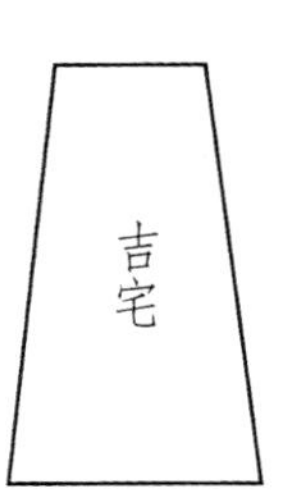

前狭后宽居之稳，富贵平安旺子孙；
资财广有人口吉，金珠财宝满家门。
(宅地前狭后宽，吉)

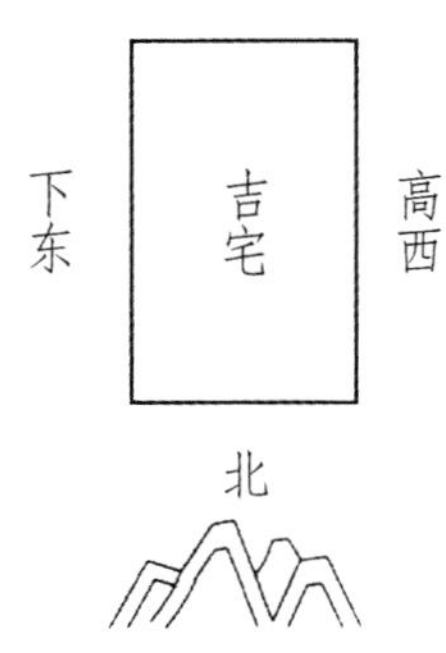

西高东下向北阳，正好修工兴盖庄；
后代资财石崇富，满宅家眷六畜强。
(西高东低、北面向山的开阔地，大吉)

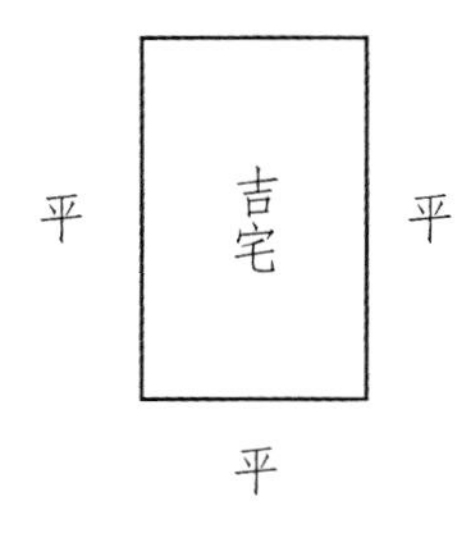

此宅方圆四面平，地理观此好兴工；
不论宫商角徵羽，家豪富贵旺人丁。
(四面皆平的地方，各种姓氏的人都可以居住，吉)

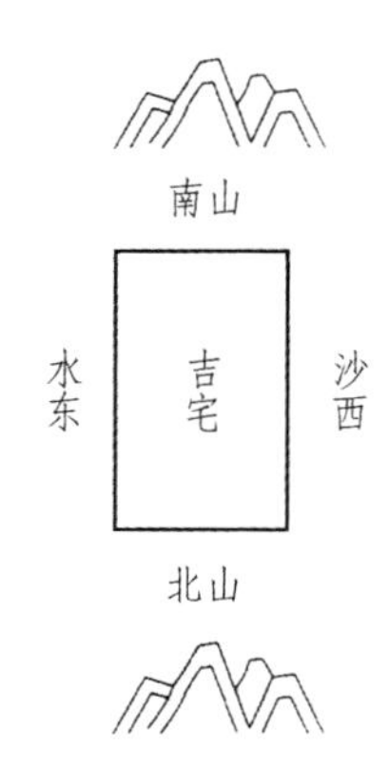

前后高山两相宜，左右两边有沙池；
家豪富贵多年代，寿命延年彭相齐。
(南、北有高山，东、西有沙池，主人富贵、长寿)

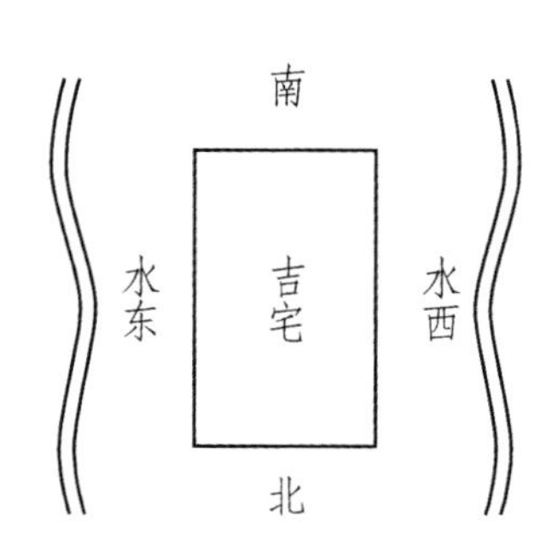

此宅左右水长渠，久后儿孙福禄齐；
禾麦钱财常富贵，儿孙聪俊胜祖基。
(宅的东、西有水渠流过，主富贵、子孙聪俊)

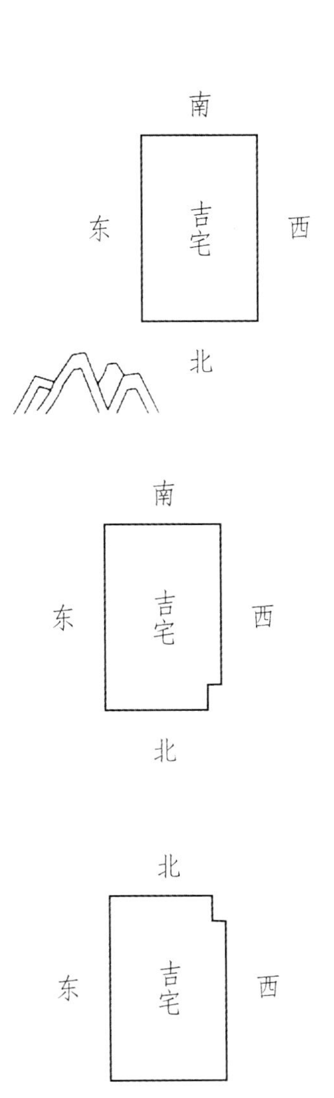

此宅后面有高岗，南下居之第一强；
子孙兴旺田蚕胜，岁岁年年有陈粮。
(东北方有丘冈，大吉)

左短右长却安然，后面夹稍前面宽；
此地修造人口吉，子孙兴旺胜田蚕。
(宅的西侧短，东侧长，人口吉，子孙旺)

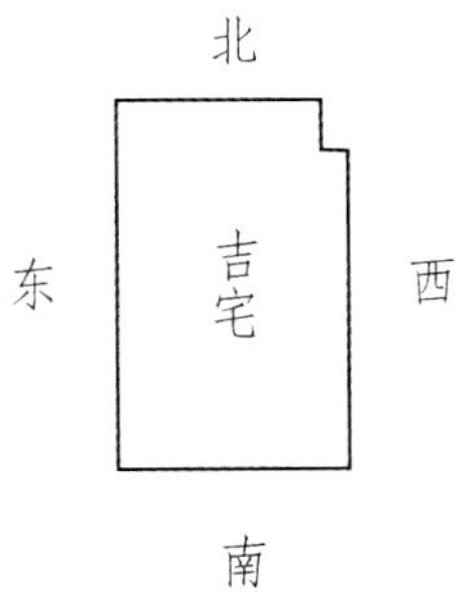

此宅右短左边长，假令左短有何妨；
后边齐整方圆吉，庶人居之出贤良。
(宅的东墙长，西墙短，北面齐整方圆，吉)

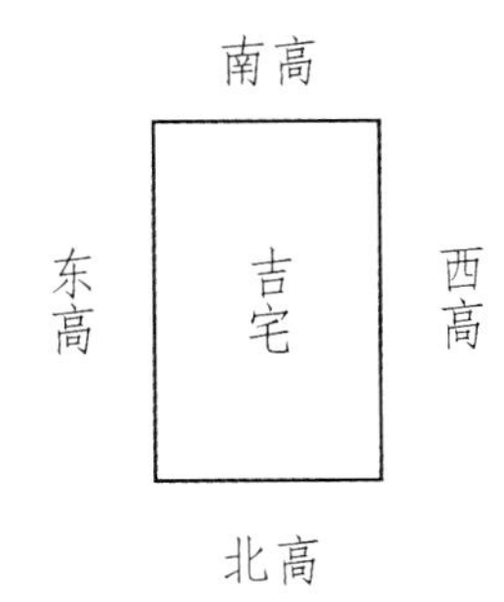

中央正面四面高，修盖中宅福有余；
牛羊六畜多兴旺，家道富贵出英豪。
(宅的四周高，吉)

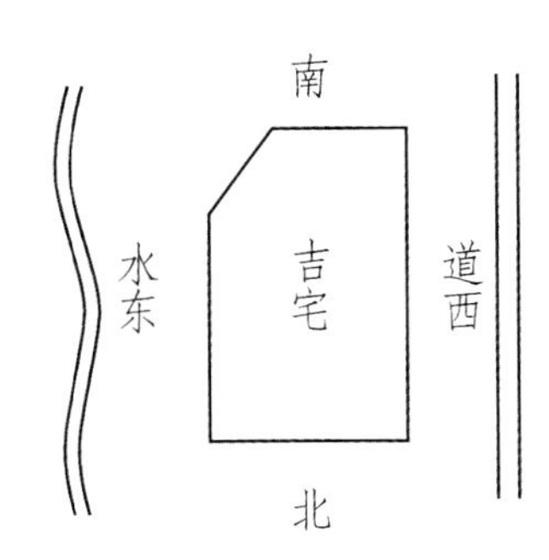

宅东流水势无穷，宅西大道主亨通；
因何富贵一齐至，右有白虎左青龙。
(宅东流水，宅西大道，吉)

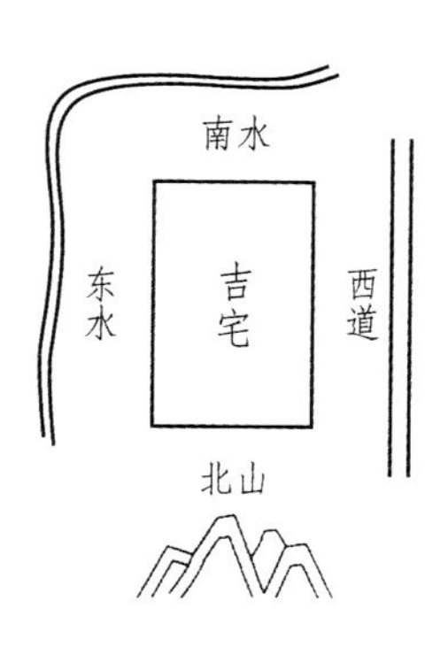

朱元龙虎四神全，男人富贵女人贤；
官禄不求而自至，后代儿孙福远年。
(东、南两面有水，北面有山，西面有道路，大吉)

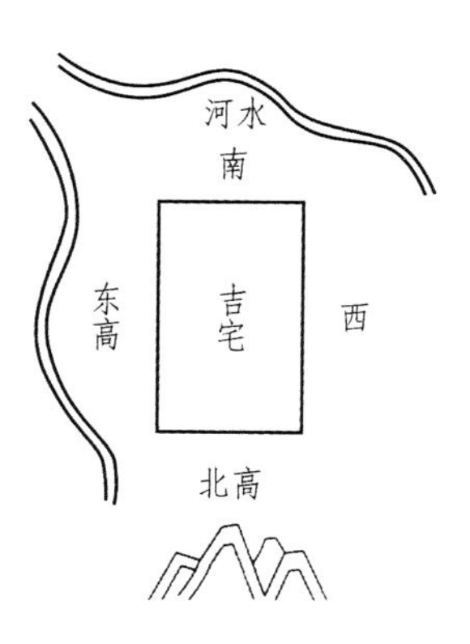

西来有水向东流，东显长河九曲沟；
后高绵远儿孙胜，禾谷田蚕岁岁收。
(有水自西流向南，东边又有水九曲转环，北面地势高，大吉)

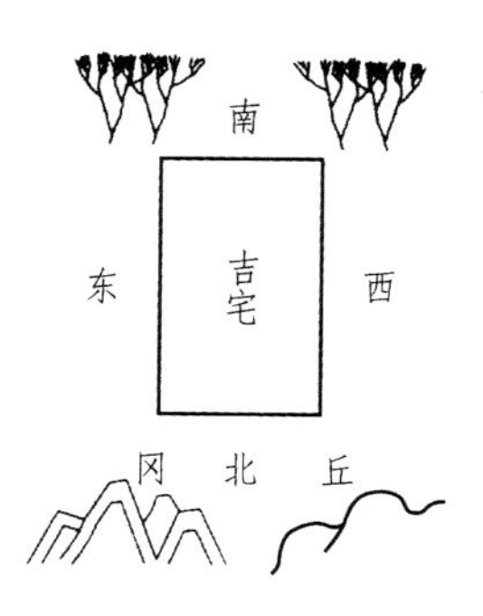

宅前林木在两旁，乾有丘阜艮有冈；
若居此地家豪富，后代儿孙贵显扬。
(宅前的林木分为两片在旁，西北有丘阜，东北有山冈，大吉)

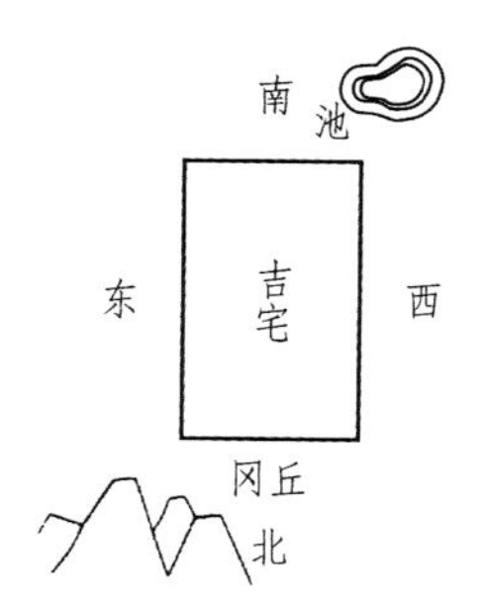

住宅西南有水池，西北丘势更相宜；
艮地有冈多富贵，子孙天赐著罗衣。
(宅西南有水池，东北有冈丘，吉)

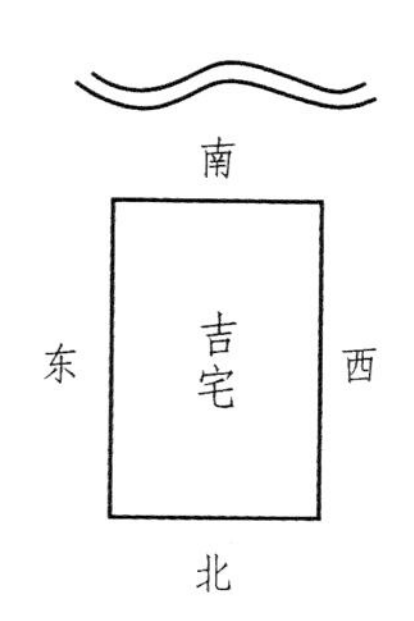

离乡迢迢是此路，儿孙出外皆发富；
若然直云不回还，定出离乡不归屋。
(宅前有曲曲弯弯的道路，主儿孙在外发富，吉)

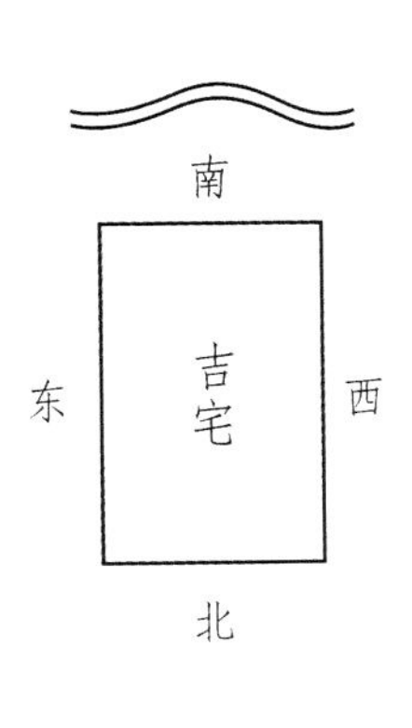

门前若有玉带水，高官必定容易起；
出人代代读书声，荣显富贵耀门闾。
(宅的南面有玉带水，吉)

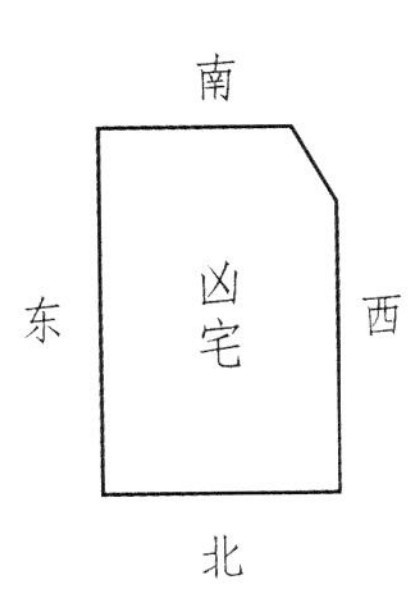

右短左长不堪居，生财不旺人口虚；
住宅必定子孙愚，先有田桑后亦无。
(宅外形右短左长，人财两虚)

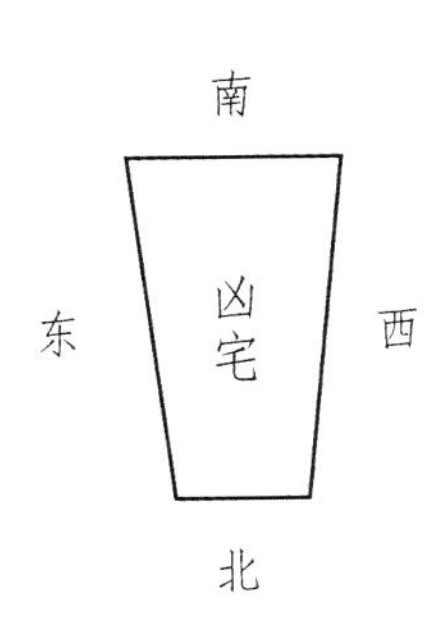

前宽后狭似棺形，住宅四时不安宁；
破尽赀财人口死，悲啼呻吟有叹声。
(宅地前宽后狭，凶)

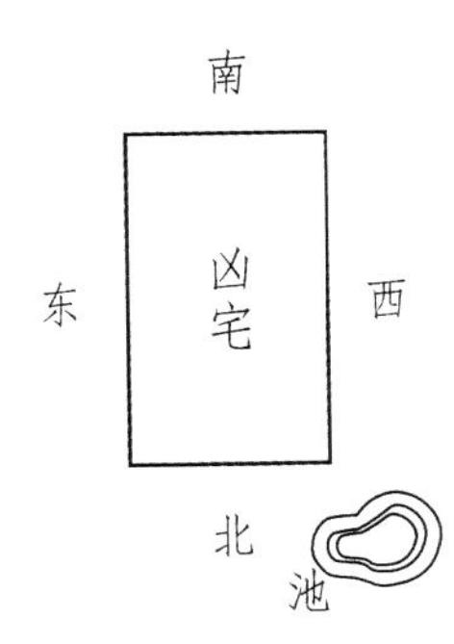

西北乾官有水池，安身甚是不相宜；
不逢喜事多悲泣，初虽富时终残疾。
(宅西北有水池，凶)

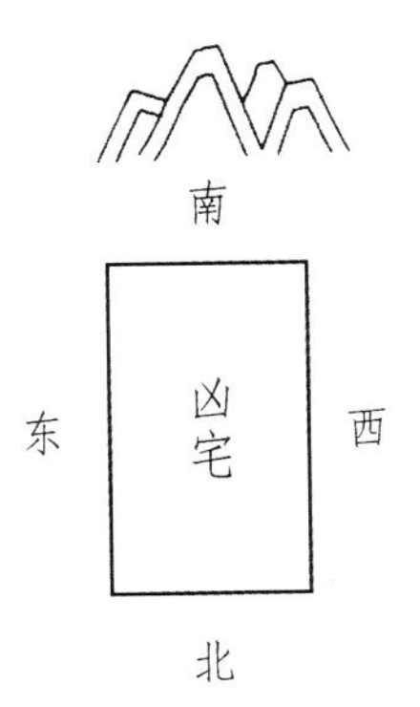

前有大山不足论，不可安庄立坟茔；
试问明师凶与吉，若居此地定灭门。
(南面有大山，凶)

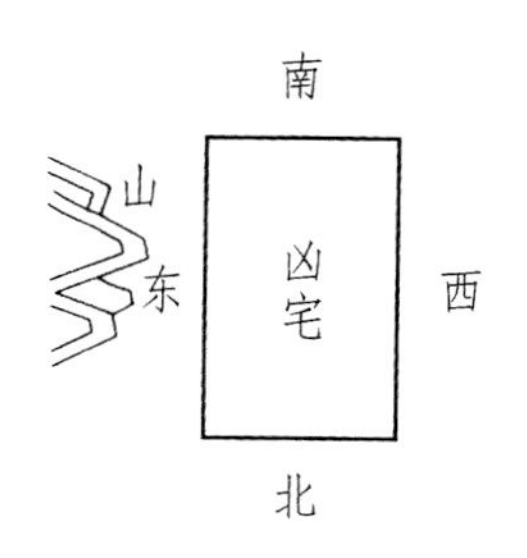

此宅东边有大山，又孤又寡又贫寒；
频遭口舌多遭难，百事先成后来难。
(宅的东边有大山，凶)

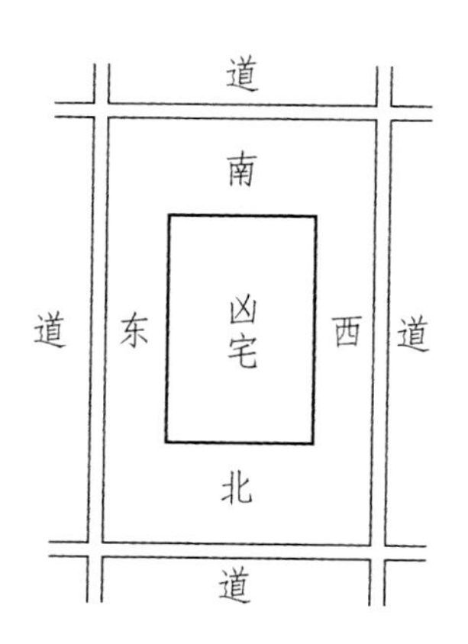

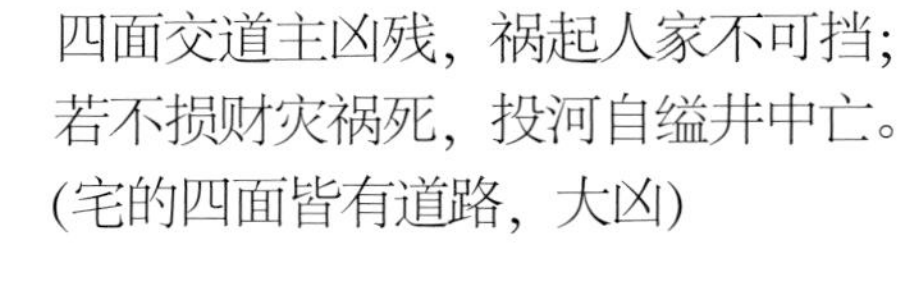
四面交道主凶残，祸起人家不可挡；
若不损财灾祸死，投河自缢井中亡。
(宅的四面皆有道路，大凶)

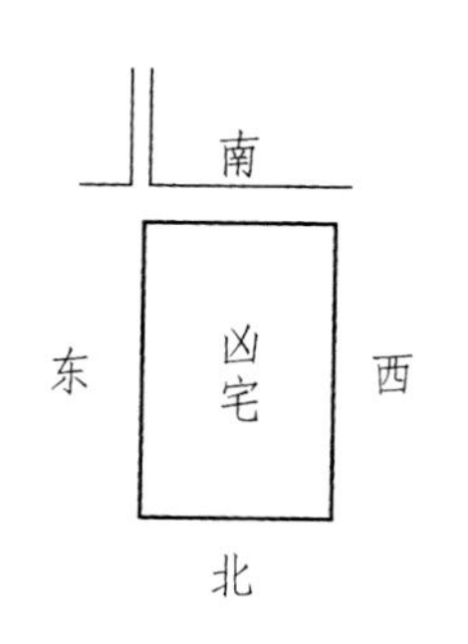

南来大路正冲门，速避直行过路人；
急取大石宜改镇，免教后人哭声频。
(南来的大路正冲门，凶，需避开一直走来的行人，必须由镇石使路口改道)

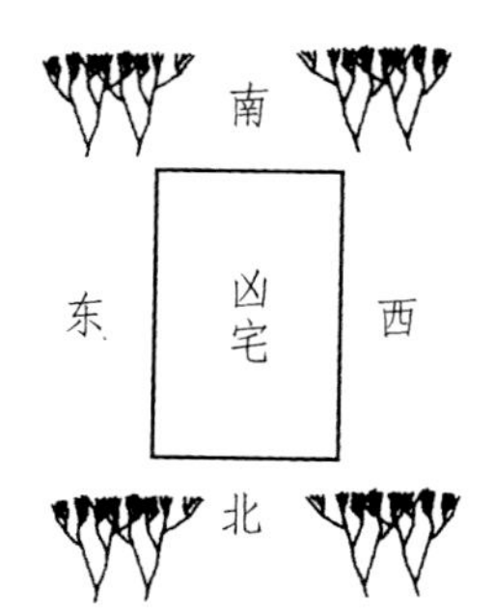

林中不得去安居，田宅莫把作丘坟；
田蚕岁岁多耗散，宅内惊扰鬼成精。
(宅的四周是林地，凶)

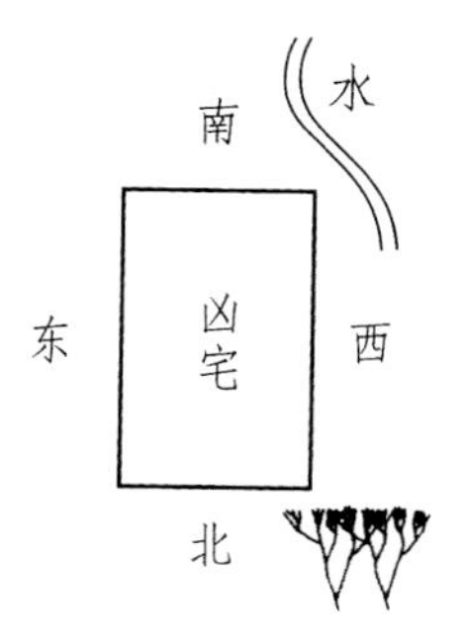

乾地林木妇女淫，沟河重见死佳人；
坤地水流妨老母，子孙后代受孤贫。
(宅的西北有林，西南流水，凶)

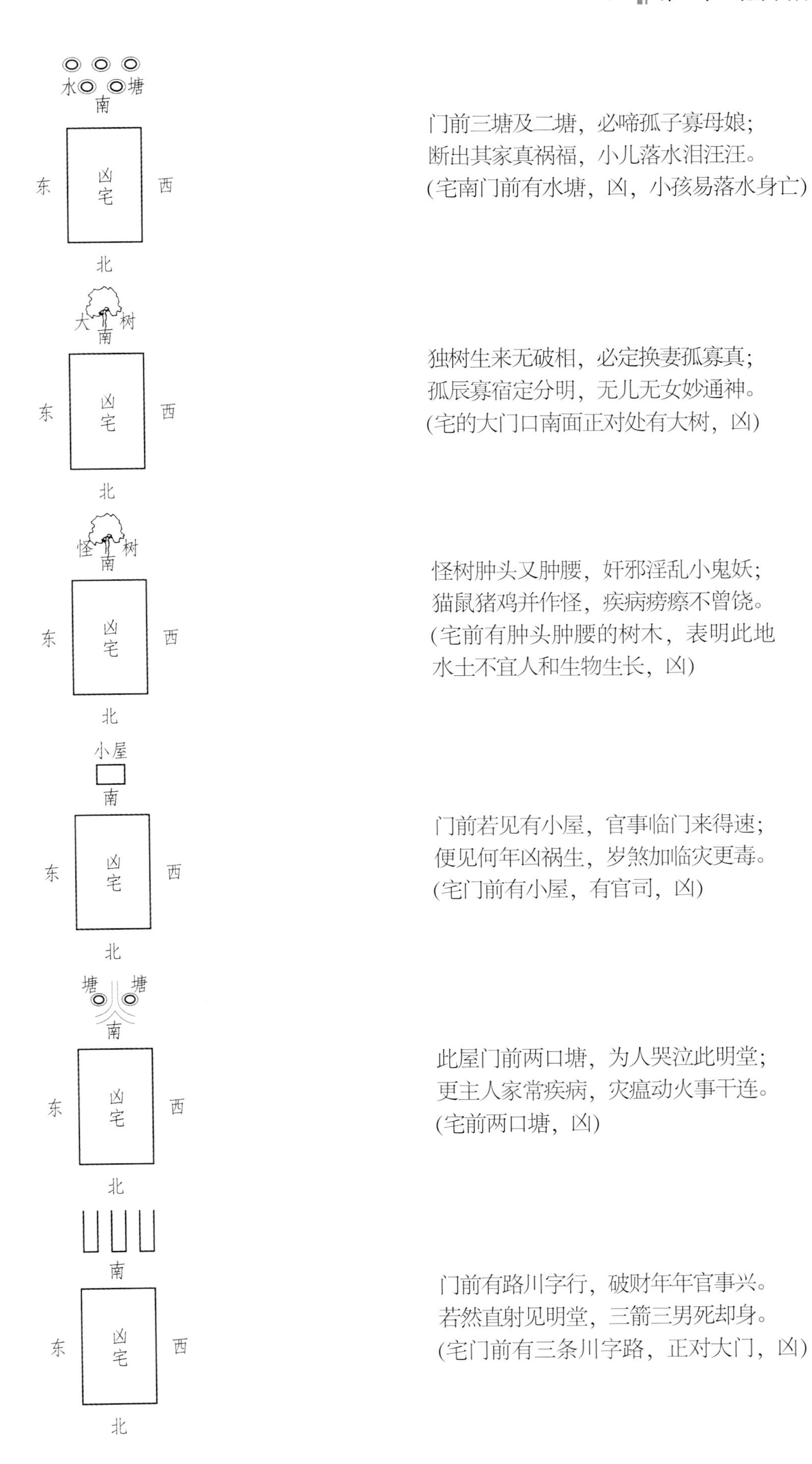

门前三塘及二塘，必啼孤子寡母娘；
断出其家真祸福，小儿落水泪汪汪。
(宅南门前有水塘，凶，小孩易落水身亡)

独树生来无破相，必定换妻孤寡真；
孤辰寡宿定分明，无儿无女妙通神。
(宅的大门口南面正对处有大树，凶)

怪树肿头又肿腰，奸邪淫乱小鬼妖；
猫鼠猪鸡并作怪，疾病痨瘵不曾饶。
(宅前有肿头肿腰的树木，表明此地
水土不宜人和生物生长，凶)

门前若见有小屋，官事临门来得速；
便见何年凶祸生，岁煞加临灾更毒。
(宅门前有小屋，有官司，凶)

此屋门前两口塘，为人哭泣此明堂；
更主人家常疾病，灾瘟动火事干连。
(宅前两口塘，凶)

门前有路川字行，破财年年官事兴。
若然直射见明堂，三箭三男死却身。
(宅门前有三条川字路，正对大门，凶)

形法中还有一些判别宅内形吉凶的标准，如“五实”“五虚”。《黄帝内经》说：“宅有五虚令人贫耗，五实令人富贵。宅大人少一虚，宅门大内小二虚，院墙不完三虚，井灶不处四虚，宅地多屋少庭院广五虚：宅小人多一实，宅大门小二实，墙院完全三实，宅小六畜多四实，宅水沟东南流五实。”此外，还有“三要”“六事”，即宅门、厅堂、居室、井、灶、路、厕、仓、碾、畜圈等布置得宜、尺寸合度为吉。这些内容也是科学的、合理的。

理法用于相宅者，以大游年法为最普遍。大游年法是阳宅相法中最基本的一种方法。其宗旨是根据天人感应观念、阴阳五行学说以及河图、洛书、九宫、八卦等宇宙图式，把天上的星宿同宅子的时空构成联系起来，分析其相生相克关系，从而作出吉凶判断。

要了解大游年法，首先要将此法涉及的中国古代哲学、宇宙观以及相关数理知识做一些简单介绍。

1.《易经》、八卦和阴阳五行学说

《易经》是我国古代最有影响的一部书，相传为伏羲、文王和孔子所作。它以乾、坎、艮、震、巽、离、坤、兑代表天、水、山、雷、风、火、地、泽八种自然现象（事物），并用阴阳两种对立势力的相互作用、相互转化、相互消长来说明万物的形成和变化。作为古代占筮之书的《易经》认为：“天下万事万物，莫不有其定数。”用现代的话来说就是：“任何事物都是按一定的相似规律在运行。”《易经》所建立的极为严密的数理结构，被专家们称为“宇宙代数学”。以《易经》理论指导自然科学研究所取得的成就古今中外尽人皆知。

阴阳学说是《易经》的灵魂，它是古代中国人的宇宙观和方法论，是人们通过对日月往来、昼夜更替、寒暖晴雨、生死更迭等自然现象的长期观察总结出来的包括朴素辩证法的一门学说。它以“—”代表阳，以“--”代表阴，并通过二者的排列组合，形成八个卦象。这八个卦象用于自然界时，“☰”代表天（乾），“☷”代表地（坤），“☳”代表雷（震），“☵”代表水（坎），“☶”代表山（艮），“☴”代表风（巽），“☲”代表火（离），“☱”代表泽（兑）。用来比拟家庭时，“☰”代表父，“☷”代表母，“☳”代表长男，“☵”代有中男，“☶”代表少男，“☴”代表长女，“☲”代表中女，“☱”代表少女。天地之间的一切事物，都可以用八卦来表征（图3－3.1）。

图3—3　风水图例

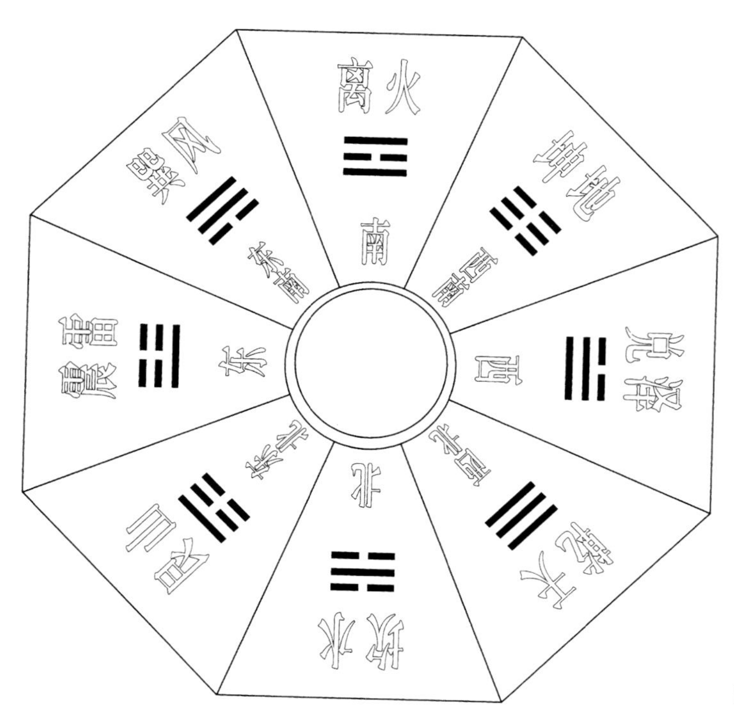

图3—3．1　后天八卦图

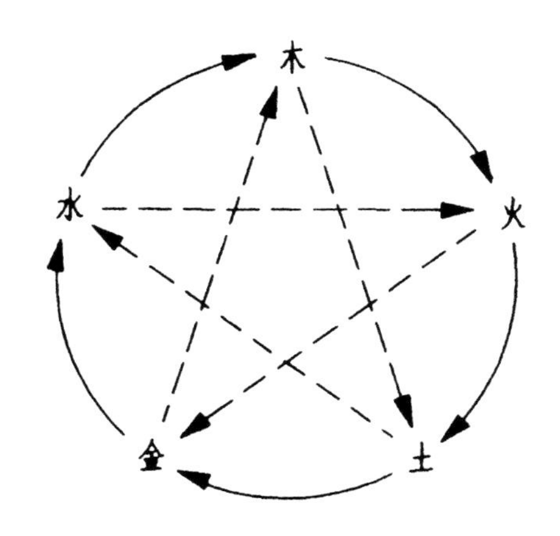

图3—3．2 五行相生相克关系图

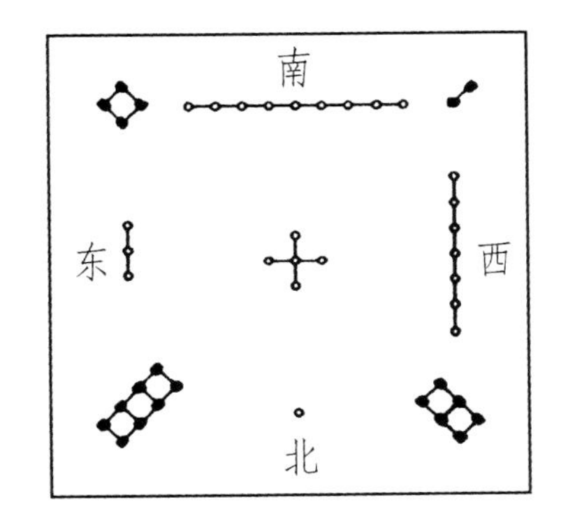

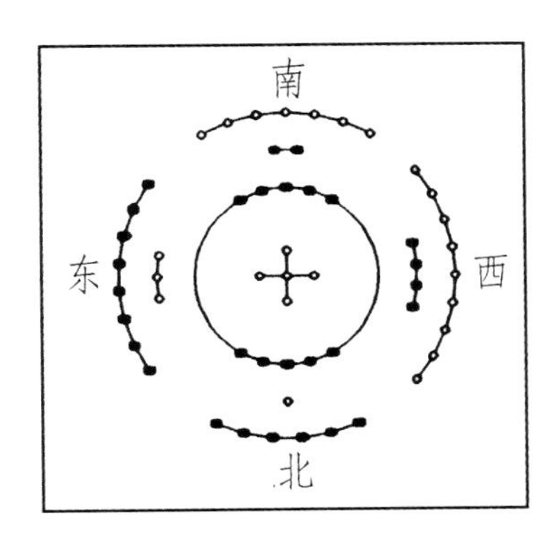

4	9	2
3	5	7
8	1	6

图3—3．3 河图、洛书、九宫图

五行原本是五种能为人所利用的物质，后经改造、完善成为系统的思想体系。《书经 · 洪范篇》这样描写五行：“……五行：一曰水，二曰火，三曰木，四曰金，五曰土。水曰润下，火曰炎上，木曰曲直，金曰从革，土爰稼穑……”五行之间有相生相克的关系。五行相生：金生水，水生木，木生火，火生土，土生金。五行相克：金克木，木克土，土克水，水克火，火克金（图3－3.2）。

2. 河图、洛书

河图、洛书是我国古代的数算之宗、数理之源，与先天、后天八卦有极为密切的对应关系。《易 · 系辞传》曰：“河出图，洛出书，圣人则之。”又曰：“天一，地二，天三，地四，天五，地六，天七，地八，天九，地十。”其中，天为阳，取奇数，地为阴，取偶数，在河图、洛书中分别用白点和黑点表示，按一定规律排列。河图的数字排列规律是：“天一以生水，地六以成之；地二以生火，天七以成之；天三以生木，地八以成之；地四以生金，天九以成之；天五以生土，地十以成之。”（图3－3.3）洛书源于河图，采方形，取龟像，并按“戴九履一，左三右七，二四为肩，六八为足”和中间是五的规律排列。将洛书的四方四隅加中间共九个方位的数字演变成九宫格，即为矩阵（图3－3.3）。

3. 七政

七政为北斗七星中的天枢、天璇、天机、天权、玉衡、开阳、摇光七颗星，又称“七曜”，再加上开阳星旁之辅星、弼星，即为“九曜”。它们在星卜中的名称依次为贪狼星（属木）、巨门星（属土）、禄存星（属土）、文曲星（属水）、廉贞星（属火）、武曲星（属金）、破军星（属金）。在相宅风水中命名为“生气”（吉星）、

“天医”（吉星）、“祸害”（凶星）、“六煞”（凶星）、“五鬼”（凶星）、“延年”（吉星）、“绝命”（凶星）。辅、弼二星在五行中属木，为中吉之唾，在宅位中居座山之位，称为“伏位”。

根据《阳宅爱众篇》的说法拟表如下（表3–1）：

表3–1

九星名称	别名	阴阳属性	五行	吉凶
生气	贪狼	阳	木	上吉
天医	巨门	阳	土	次吉
延年	武曲	阳	金	次吉
绝命	破军	阴	金	大凶
五鬼	廉贞	阴	火	大凶
六煞	文曲	阴	水	次凶
祸害	禄存	阴	土	次凶
左辅	左辅	阴	木	小吉
右弼	右弼	不定	不定	不定

大游年的具体做法如下：

（1）将宅全按九宫格划分，并按后天八卦标出方位。

（2）确定主房的坐宫卦位，称为“伏位”(辅、弼二星所占之位即为伏位)，然后由伏位，循八卦方位顺序沿顺时针方向按照大游年歌诀顺布辅、弼二星之外的其他七星位置，使之与其他宫卦相对应。关于伏位，再一种方法是定门的位置，以宅门之坐宫卦位为伏位，然后依照大游年歌诀分布七星宫卦。所排相同。

（3）根据九星临宫情况，按星的吉凶及阴阳五行生克关系判定宅中各宫位的吉凶。

（4）依吉凶程度确定宅舍形势、各房高卑大小及层数等。一般吉位应建高大房屋，凶位建低小房屋。

（5）依吉凶确定房屋功能。吉地宜做主房、厅堂、宅门，灶亦应处吉位。凶位可安排仓房、厕所等。

大游年歌诀共八句，每句八个字，第一个字为伏位卦名，其余七个字为七星的星名缩写，八句歌诀按卦序组成。兹将《阳宅十书》所载大游年歌诀转录如下：

乾六天五祸绝延生
坎五天生延绝祸六
艮六绝祸生延天五
震延生祸绝五天六
巽天五六祸生绝延
离六五绝延祸生天
坤天延绝生祸五六
兑生祸延绝六五天

以大游年歌顺布七星，可反映出伏位的宫卦与其他七个宫卦因八卦数理关系和阴阳五行生克关系而产生的七种不同的吉凶结果，这样便演化出八种宅型（图3–3.4）。

北京四合院的方位多为坐北朝南，即所谓“坐坎朝离”。如以北房为伏位，则北房、南房、东房、东南方向均为吉位。如以东南角大门为伏位，所排的结果与此相同。联系到形法对四合院的判别可以看到，不论形法还是理法，都认为坐北朝南的院落是最理想的格局（图3–3.5巽门坎宅的九星分布和宅院布局）。

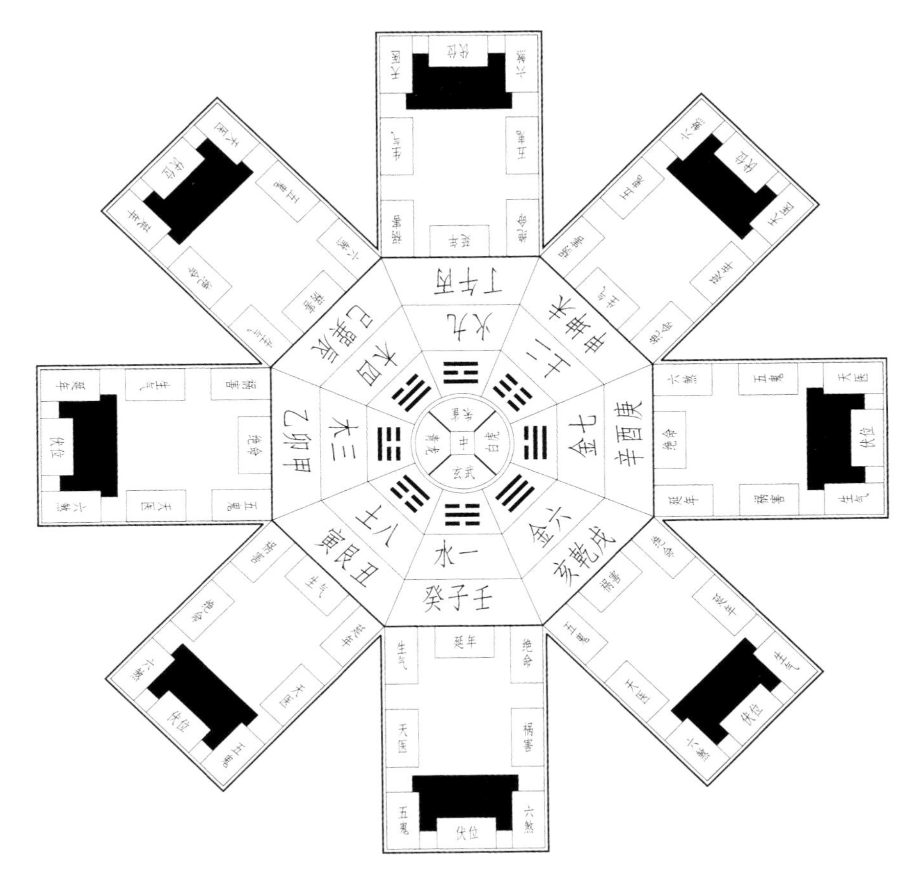

图 3—3.4 按大游年歌诀演化出的八种宅位示意

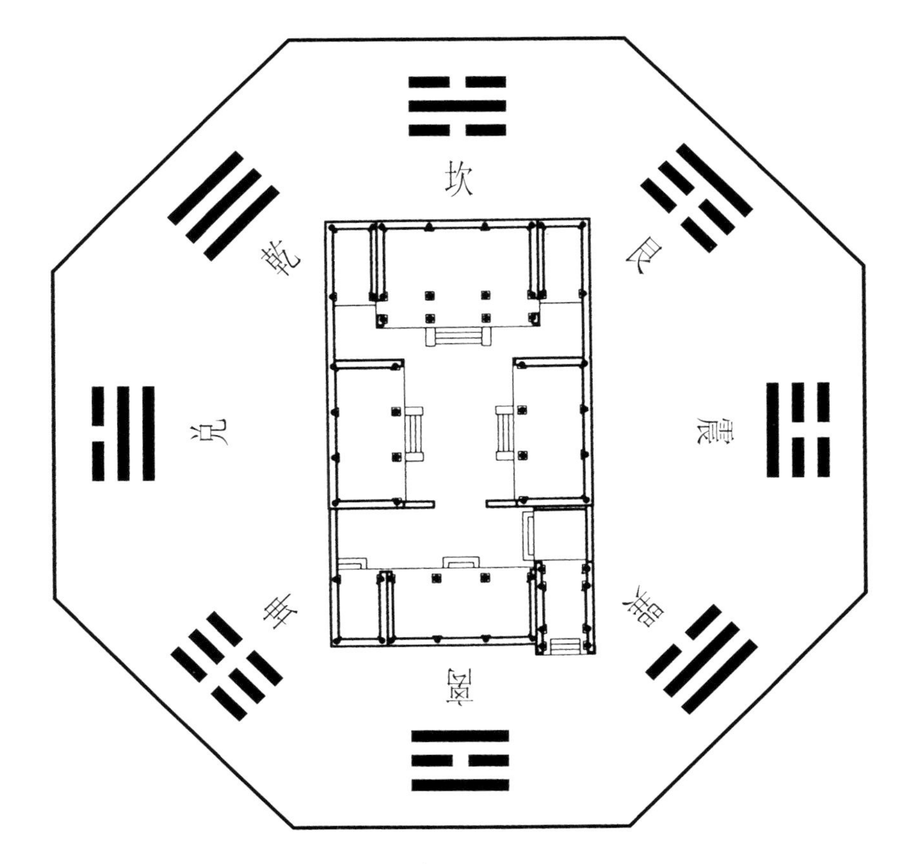

图 3—3.5 巽门坎宅的九星分布和宅院布局

关于风水还有一点要说及的，就是关于门光尺的应用。

门光尺，又名“门尺”“鲁班尺”“八字尺”，是用来确定门、窗、床等器物尺寸、判定吉凶的专用尺。门光尺长度为1.44营造尺（1营造尺=320mm），折合公制460.8mm，宽55mm，厚13mm，正反面、侧面均标有刻度和文字（图3–4）。

门光尺一尺分为八寸(每寸合营造尺1.8寸，折合公制为57.6mm)。分别标有财、病、离、义、官、劫、害、吉，或者标有贵人、疾病、离别、义顺、官禄、劫盗、伤害、福本。每一寸又分为五小格(可称分)，每格合公制15.2mm。门光尺的另外一面对应一至八寸标写星宿名，依次为贪狼星、巨门星、禄存星、文曲星、廉贞星、武曲星、破军星、辅弼星。星名的上方居中标注有一白木、二黑土、三碧土、四绿水、五黄火、六白金、七赤金、八白木。这之中的三个字，表示三个内容：“一”是自然数序，同时含有九宫的“一宫”之意，表示方位为北，在八卦中居坎位；“白”是九宫飞白的颜色，是用色彩表示吉凶的又一种风水内容；第三字“木”是贪狼星的五行属性。余同此。星名两侧是有关吉凶内容的判词。

从门光尺上所标注的吉凶尺寸看，八寸中的第一、四、五、八为吉寸，二、三、六、七为凶寸。确定门的尺寸时，只要将尺寸定在吉寸的范围内便可。如定财门（又称富贵门），则一尺零一分至五分、二尺零一分至五分、三尺零一分至五分……均为财门；如定“义顺门”，则一尺三寸一分至五分、二尺三寸一分至五分、三尺三寸一分至五分……均为“义顺门”。门的四种吉尺寸中，“义顺”“官禄”二种吉门尺寸一般用于官邸、衙署、寺庙等建筑，平民百姓多用“财门”“吉（福本）门”。

对大门、隔门而言，吉尺寸是指门口的里皮尺寸。如街门门口的净高、净宽，都应合于吉尺寸。如果是对开的隔扇门，则指两扇门对开的宽度和高度。如隔扇外有帘架时，则帘架里皮尺寸应为吉尺寸（下面自下槛上皮，上面至帘架横陂下皮）。如帘架内有风门，则风门的尺寸应为吉尺寸。

这里还有一点要说明的，就是关于门光尺的准确性问题。由于风水书所载及工匠间所传的门光尺多为辗转传抄的复制品，其中错处不少。本章所举门光尺二例，就有不少明显的错误。因苦于无处寻找“正根”，对这些错误无法更正，只能以讹传讹。对于这个问题，对风水研究有素的何俊寿先生曾列一表，名为“门光尺、营造尺与公制尺换算表”，表中，对吉凶尺寸都做了标注，这就为我们使用门光尺排尺寸提供了极大的方便，从而避免了因门光尺图中的错误导致实际中的吉凶错位。笔者特将此表转录如下，以供读者参照使用（表3–2）。

研究传统住宅，必须要涉及风水。但本章所谈的风水问题，仅是一些皮毛，而且都是前人研究的成果。由于本人水平、章节内容及书稿篇幅所限，不可能谈得更多更深。

风水学说的内容，都是历史的积淀，其中有许多科学合理的成分，也有不少源于巫术的迷信内容。但在人类社会的发展中巫术和迷信都是不可避免的。英国著名人类学家马林诺夫斯基曾精辟地指出：“无论有多少知识和科学能帮助人满足他的需要，它们总是有限度的。”“人事中有一片广大的领域，非科学所能用武之地。它不能消除疾病和腐朽，它不能抵抗死亡，它不能有效地增加人和环境间的和谐，它更不能确立人和人之间的良好关系。这领域永久是在科学支配之外，它是属于宗教的范围。”可见，巫术是在科学支配之外，人们与自然及命运抗争的一种手段。人们通过开斜道、安假门、放镇物等迷信手段，可以增强抵抗灾难的信心，获得一种必需且有益的生理功效，达到类似“心理暗示”的效果，这是巫术和迷信能够长期存在的社会根源。历史地、客观地看待风水学中一些迷信成分，是我们应取的正确态度。

图 3—4 门光尺图二例

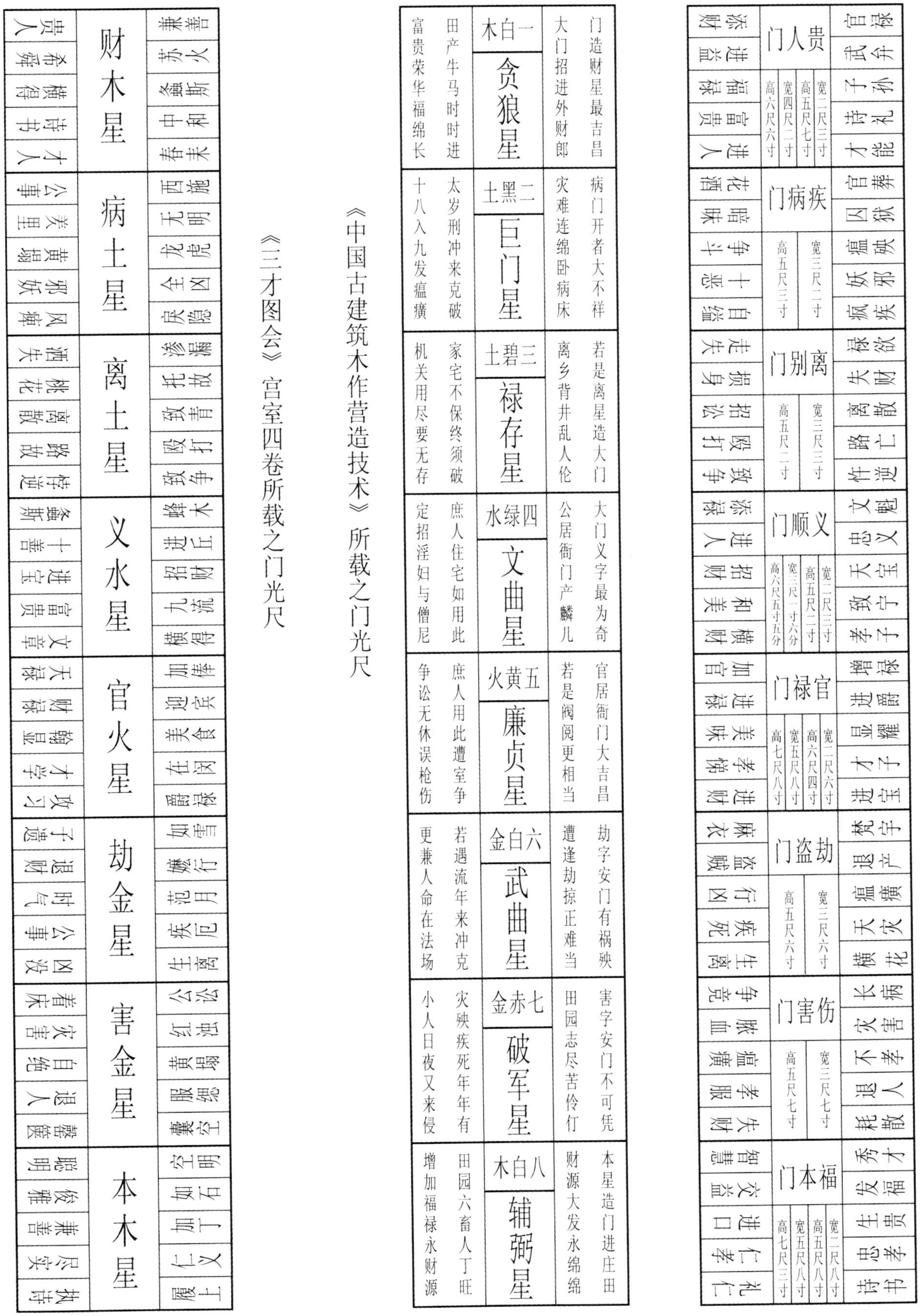

《三才图会》宫室四卷所载之门光尺

星	上	下
财木星	兼善、苏火、螽斯、中和、春来	贵人、希舜、横得、诗书、才人
病土星	西施、无明、龙虎、全凶、宗隐	公事、差田、黄煽、邪妖、风瘫
离土星	渗漏、托故、致青、殴打、致争	流失、桃花、离散、路故、忤逆
义水星	蜂木、进丘、招财、九流、横得	螽斯、十善、进宝、富贵、文章
官火星	加俸、迎宾、美食、在闵、爵禄	天禄、财禄、翰显、才学、攻习
劫金星	如雪、媵行、范月、疾厄、生离	子遗、退财、时气、公事、凶殁
害金星	公讼、红浊、黄煽、服缌、囊空	看床、灾害、自绝、退人、蹉跎
本木星	空明、如石、加丁、仁义、履上	聪明、俊雅、兼善、尽实、执诗

《中国古建筑木作营造技术》所载之门光尺

星	诗
一白木 贪狼星	门造财星最吉昌，大门招进外财郎。田产牛马时时进，富贵荣华福绵长。
二黑土 巨门星	病门开者大不祥，灾难连绵卧病床。太岁刑冲来克破，十八入九发瘟癀。
三碧土 禄存星	若是离星造大门，离乡背井乱人伦。家宅不保终须破，机关用尽要无存。
四绿水 文曲星	大门义字最为奇，公居衙门产麟儿。庶人住宅如用此，定招淫妇与僧尼。
五黄火 廉贞星	官居衙门大吉昌，若是阀阅更相当。庶人用此遭室争，争讼无休误枪伤。
六白金 武曲星	劫字安门有祸殃，遭逢劫掠正难当。若遇流年来冲克，更兼人命在法场。
七赤金 破军星	害字安门不可凭，田园志尽苦伶仃。灾殃疾死年年有，小人日夜又来侵。
八白木 辅弼星	本星造门进庄田，财源大发永绵绵。田园六畜人丁旺，增加福禄永财源。

门	上	尺寸	下
贵人门	官禄、武弁、子孙、诗礼、才能	宽二尺三寸、高五尺七寸、宽四尺二寸、高六尺六寸	添财、进益、福禄、富贵、进人
疾病门	官葬、囚狱、瘟殃、妖邪、疯疾	宽三尺二寸、高五尺三寸	花酒、暗昧、争斗、十恶、自缢
离别门	禄欲、失财、离散、路亡、忤逆	宽三尺三寸、高五尺二寸	走失、损身、招讼、殴打、致争
顺义门	文魁、忠义、天宝、致宁、孝子	宽二尺三寸、高五尺二寸、宽三尺一寸六分、高六尺五寸五分	添禄、进人、招财、和美、横财
官禄门	增禄、进爵、显耀、才子、进宝	宽二尺六寸、高六尺四寸、宽五尺八寸、高七尺八寸	加官、进禄、美味、孝悌、进财
劫盗门	梵宇、退产、瘟癀、天灾、横花	宽三尺六寸、高五尺六寸	麻衣、盗贼、行凶、疾死、生离
伤害门	长病、灾害、不孝、退人、耗散	宽三尺七寸、高五尺七寸	争竞、脓血、瘟癀、孝服、失财
福本门	秀才、发福、生贵、忠孝、诗书	宽二尺八寸、高五尺八寸、宽五尺八寸、高七尺三寸	智慧、交益、进口、仁孝、礼仁

表3－2　门光尺(鲁班尺)、营造尺与公制尺换算表

单位：门光尺　尺
营造尺　寸
公制尺　mm

门光尺	营造尺	公制尺	门光尺	营造尺	公制尺	门光尺	营造尺	公制尺	门光尺	营造尺	公制尺	门光尺	营造尺	公制尺	门光尺	营造尺	公制尺
○ 0.1	1.8	57.6	○ 3.1	45.0	1440.0	○ 6.1	88.2	2822.4	○ 9.1	131.4	4204.8	○ 12.1	174.6	5587.2	○ 15.1	217.8	6969.0
0.2	3.6	115.2	3.2	46.8	1497.6	6.2	90.0	2880.0	9.2	133.8	4262.4	12.2	176.4	5644.8	15.2	219.6	7027.2
0.3	5.4	172.8	3.3	48.6	1555.2	6.3	91.6	2937.6	9.3	135.6	4320.0	12.3	178.2	5702.4	15.3	221.4	7084.8
○ 0.4	7.2	230.4	3.4	50.4	1612.8	6.4	93.6	2995.2	○ 9.4	136.8	4377.6	○ 12.4	180.0	5760.0	○ 15.4	223.2	7142.4
○ 0.5	9.0	288.0	○ 3.5	52.2	1670.4	○ 6.5	95.4	3052.8	○ 9.5	138.6	4435.2	○ 12.5	181.8	5817.6	○ 15.5	225.0	7200.0
0.6	10.8	345.6	3.6	54.6	1728.0	○ 6.6	97.2	3110.4	9.6	140.4	4492.6	12.6	183.6	5857.2	15.6	226.8	7257.6
0.7	12.6	403.2	3.7	55.8	1785.6	6.7	99.0	3168.0	○ 9.7	142.2	4550.4	12.7	185.4	5932.6	15.7	228.6	7315.2
○ 0.8	14.4	460.8	○ 3.8	57.6	1843.2	○ 6.8	100.8	3225.6	○ 9.8	144.0	4608.0	○ 12.8	187.2	5990.4	○ 15.8	230.4	7372.8
○ 1.1	16.2	518.4	○ 4.1	59.4	1900.8	○ 7.1	102.6	8283.2	○ 10.1	145.5	4665.6	○ 13.1	189.0	6048.0	○ 16.1	232.2	7430.4
1.2	18.0	576.0	4.2	61.2	1958.4	7.2	104.4	3340.8	10.2	147.6	4723.2	13.2	190.8	6105.4	16.2	234.0	7488.0
1.3	19.6	633.6	4.3	63.0	2016.0	7.3	106.2	3398.4	10.3	149.4	4780.8	13.3	192.6	6163.2	16.3	235.8	7545.6
○ 1.4	21.6	691.2	○ 4.4	64.6	2073.6	○ 7.4	108.0	3456.0	○ 10.4	151.2	4838.4	○ 13.4	194.4	6220.6	○ 16.4	237.6	7603.2
○ 1.5	23.4	748.8	○ 4.5	66.6	2131.2	○ 7.5	109.6	3513.6	○ 10.5	153.0	4896.0	○ 13.5	196.2	6278.4	○ 16.5	239.4	7660.8
1.6	25.2	806.4	4.6	68.4	2188.8	7.6	111.6	3571.2	○ 10.6	154.8	4953.6	13.6	198.0	6336.0	16.6	241.2	7718.4
1.7	27.0	864.0	4.7	70.2	2246.4	7.7	113.4	3628.8	10.7	156.6	5011.2	13.7	199.8	6369.6	16.7	243.0	7776.0
○ 1.8	28.8	921.6	○ 4.8	72.0	2304.0	○ 7.6	115.2	3685.4	○ 10.8	158.4	5068.8	○ 13.8	201.6	6451.2	○ 16.6	244.6	7833.6
○ 2.1	30.6	979.2	○ 5.1	73.8	2361.6	○ 8.1	117.0	3744.0	○ 11.1	160.2	5126.4	○ 14.1	203.4	6508.8	○ 17.1	246.6	7891.2
2.2	32.4	1036.0	5.2	75.6	2419.2	8.2	118.8	3801.6	11.2	162.0	5184.0	14.2	205.2	6566.4	17.2	248.4	7948.8
2.3	34.2	1094.4	○ 5.3	77.4	2476.8	8.3	120.6	3859.2	11.3	163.6	5241.6	○ 14.3	207.0	6624.0	17.3	250.2	8006.4
○ 2.4	36.0	1152.0	○ 5.4	79.2	2534.4	○ 8.4	122.4	3916.8	○ 11.4	165.6	5299.2	○ 14.4	208.8	6681.6	○ 17.4	252.0	8064.0
○ 2.5	37.8	1290.6	○ 5.5	81.0	2592.0	○ 8.5	124.2	3974.4	○ 11.5	167.4	5356.8	○ 14.5	210.6	6379.2	○ 17.5	253.8	8121.6
2.6	39.6	1267.2	5.6	82.8	2649.6	8.6	126.0	4032.0	11.6	169.2	5414.4	14.6	212.4	6796.8	17.6	255.6	8179.2
2.7	41.4	1324.8	5.7	84.6	2707.2	8.7	127.8	4089.6	11.7	171.0	5472.0	14.7	214.2	6854.4	17.7	257.4	8236.8
○ 2.8	43.2	1382.4	○ 5.8	86.4	2764.8	○ 8.8	129.6	4147.2	○ 11.8	172.8	5529.6	○ 14.8	216.6	6912.0	○ 17.8	259.2	8294.4

本表转引自何俊寿《门光尺析证》(原载《古建园林技术》总44期)

第四章
北京四合院的空间关系

北京四合院经历千百年的发展演变，形成了比较固定的、程式化的空间模式。关于这一点，我们通过对其基本格局的分析，已经有了初步了解。

建筑的空间形式主要是由功能决定的。作为中国人世代居住的四合院，它在空间形式和功能之间存在着一种什么联系呢？要弄清这一点，我们须先分析四合院的功能。

一、四合院的功能

根据中国人传统的生活方式、风俗习惯、纲常伦理和道德准则，对四合院的功能要求应有以下几个方面。

1. 居住功能

居住功能是四合院的主要功能。传统的四合院是由一家一户单独居住的。在尊卑有别、长幼有序的封建社会，家庭成员及其居住的房子是有严格的秩序的。宅院中最大最好的正房，要由家庭中年纪最长、权威最高的家长居住。家长居室的明间称为堂屋，它既是家长的起居之所，也是全家人聚集的中心。家庭中的子女、儿孙，分别住在厢房和其他配属房中。未出阁的香闺女子由女仆陪伴住在后罩房或后罩楼内。家中的男仆通常是不能进入内宅的，只能住在外院。在封建社会，这种居住方面的严格秩序是任何人都不能打破的。

2. 读书学习的功能

重视教育是中国人的优良传统，读书学习是人生的重要内容。过去古人读书，是将教书先生请到家里，在家中设私塾。家中的适龄儿童均可入塾就读。在启蒙教育阶段，由先生统一教读；到深造阶段，学生则在各自的书房中独自攻读，即所谓“十年寒窗”。即便是功成名就的成年人，也需要经常读书。读书需要有安恬宁静的环境。在带有私家园林的宅院中，书房多设在花园之中，北京某宅以“西园翰墨”命名的私园，就是以读书为主要功能的幽雅小园。没有私园的宅院一般都将书房设在相对安静的地方。

3. 社会交往的功能

社会交往是人生不可缺少的内容。古代的公共社交场所不多，社交活动多以朋友之间登门造访的形式在家中进行，因此，每家都设有客厅。客厅一般设在外院的倒座房（南房）内。大型住宅中，也有将客厅设在中路的实例。客至后，由仆人将宾客让至客厅内等候，或主人在客厅内恭候宾客的到来。一般的人家多将客厅和书房合为一体，一室二用。有客时待客，无客时读书。

4. 团拜、请安、开会、议事的功能

中国封建社会的豪门大户，家庭人口一般都很多，三世同堂、四世同堂乃至五六世同堂者不乏其例，人口一般二三十人、五六十人，多者可达百人以上（除主人外，还有仆人、奴婢）。家庭（族）中遇有重大事情，需由家长（或族长）召开会议，或商议家事，或褒扬孝悌、惩治忤逆、行使家规。进行这种活动的场所，或在家长的堂屋，或在议事的厅堂。这种厅堂一般设在中轴线上，平时可做接待之用（参见图2–10.2、2–10.3）。

5. 祭神、祭祖的功能

族权和神权，是维系封建家族统治的主要权力支柱。为体现神权、族权，在大型住宅中一般都有专设的祠堂，里面供奉祖宗牌位及佛像神龛，设有跪拜祭祖的场所。这种设族长、建宗祠的制度，在我国南方尤其盛行。有些地区将宗祠建得高大庄严，成为宅院中最重要的建筑。

6. 维持纲常伦理和家族秩序的功能

作为封建社会组成细胞的封建家庭，在维护封建的纲常伦理及社会秩序方面具有特殊作用。而作为封建家庭载体的四合院住宅，自然也就成了维系这个制度的物质因素。传统的北京四合院多采用对称格局，具有明显突出的中轴线。这种严整对称的格局，营造出一种庄严肃穆的气氛，使中轴线上高大的正房具有压倒一切的威严感，使其他房间退位到附属、仆从地位，以烘托居住在正房内的家长或族长的无上权威。这种由空间造成的精神上的威慑感，正是封建制度和秩序所需要的。

7. 体现风水学说的功能

风水学说对中国的传统建筑有着广泛深刻的影响，这一点在前文已经叙及。按照风水学说，处于“坐坎朝离”方位的四合院，其北房、南房、东南和东房均属吉位，可建高大房屋。四合院以北为尊的严谨对称的格局，与风水要求正好相合。建筑合乎风水学说，可给人一种精神上的安定感，可满足人们心理上祈福消灾的要求，获得必需而有益的功效。

以上是我们对传统四合院建筑在功能方面的简略分析。结合这些内容，我们再来分析四合院建筑的空间模式与功能之间的内在联系。

二、四合院建筑的空间关系

传统四合院建筑之间的空间关系，可以用两句话来概括，即“出入躲闪”“高低错落”。其中，“出入躲闪”是讲平面关系，“高低错落”是讲立面关系。

1. 建筑平面中的出入躲闪关系

（1）*正房与厢房的关系*　正房是宅院中的主房，是家长居住的地方，在风水说中，居于吉位，在朝向上坐北朝南，是最佳位置。这就要求正房的体量最大，是全院建筑的中心。而处于院落东西两侧的厢房，无论位置、朝向、居住人的地位都处在次要和陪衬位置。因此，它们的开间、进深尺寸必须要小于正房。在院落宽度允许的条件下，厢房的前檐还要尽量躲闪到正房山墙轴线之外，以确保露出正房的前脸，使正房处在足够显要、突出的位置。此谓：厢房躲闪，突出正房。

（2）*正房与耳房的关系*　耳房是正房的陪衬。从平面看，其前檐比正房大大向后收缩。如果正房两侧有抄手游廊的话，那么耳房还要退至抄手游廊之后，并与抄手游廊之间留出空隙以利采光、排水。耳房不仅在开间、进深上小于正房，而且也应小于厢房。此谓：耳房退够，突出正房。

（3）*厢房与厢耳房的关系*　稍加讲究的四合院，在厢房南侧还要设置厢耳房一至二间。厢耳房是厢房的陪衬，它的尺寸不仅小于厢房，而且还要小于正房的耳房，甚至在建筑形式上有时也只能采取平台形式，称为“盝顶”。设置厢耳房，主要是为了利用抄手游廊与厢房南侧相接后余出的空间。由于厢耳房的尺度很小，它只能用做厨房、车房或女仆休息用房。

（4）垂花门与抄手游廊的关系　垂花门是宅院的二门，是沟通内外宅的通道，又是东西两侧抄手游廊的交汇之点。如果只考虑通道的功能，那么垂花门在平面、立面上的尺度完全可以与抄手游廊相同。但垂花门处在院落的中轴线上，位置显赫，在风水说中也占在吉祥位置，而且又是作为二门存在，所以，它的开间、进深尺寸都要大于两侧的游廊，使它处在一个非常突出的地位，相对于垂花门的游廊则处在明显的陪衬地位。此谓：游廊为衬，突出二门。

（5）宅门与倒座的关系　一般宅院的宅门位于东南角的巽位，与倒座房串联相接，是倒座房的一部分。但宅门的进深往往大于倒座房，处在一个比较突出的位置。这种关系也是由功能决定的。中国人历来对宅门十分重视，它不仅是宅人出入的要塞，也是主人社会地位、经济地位的重要标志。在风水说中，它位于生方，处在吉位，这些都要求它外形显要，位置突出。

从以上诸种关系可以看出，四合院中正房和厢房之间，是以正房为中心，厢房要躲开正房，闪在两旁；正房与耳房之间，耳房是正房的陪衬，它在平面上要退足尺寸，将空间让给正房以及它的附属物抄手游廊；厢耳房作为厢房的陪衬，尺度更小，而且要让开游廊。它们基本是按照正房→厢房→正房耳房→厢房耳房这样的顺序来确定房间的地位和尺寸的。在四合院的外宅，则要突出宅门和垂花门的地位，宅门要得到充分的强调，倒座应处在衬托的位置；垂花门也应得到充分的强调，游廊要处于陪衬的位置（图4–1）。

2. 建筑立面中的高低错落关系

凡是在平面中得到突出和强调的建筑，在立面中也应得到相应的突出和强调。这种突出和强调，一般是通过尺度的增减来实现的，主要有以下几种措施。

（1）提高台明高度以突出主要建筑　中国传统建筑都是建在台基上面的，住宅建筑也是如此。台基露出地面部分称为“台明”。在一进或二进院落的四合院中，正房的台明是最高的，它要比厢房高出一步台阶（一步台阶高度为4寸，约合130mm）。垂花门、大门的台明可与正房相同。厢房、抄手游廊及倒座房的台明比正房低一步台阶。这种提高台基的方法，可使正房、宅门、垂花门在立面上得到一定程度的强调。

（2）加大柱高尺寸以突出主要建筑　加大柱子的高度，也是突出主要建筑的手段之一。在一般情况下，正房的面宽大于厢房，按权衡比例折算后，正房的柱高也会大于厢房。但有时为了突出正房，还要另外再加大正房的柱高，使之达到恰当的高度。与正房相比，厢房的柱高就要矮得多。正房两侧的耳房，其柱高不仅要低于正房还应低于厢房。同理，作为厢房陪衬的厢耳房，柱高就更矮一些。至于增减尺寸的多少，应视院落空间的宽窄、房屋开间的大小等客观因素，根据环境要求而定，不宜一概而论，要以满足功能和视觉感受要求为原则。

倒座房的柱高可与厢房相等或相似。宅门柱高尺寸应大于倒座房。垂花门柱高应远远大于游廊，可略同于厢房。

图4—1　四合院平面中的出入躲闪关系

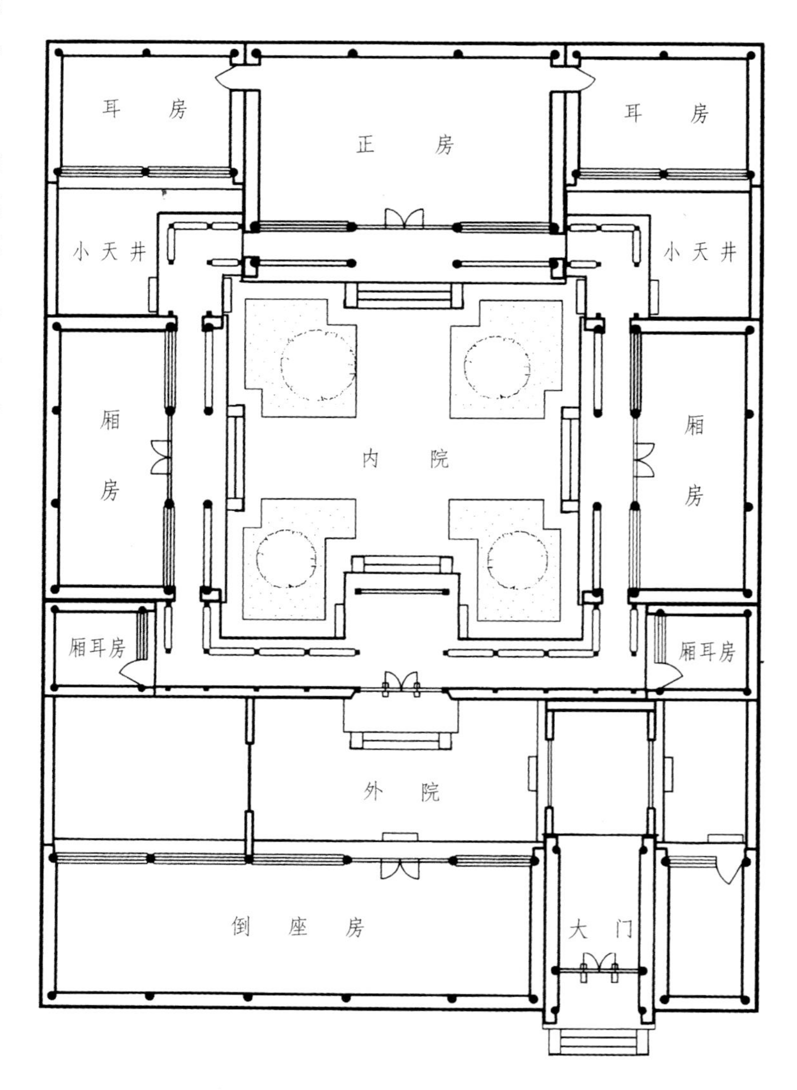

（3）进深尺寸对屋顶高度的影响　四合院是坡屋顶建筑。在坡度相同的情况下，房屋的进深越大，屋顶的高度就越高。在四合院建筑中，正房的进深最大，因此，正房屋顶的高度也就最高。宅门的进深大于倒座房，垂花门的进深大于抄手游廊，也使它们的屋顶尺度相应增高。

（4）举折的因素　了解中国传统建筑的人都知道，屋顶各部分的不同举折，构成了屋面优美的曲线。屋顶举折与房屋进深的大小以及步架的多少有直接关系。如一幢进深4m的房子，进深方向分四步的话，可采用五举、七举；如一幢进深6m的房子，进深方向分六步，则可采用五举、七举、九举（或五举、六五举、八举）。显然，后者屋面比前者要高大陡峻。如果进深再大，步架再多，屋顶就会更陡。这也是造成正房高大的因素之一（图4–2）。

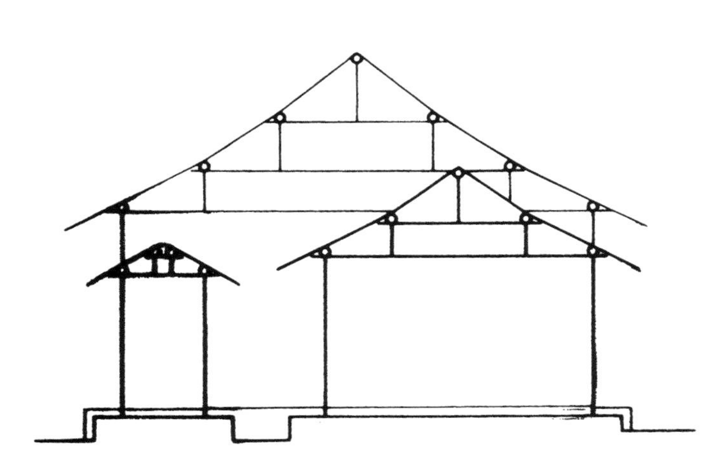

图4—2　台明、柱高、进深、举折对建筑尺度的影响

综合以上各种因素，我们可以看出，正房在台基、柱高、进深、举折诸方面都占有绝对优势，所以体量最大，最显赫；其他厢房、耳房、厢耳房，都按不同的因素，构成不同的高度和体量；大门高于倒座房，垂花门高于游廊，使得这两道门也得到了充分的突出。在一座宅院中，正房尺度最高，体量最大，形成建筑群体的中心。正房高于（大于）厢房，厢房高于（大于）耳房，宅门高于倒座房，垂花门高于游廊，构成主次分明、尊卑有序的空间格局。无论从平面布局还是从立面尺度的构成去考察，都会使人感受到一种固定的、井然有序的空间组合模式；而这种组合模式又给人一种高低参差、错落有致的优美的韵律感（图4–3）。

图4—3　四合院各建筑的高低错落关系

关于四合院建筑的空间关系，还有一点应说及的，就是院落的构成以及室内外空间的联系与沟通。

四合院是以庭院为中心的内院式住宅，院落是四合院住宅中必不可少的定义性元素，没有院落就不称其为四合院。

北京四合院的院落是沿中轴划分成若干层次的，北京人称之为“几重院落”或“几进院落”。现在我们以三进院落的中型四合院为例，来分析院落的空间构成。

由宅门进入宅院之后，在倒座房和垂花门之间是第一进院落。明清四合院的第一进院落称为外院，又称外宅。它是主人接待宾客、进行社交活动的场所。外院的建筑功能主要有门道、门房（传达室、门卫）、客厅（可兼作书房）、库房、厕所等。院落呈东西狭长状。院落北端在中轴线位置设二门（垂花门）通向内院。第一进院落是宅院的起点，但不是中心。

第二进院落是宅院的中心，它由坐北朝南的正房、东西厢房、与外院相通的垂花门及联系各房的抄手游廊围合而成，开阔而宽敞。第二进院落在宅院中有着十分重要的地位和作用，它是住宅中多功能共享空间，院内各房间的采光、通风、联系、交通、人们的户外活动、儿童的玩耍游戏、冬日晒暖、夏天纳凉、赏花观鱼、弦歌诵乐及至逢红白喜事时招待亲朋都要在内宅的大院里进行。北京四合院这种宽敞的内院不仅补充了室内空间的不足，而且提供了一个家庭露天活动的场所。它给家庭生活带来了诸多方便和舒适，增添了许多情趣和交流。这种露天活动场所的优点，是其他任何形式的住宅都无法相比的。

在第二进院落中，还有两处空间特别值得提及，这就是由耳房、厢房、抄手游廊和院墙围成的两处小院落，称为“小天井”。小天井的面积不大，与中心院落既分隔又联系，但它们位处一隅，安谧幽静，另有一番情趣和功用，是内宅院落的重要组成部分。

第三进院落是由后罩房（或后罩楼）及两侧院墙围合起来的狭长空间。它的南面是正房和耳房，东西两侧一般不设厢房，院落的进深不大，是居住在后罩房内人们的室外活动空间。第三进院落和第二进院落通常由位于东耳房一侧的通道相联系（参见图4—1）。

谈及院落的构成和空间，还有一点要说及的，就是廊子的作用。讲究的四合院，内宅的各房都有廊，这些外廊作为室内外的过渡部分，起着联系、沟通内外空间的作用。廊内通常都安有坐凳，天气晴和之时，宅人可以在廊内安坐小憩、赏花观鱼、戏鸟听蝉、悠然自乐。正房、厢房的外廊由抄手游廊串连起来，并与垂花门相接，在内宅形成一个环形交通系统。这种设施为居住者带来了极大的方便，尤其在雨雪天气或烈日当头时，人们可以通过环廊达于各个房间而避免日晒雨淋。四合院的这种优点，也是其他形式的住宅难以相比的（参见图4—1并见图4—4）。

由正房、厢房、倒座、耳房、罩房、游廊、宅门和院落组成的北京传统的四合院，除去满足一家人的生活功能外，还是维护封建社会礼制的工具。晋人阮籍在《乐论》里说：“尊卑有分，上下有等，谓之礼……车服、旌旗、宫室、饮食、礼之具也。”在封建社会，包括车马、衣服、饮食在内的一切，都是礼制的工具，作为居住用的“宫室”，自然就更是礼制的工具了。四合院住宅建筑的空间关系，无处不体现出封建社会的等级和秩序，这是它的功能的核心所在。由这种功能决定的空间，自然就是主从分明、尊卑有别、严格对称、秩序井然了。

封建制度消亡以后，传统四合院依然存在。但今天的四合院已不再是封建社会的“礼之具”。它留给人们的是安恬舒适的居住环境和令人回味无穷的居住文化。

图4—4　沟通室内外空间的外廊

第五章
四合院建筑及其构造

当我们了解了四合院的历史、格局、风水讲究和空间构成之后，需要进一步探究的便是它那令古今中外无数人为之倾倒的建筑和装饰了。

仅就四合院建筑的型制等级和建筑形式而言，它在中国传统建筑中是最基本、最朴素、最简单的——以硬山式建筑为主的建筑外形，以砖石材料本色构成的青灰色调，简单素雅的油饰和彩绘。这些与豪华的宫殿建筑和精美的园林建筑相比，显得是那样的简朴和平淡。但当这些简单朴素的建筑被赋予艺术的装点、文化的内涵和丰富的民俗内容之后，展现在我们面前的则是一个具有神奇魅力的建筑文化艺术世界。

第一节　宅门、影壁、上马石和拴马桩

一、宅门的等级、种类和构造

宅门是住宅的出入口，是宅院的门面。中国人历来重视宅门的作用。历代统治者都把门堂制度看做是封建等级制度的重要内容，并对各种门堂的建制做出具体而严格的规定，使宅门从建筑的规模、形式、装修色彩、建筑材料的使用等各个方面都划分出森严的等级，从而使宅门成了宅主人社会地位和经济地位的重要标志。这种观念渗透到社会生活领域又派生出“门第”“门阀”“门派”“门户”等各种复杂的等级观念，深刻地影响着人们的生活。浸透在中国传统住宅文化中的风水学说，也格外重视宅门的作用。在确定建筑的吉凶时，首先将宅门定在坐宫卦位，使之处于吉祥位置。

北京四合院住宅的宅门，从建筑形式上可以分为两类：一类是由房屋构成的屋宇式门；另一类是在院墙合拢处建造的墙垣式门。北京四合院的屋宇式门主要有广亮大门、金柱大门、蛮子门和如意门。

1. 广亮大门

广亮大门是仅次于王府大门的宅门，是具有相当品级的官宦人家采用的宅门形式。广亮大门位于宅院的东南角，一般占据倒座房东端第二间的位置。它的进深方向的尺度明显大于倒座房，显得非常突出。广亮大门的台基高于倒座房的台基，柱高也明显高于倒座房，从而使它的屋面在沿街房屋中突兀而起，格外显赫。

广亮大门门庑的木构架一般采用五檩中柱式，平面有六根柱子，分别是前后檐柱和中柱。中柱延伸至屋脊部分直接承托脊檩，并将五架梁切分为双步梁和单步梁。这种做法可以利用短料，节省长材。

广亮大门的门扉设在门庑的中柱之间，由抱框、门框、余塞、走马板、抱鼓石（或门枕石）、板门等组成。门扉居中，使得门前形成较大的空间，使大门显得宽敞而亮堂，这可能就是广亮大门名称的由来。大门的外檐柱间，檐枋之下安装雀替。这一构件既有装饰功能，又能代表大门的规格等级。

广亮大门的装饰也很讲究，门庑山墙墀头的上端，有向外层层挑出的砖檐，称为“盘头”。盘头通常由四层砖料组成，砍磨加工成半混、炉口、枭的形式叠涩挑出，构成优美的曲线。盘头之上的两层砖料，称做“拔檐”，再上是向外斜出的方砖，称为“戗檐”。广亮大门墀头部分的戗檐砖可做得素朴无华，也可以做出精美的雕刻。

抱鼓石和门簪，也是广亮大门着意装饰的部位。抱鼓石是安装在大门的下槛下面的石构件。其门槛以内部分呈方形，上面装有铸铁海窝，做承接门扇之用，称为门枕石。门枕石的外

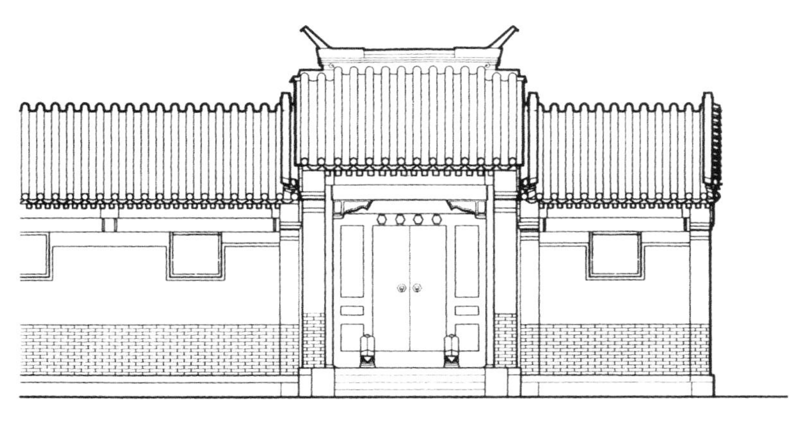

沿街立面图

剖面图

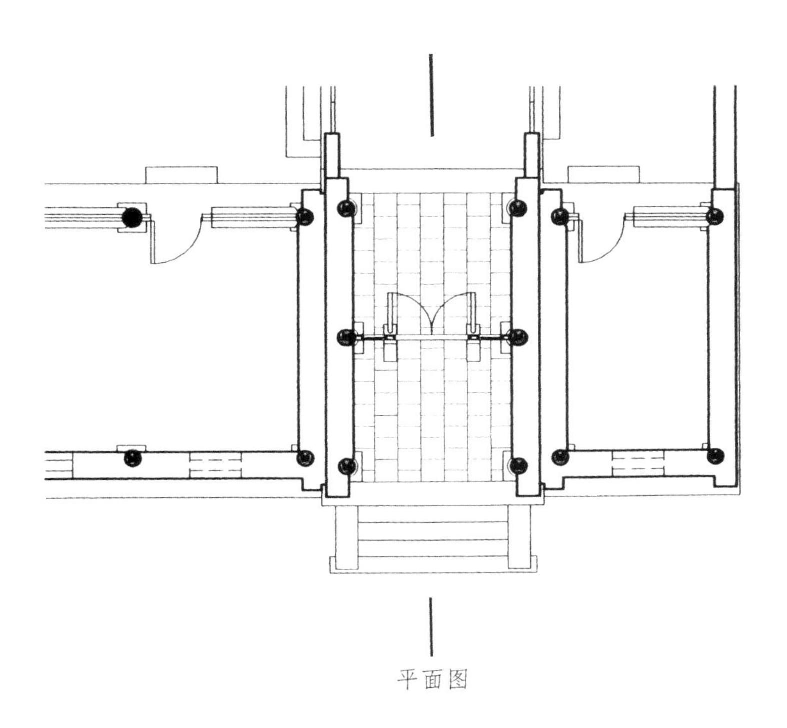

平面图

图 5—1—1　广亮大门平面、立面、剖面图及实例

侧打凿成圆鼓形状，其上镌刻卧狮兽面和其他吉祥图案。门簪是用来锁合中槛和边楹（俗称门龙）的木质构件，因其形状和功能类似旧时妇女固定发髻的簪子，故名门簪。门簪的头部呈六边形，一组四只，在迎面刻“平安吉祥”或“福禄寿喜”等吉辞，也可雕刻牡丹、荷花、菊花、梅花等四季花卉（图5—1—1）。

广亮大门多用于相当品级地位的官宦人家，大门的色彩、装饰受到比较严格的限制，一般不施华丽的彩画，仅做适当的点缀。有的广亮大门在山墙墀头两侧做两块反八字影壁（又称撇山影壁），使大门前面形成一个小广场，更显出广亮大门的气派。这种做法在实际中也很常见（图5—1—2）。

2. 金柱大门

金柱大门是型制上略低于广亮大门的一种宅门，也是具有一定品级的官宦人家采用的宅门形式。金柱大门与广亮大门的区别主要在于，门扉是设在前檐金柱之间，而不是设在中柱之间，并由此而得名。这个位置，比广亮大门的门扉向外推出了一步架（约1.2～1.3m），因而显得门前空间不似广亮大门那样宽绰。金柱大

图 5—1—2　带反八字影壁的广亮大门

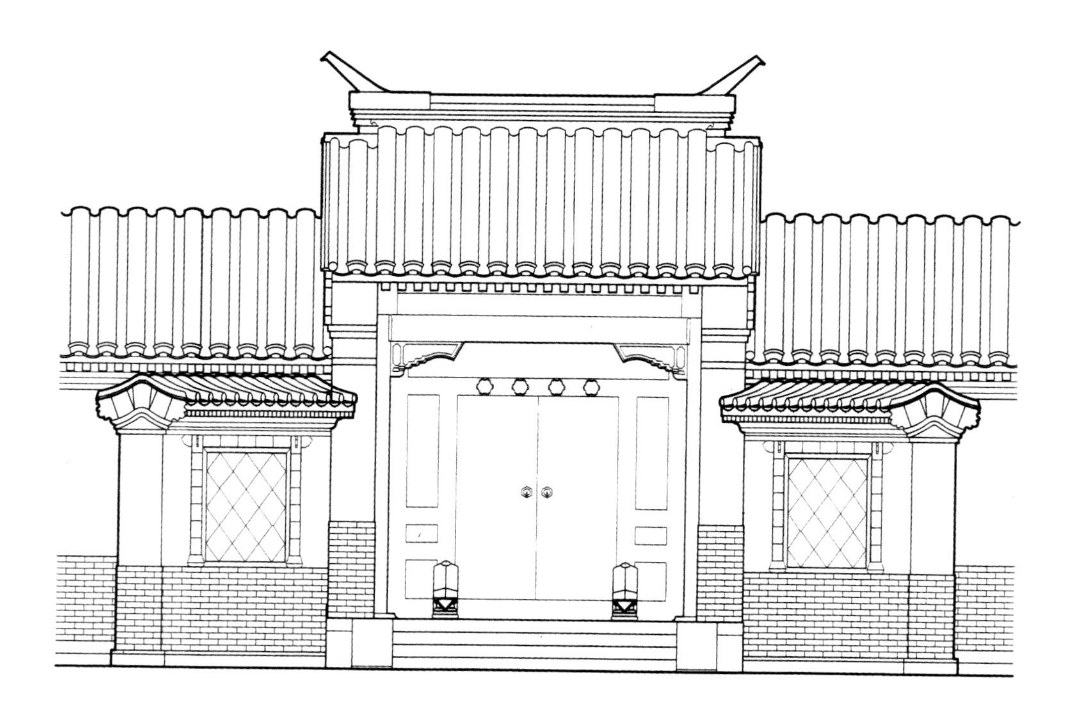

带反八字影壁的广亮大门立面

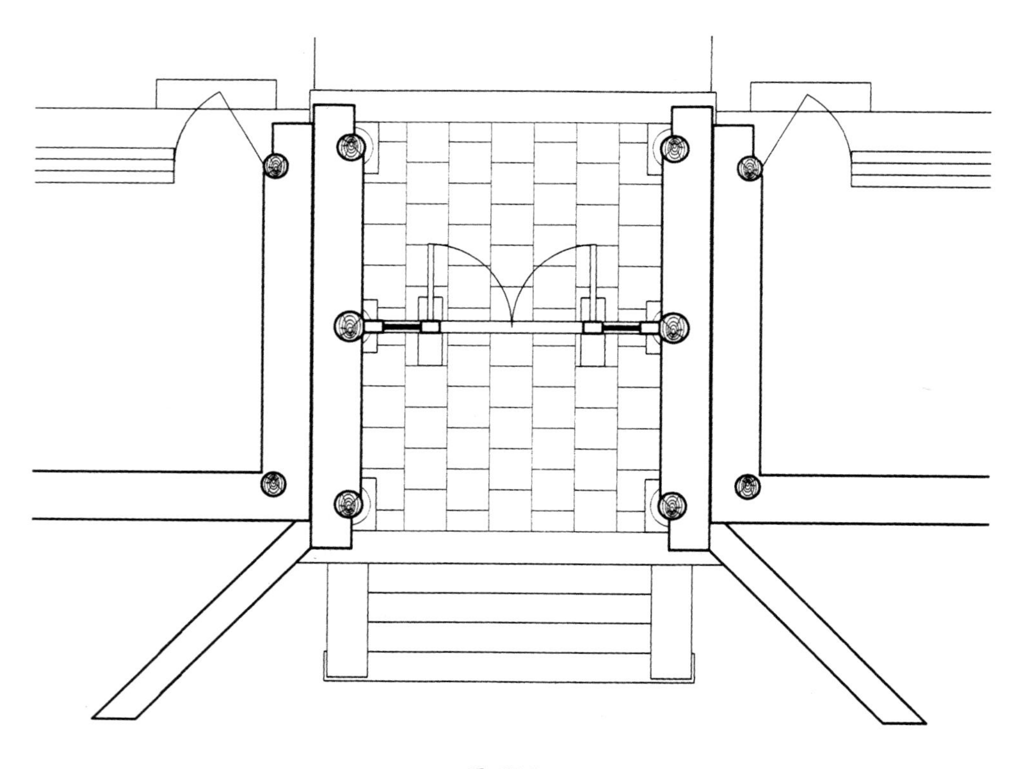

平面图

门的木构架一般采取五檩前出廊式，个别采取七檩前后廊式，平面列三排或四排柱子，即前檐柱、前檐金柱（后檐金柱）、后檐柱。金柱之上承三架梁或五架梁，檐柱、金柱间施穿插枋或抱头梁。

同广亮大门一样，金柱大门外檐檐枋之下也施雀替作为装饰（图5-1-3）。

3. 蛮子门

蛮子门是将槛框、余塞、门扉等安装在前檐檐柱间的一种宅门，门扉外面不留容身的空间。这种宅门从外表看来，不如广亮大门和金柱大门深邃气派。至于它名称的由来，更无确据可考。有一种说法是，到北京经商的南方人为安全起见，特意将门扉安装在最外檐，以避免给贼人提供隐身作案条件，并因此而得名为蛮子门。

蛮子门的型制低于广亮大门和金柱大门，是一般商人富户常用的一种宅门形式，其木构架一般采取五檩硬山式，平面有四根柱，柱头置五架梁。宅门、山墙、墀头、戗檐处做砖雕装饰，门枕抱鼓石或圆或方并无定式（图5—1—4）。

4. 如意门

如意门是北京四合院采用最普遍的一种宅门形式。它的基本做法是在前檐柱间砌墙，在墙上居中留一个尺寸适中的门洞，门洞内安装门框、门槛、门扇以及抱鼓石等构件。如意门洞的左右上角，有两组挑出的砖制构件，砍磨雕凿成如意形状（一称“象鼻枭”）。门口上面的两只门簪迎面也多刻“如意”二字，以求“万事如意”，这大概就是如意门名称的由来。

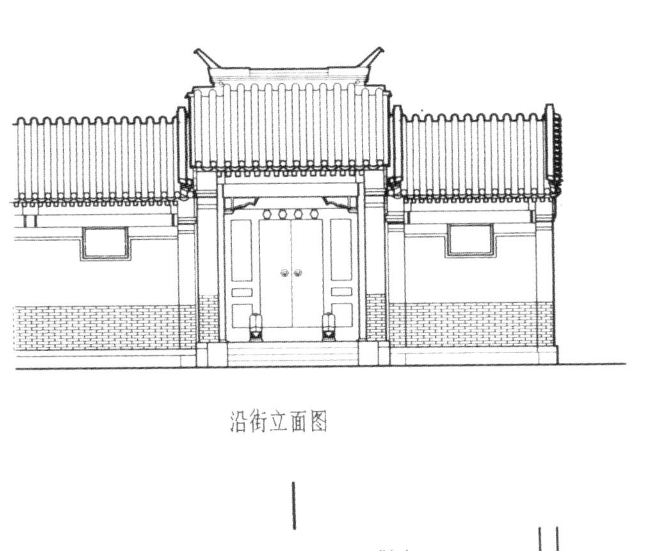

沿街立面图

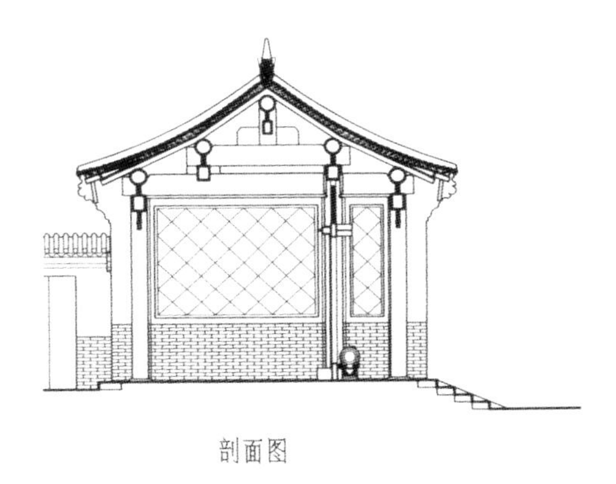

剖面图

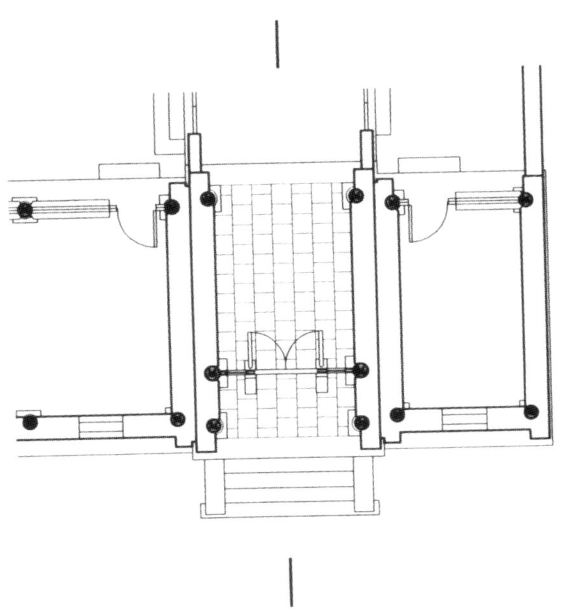

平面图

图5—1—3 金柱大门平面、立面、剖面图及实例

图 5—1—4 蛮子门平面、立面、剖面图及实例

图 5—1—4.1 带反八字影壁的蛮子门

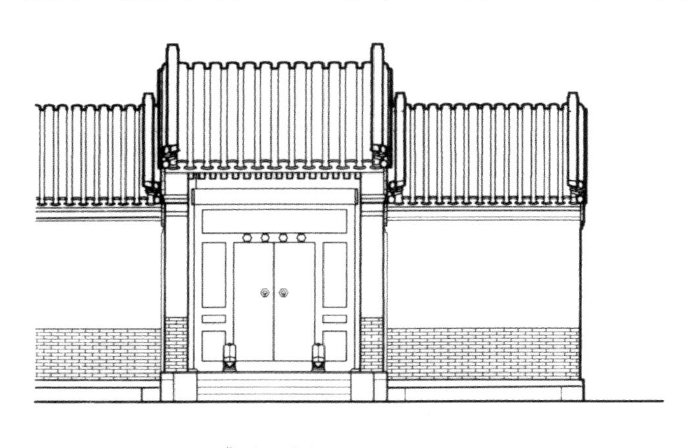

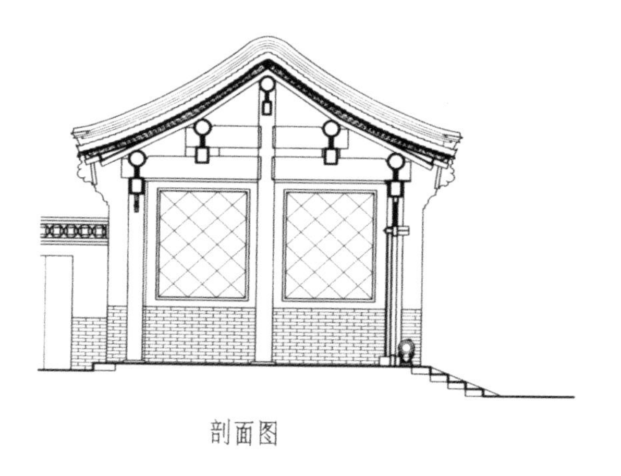

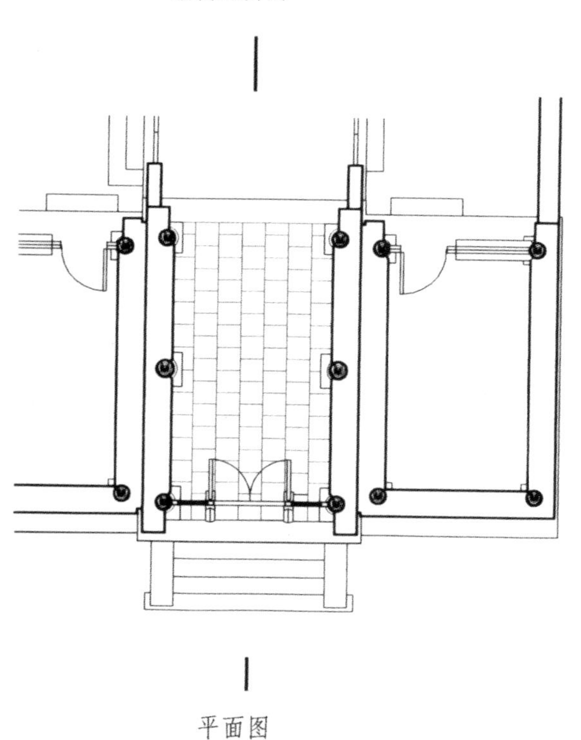

图 5—1—4.2 平面、立面、剖面图

图 5—1—4.3 常见的蛮子门

如意门这种宅门形式，多为一般百姓所采用，其型制虽然不高，但不受等级制度限制，可以着意进行装饰。它既可以雕琢得无比华丽精美，也可以做得十分朴素简洁，一切根据主人的兴趣爱好和财力情况而定。做得讲究的如意门，在门楣上方要做大面积的砖雕，砖雕多采用朝天栏杆形式，它的部位名称由下至上依次为挂落、头层檐、连珠混、半混、盖板、栏板望柱。在这些部位，依主人的喜好或传统装饰内容，分别雕刻花卉、博古、万字锦、菊花锦、竹叶锦、牡丹花、丁字锦、草弯等图案。如果房主的财力不够或偏爱素雅，则可做素活，或只加少许雕饰作为点缀。更简朴者，还可用瓦片摆出各种图案。总之，形式多样，不一而足，充分体现了如意门头装饰的随意性（图5－1－5）。

图 5—1—5 各种档次的如意门举例

图 5—1—5.1 典型的如意门

图 5—1—5.2 较朴素的如意门

图 5—1—5.3 雕刻华丽的如意门

如意门的构架多采用五檩硬山形式，平面有四或六根柱。两根前檐柱被砌在墙内不露明，柱头以上施五架梁或双步梁。如意门区别于其他宅门的地方，是前檐柱间的门墙以及它的构造和装饰。如意门的门口，要结合功能需要和风水要求确定尺寸，一般宽0.9～1m，高约1.9m（指门口里皮净尺寸）。民间有“门宽二尺八，死活一起搭”的说法，是指二尺八的宽度已能满足红白喜事的功能需求。这个尺寸也正好合乎门尺中“财门”的尺度要求。如意门的门楣装饰，无论是采取冰盘檐挂落形式，还是采取其他形式，它上面的砖构件都要接近檐椽下皮，将檐檩挡在里面，不使露明，以突出砖活的完整性（图5—1—6）。

5. 小门楼

小门楼是墙垣式门的一种。这种门比起屋宇式门，等级要低得多。它是三合院或一进院落的四合院等小型住宅所采用的宅门形式。小门楼是纯砖结构，构造比较简单，主要由腿子、门楣、屋顶、脊饰以及门框、门扉等构成，装饰一般都比较简洁朴素，但也有为数不少的豪华式小门楼，在门楣发一遍施砖雕，于小巧中见华丽，亦不失为宅门中的精品（图5—1—7）。

6. 西洋式宅门

西洋式宅门在北京四合院中采用得也很普遍，它是清代中期以后，西方建筑文化传入中国，并与中国传统建筑文化融合事例的结晶。这种门所在位置与其他宅门没有差别，只是采取了西洋式建筑的形式。西洋式宅门的一般构造是单开间，两端为砖柱，砖柱间是砖墙，在砖墙上居中留出大小适中的门洞。这种做法和如意门十分

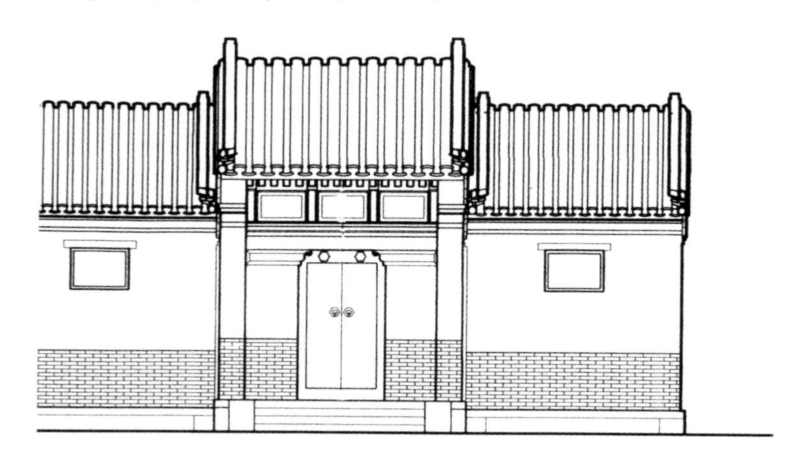

沿街立面图

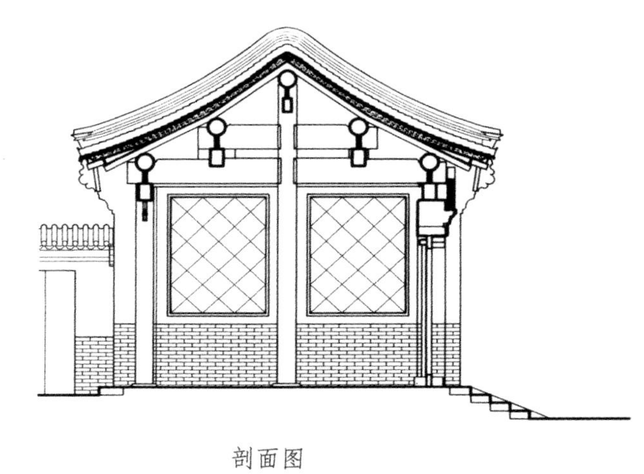

剖面图

平面图

图 5—1—6　如意门平面、立面、剖面图

图5—1—7 小门楼平面、立面图及实例

图5—1—7.2 较典型的小门楼（随墙门）

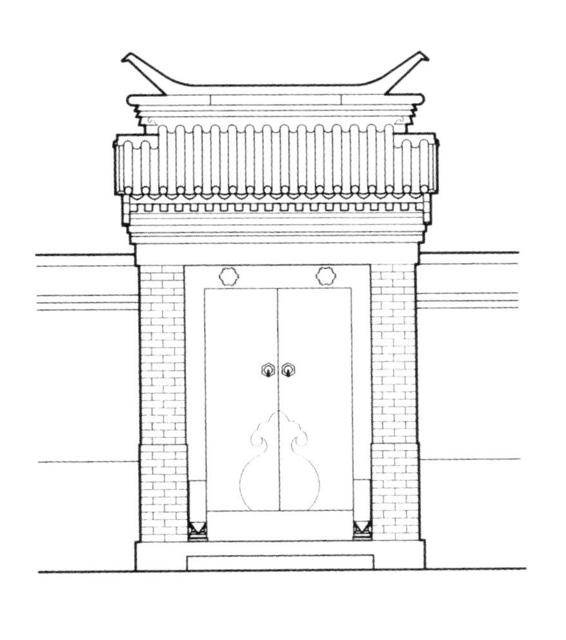

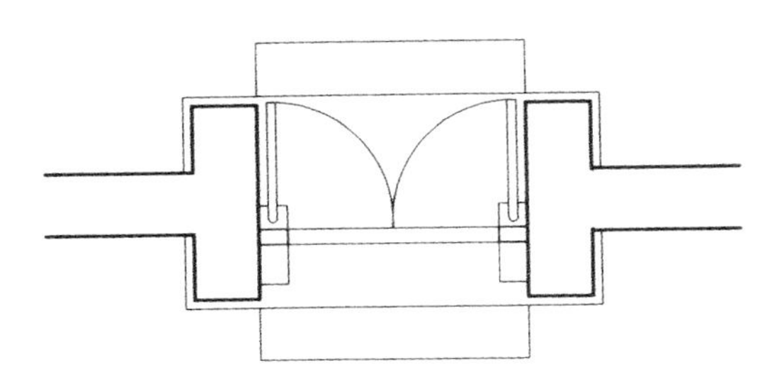

图5—1—7.1 小门楼平面、立面图

图5—1—7.3 常见的随墙门

相似。砖柱上一般有二重或三重冰盘檐向外挑出，将砖柱分为二段或三段。其中，下面两重冰盘檐与柱间砖墙上的冰盘檐贯通一气，形成两道装饰线。装饰线之间为横额。砖柱呈冲天柱式，中间顶墙做成阶梯状或其他形状（图5-1-8）。西洋式宅门的门框、门扉等构造、做法与其他宅门相同，依旧采用中国传统式样。这类“入乡随俗”的西洋式宅门，都不同程度地掺入了中国传统建筑的零件，是中西合璧的“混血儿”。

除去以上六种宅门之外，实例中还有各种不同形式的门，如大车门、栅栏门、半间门、小窄门等等。这些形形色色的门，除有特殊功能者外，都是以上六种宅门的简化或变形，其构造与以上宅门类似，故不再逐一详述（图5−1−9）。

图5—1—8　西洋式宅门举例

图5—1—8.1　普通的西洋门

图5—1—8.2　常见的西洋门

图 5—1—9 其他形式的宅门或栅栏门举例

图 5—1—9.1 做成随墙门形式的屋宇式门

图 5—1—9.2 残破的大车门

二、影壁的种类和构造

影壁是设在四合院宅门内外的装饰墙面。它的作用在于遮挡宅门内外杂乱呆板的墙面或景物，美化宅门出入口的环境，使人们在进出宅门时，迎面首先看到的是垒砌考究、雕饰精美的壁面和镶刻在上面的吉辞颂语，从而获得视觉上和心理上的良好感受。

用于四合院的影壁有三种。第一种是位于宅门里面，呈一字形迎门而设的影壁，又称照壁。这种影壁一般设在四合院东厢房南山墙位置。单独建造的，称为独立影壁；镶砌在厢房山墙之上与山墙连为一体的，称为坐山影壁。第二种是设在宅门外边的影壁，这种影壁坐落在宅门对面，与宅门有胡同相隔，常见有两种形式：平面呈一字形的一字影壁和平面呈︿形的八字影壁。这种影壁或与对面宅院墙壁隔开一段距离，或倚砌于对面院墙之外，主要用于遮挡对面房屋和不甚整齐的房角檐头，给从宅门内出来的人以整齐美观的愉悦感觉。还有一种影壁，斜置于宅门前脸的山墙墀头两侧，与宅门成60°或45°夹角，平面成八字形，称做“反八字影壁”或“撇山影壁”。这种反八字影壁，除去上述功能之外，还有拓宽门前空间、衬托宅门的作用（图5–1–10）。

影壁作为独立的建筑物，也有它的建筑形式。用于民居的影壁，其外形主要有硬山式和悬山式两种。

图 5—1—10　宅门内外的几种影壁

图 5—1—10.1　宅门对面的影壁

图 5—1—10.2
宅门对面的大型一字影壁

图 5—1—10.3
宅门外刻有吉辞的影壁

图 5—1—10.4
宅门内的独立影壁

图5—1—10.5　从门口看内影壁

图5—1—10.6　宅门内的坐山影壁

硬山式影壁的外形如同一间硬山式建筑。在正立面上，是三段式的构成：下碱（相当于硬山建筑的台基部分）、墙身（相当于硬山建筑的构架部分）、屋面（相当于硬山建筑的屋面部分）。这种对硬山建筑的模仿（或称影壁与对应建筑形式的同构关系），从它的墙身部分看得更加明显。硬山影壁的壁心部分有马蹄磉、砖柱子、大枋子、线枋子、耳子和三岔头等构件。这些构件，分别模仿了木构硬山建筑的柱顶石、檐柱、檐枋、槛框以及三岔头、箍头枋等构件，而位于砖柱子外侧的撞头（墙体）部分，正好相当于硬山建筑的山墙墀头（图5–1–11）。

悬山影壁也是如此。从立面看，它与悬山式建筑完全同构：影壁的最下面是下碱，相当于悬山建筑的台基；中间部分的马蹄磉、砖柱子、大枋子、线枋子等，相当于悬山建筑的柱顶石、柱子、额枋、槛框；上部的屋面向两山挑出的形状和构成，也与悬山建筑毫无二致，三岔头、燕尾枋、博缝板应有尽有。悬山式影壁两边不做撞头，这与悬山建筑一般不设墀头有直接的对应关系（图5–1–12）。

前面我们谈到了影壁的三段式构成，即下碱、墙身、屋面。现在，我们分别谈谈这三个部分的构造。

1. 下碱部分

影壁的下碱相当于建筑的台基。台基通常有两种做法：一种是直方形台基，一种是须弥座式台基。硬山式影壁的下碱多采取直方形。这种选择与同构建筑的等级有对应关系。下碱的宽度按影壁满宽尺寸（本书正文所使用的单位“尺”“寸”，均为营造尺或营造寸。清代每1营造尺=320mm，每1营造寸=32mm——编者），高度约占影壁全高的1/4，影壁厚度的确定应视其体量大小，通常在二尺左右（约60cm）。悬山式影壁的下碱一般为须弥座式。须弥座与影壁等宽，高度约为影壁全高的1/4。须弥座各部位的分法与构造，同一般做法。用于民宅的影壁须弥座多用砖料摆砌，很少见有用石料的。

2. 墙身部分

前面已略微谈到了影壁墙身部分的构成。其中，最主要的部分是圈在线枋子范围以内的墙心，称为影壁心。影壁有硬心和软心两种做法。

硬心做法是在心内贴砌斜置的方砖，采用磨砖对缝干摆做法，细致而讲究。为使贴出的砖块整齐对称，影壁心的长宽尺寸一定是方砖对角线尺寸的倍数，如墙心横向要摆6块方砖，竖向摆4块方砖，那么，影壁心的宽、高就应是砖块对角线尺寸的6倍和4倍。方砖砖块的大小要根据所用方砖的型号尺寸而定，如尺二方砖，毛料尺寸为384mm见方，经砍磨加工后，最大尺寸可落到350mm左右，对角线尺寸则为490mm左右。假定影壁心横向摆放6块，竖向摆放4块，则影壁心尺寸应为2940mm×1960mm。在设计施工中，一般

是先定出影壁心尺寸，然后确定砖块的大小。仍以上述情况为例，假设影壁心尺寸定为宽2400mm、高1600mm，横竖向分别摆放6块和4块方砖，则方砖的对角线长应为400mm，边长应为283mm。确定影壁心的尺寸，既要考虑影壁的总尺寸和影壁心与总尺寸的关系，又要考虑砖块的分格要求和摆砌要求，有时需要砖块坐中，有时可以缝子坐中。影壁心内有雕刻时，还要考虑雕刻内容、范围和砖块划分的关系，既要美观合理，又要方便施工。但不管如何分块，影壁心内不应出现除1/2、1/4块（斜裁）之外的零碎砖块，也应避免出现上下、左右不对称的情况，特别是不能出现左右不对称的情况。

图5—1—11 硬山式影壁及其细部构造

图5—1—11.1 硬山式独立影壁

图5—1—11.2 硬山式影壁山面细部

图5—1—12 悬山式影壁及其细部构造

图5—1—12.1 悬山式独立影壁山面细部

图5—1—12.2 悬山式坐山影壁正细部

图5—1—13 影壁心构造、做法

影壁心有素作和雕花等不同做法。素作仅贴方砖于墙面，素雅大方；雕花要在中心和四角刻出传统的装饰图案，为中心花和岔角花，雕刻内容根据房主人的兴趣爱好而定。也有在影壁中心雕刻砖匾的，上面题刻“平安”“吉祥”“吉迪”“鸿禧”等吉辞（图5—1—13）。

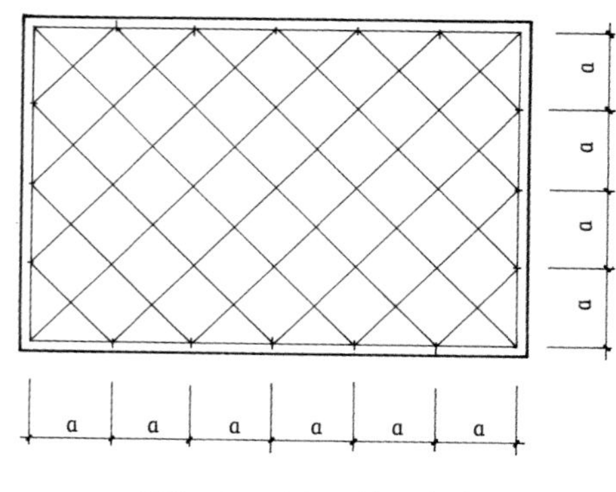

图5—1—13.1 缝子坐中

软心做法的影壁心，是在壁心内砌砖墙，然后在表层抹白灰。这是一种低等级的做法，一般用于小型住宅。影壁的软心做法与影壁自身的等级也是配套的，采用软心做法的影壁，其立面形式也与前述“硬山影壁”和“悬山影壁”不同，由于这种影壁的立面形式与民房中的平台建筑相近似，我们不妨称它为“平台式影壁”。

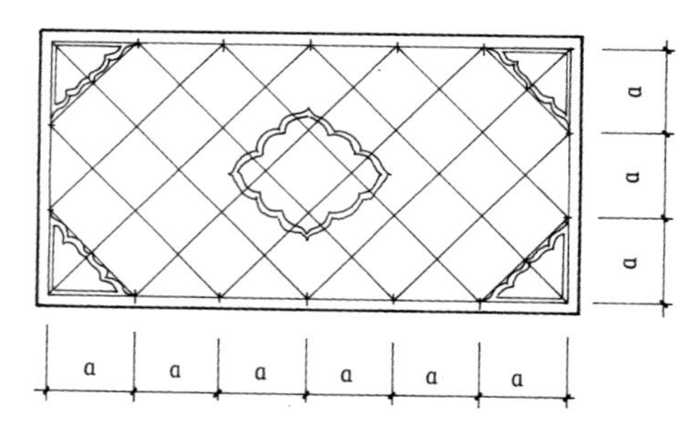

图5—1—13.1 砖块坐中

平台式影壁的立面也由三部分构成，即下碱、墙身和墙顶。下碱为直方形，墙面的砌筑多用“淌白”做法。墙身没有马蹄磉、砖柱子等模拟构件，墙心部分凹进墙体8～12mm，四角做直角或做海棠线角。墙体上方，第一层拔檐至第二层拔檐之间为墙顶，墙顶的中段与墙心对应部分常做些花瓦图案装饰，墙顶形状与平台房屋顶上的女儿墙十分相近。平台式影壁虽然简陋，其壁心内有时也镶砌砖匾，上刻“吉祥”“平安”等吉辞（图5—1—10）。

3. 屋面部分

影壁的屋面，由于与硬山或悬山建筑的屋面存在同构关系，因而在做法上也有两种情况。一种是仿照过垄脊屋面的做法，影壁屋面正中不做高起的正脊。这种做法多用于独立影壁。贴砌在山墙上的座山影壁，屋面多采用带正脊的做法，沿屋脊部分做出清水脊的形状。影壁屋面一律采用小号青瓦而不能用板瓦。

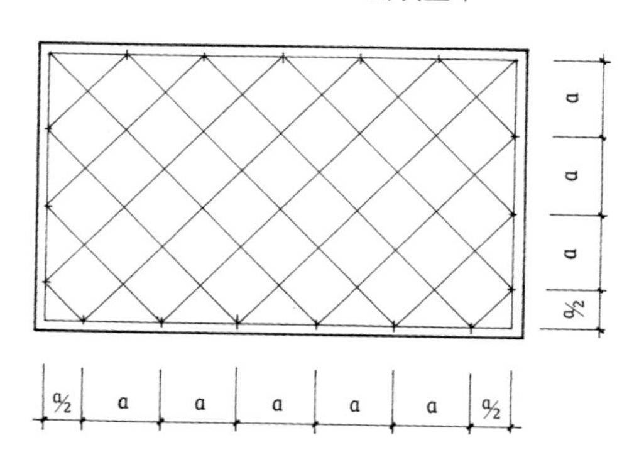

图5—1—13.3 砖块、缝子均不坐中

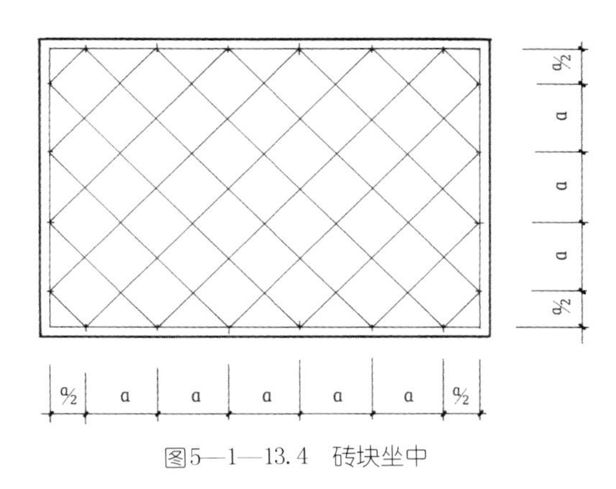

图5—1—13.4 砖块坐中

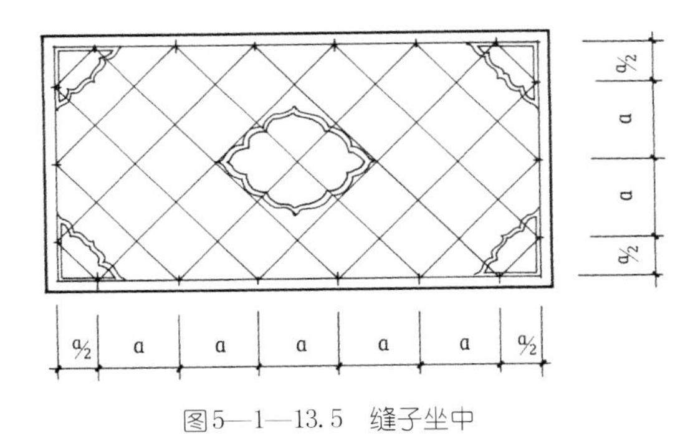

图5—1—13.5 缝子坐中

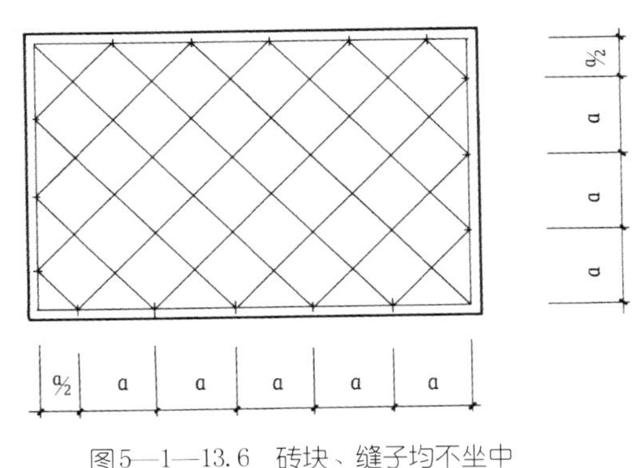

图5—1—13.6 砖块、缝子均不坐中

图5—1—14 上马石

图5—1—14.1 带雕刻的上马石

三、上马石和拴马桩

上马石和拴马桩是过去四合院宅门外必备的设施，虽不属建筑范畴，但与人们的生活息息相关，特别是它们蕴含的人文知识和历史风情，令人饶有兴趣。

在汽车和自行车成为主要交通工具之前，人们的交通工具主要是轿子、马车和马匹。“武官骑马、文官坐轿”是明清乃至更远的古代的主要交通方式。特别是在清代，满洲贵族年纪未满60岁的都善骑马，因此，骑马成为人们主要的交通手段。

人们上马时，需要蹬在一块石头上，下马时也需要有石阶相助，于是就产生了固定在住宅门外的上马石。

上马石是呈阶梯状的石构件，高约2尺（约合65cm），宽约2尺，长（包括第一步踏跺在内）约4.5尺（约合145cm），一般都用青白石打造而成。上马石有素作和雕刻两种做法，素作的上马石就是一块阶梯状的石头；经过雕刻的上马石下部刻出圭脚形状，上面刻成搭袱子（包袱）形状，包袱上面刻锦纹和吉祥图案。上马石一般是成对设置，安放在宅门两侧，专供上下马之用（图5-1-14）。

如果客人骑马来访，下马后需有拴马的地方，这便出现了拴马桩。

拴马桩常见有两种。一种是在倒座房临街的后檐墙上留出约15cm×15cm大小的洞口，洞口位置正对后墙内的木柱，在木柱上面安装铁环一只，马匹的缰绳就系在铁环上。供拴马的洞口一般事先用石块打凿而成，讲究者还在里口刻上花饰加以美化，砌墙时，将石块砌在对应位置。

还有一种拴马桩，是将一块约6寸见方的石柱埋入地下，露出地面约4尺左右，端头略作雕刻，以供拴马之用；也有采用宽约1.5尺、厚约4寸的块石，部分埋入地下，露出地面2～3尺，其上刻出穿缰绳用的“鼻梁儿”，称做拴马桩或“拴马碑”（图5-1-15）。

小小的上马石和拴马桩，影射出一百多年前老北京人生活的历史风情画。

图5—1—14.2 素作上马石

图5—1—14.3 上马石的另一种形式

图5—1—14.4
雕刻精美的上马石

图5—1—14.5 上马石的正面雕刻

图5—1.15 拴马桩

图5—1—15.1 镶嵌在倒座房后檐墙上的拴马桩（拴马环）

图5—1—15.2 拴马桩（拴马环）细部

图5—1—15.3　拴马桩的一种形式

图5—1—15.4　简易的拴马桩

第二节　垂花门、屏门、看面墙和抄手游廊

一、垂花门和屏门

垂花门是四合院住宅的二门。它是分隔、沟通内外宅的一道门，坐落在院落的中轴线上，在外院和内院之间。垂花门适用于二进以上的院落，一进院落的四合院不设垂花门。

在礼法森严的封建社会，成年男女是不能随便接触的，他们的活动范围也有严格限制。《事林广记》中说："男治外事，女治内事。男子昼无故不处私室，妇人无故不窥中门，有故出中门必拥蔽其面。"民间也有闺房小姐"大门不出，二门不迈"的清规。垂花门就是这样一道区分内外宅的二门。垂花门以外算作外宅，可以接待外来客人，垂花门以内是内宅，不允许外人进入，家中的男仆也不能擅入，即所谓"宾不入中门"。被称为是封建社会一面镜子的清代著名小说《红楼梦》中有两段发生在垂花门前的情节描写，颇能说明问题：林黛玉初到贾府，进了荣国府门以后，"另换了四个眉清目秀的十七八岁的小厮上来抬着轿子，众婆子步下跟随，至一垂花门前落下，那小厮俱肃然退出……"。该书第二十四回说贾芸来访宝玉，在书斋等候，宝玉的贴身小厮焙茗给他向宅内传话，站在垂花门外等了半天，见丫环小红出来，才托她给带了个信儿进去。这两段描写，把内外宅的森严分隔勾画得一清二楚。

垂花门是单开间悬山建筑，体量不大，开间尺寸八尺至一丈（约合2.5～3.3m），进深略大于面宽。其主梁前端穿过前檐柱并向外挑出，形成悬臂梁的形式，在挑出的梁头之下，各吊一根悬空短柱，柱头有雕刻精美的花饰，十分美观别致，垂花门也因此而得名。

图5—2—1　垂花门、抄手游廊、看面墙平面、正立面、背立面图及各种垂花门举例

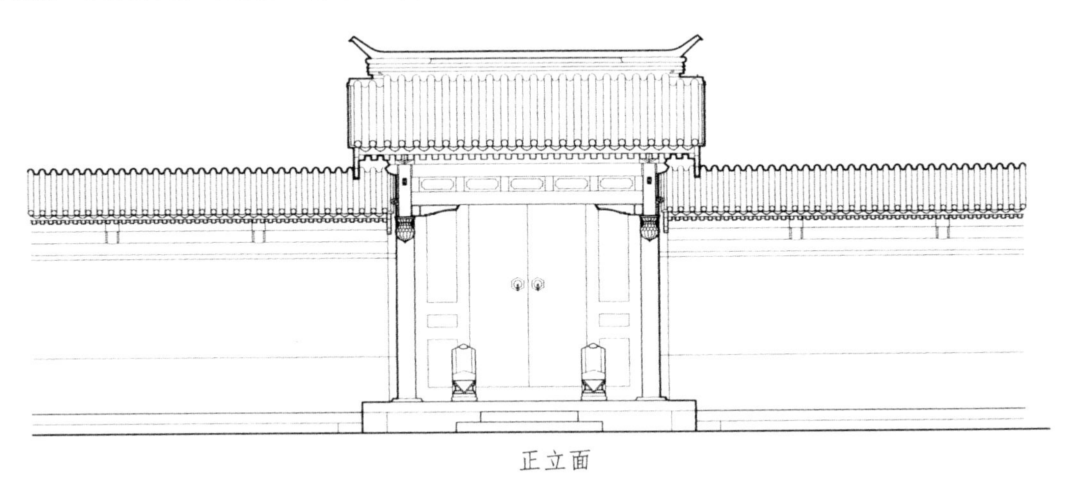

正立面

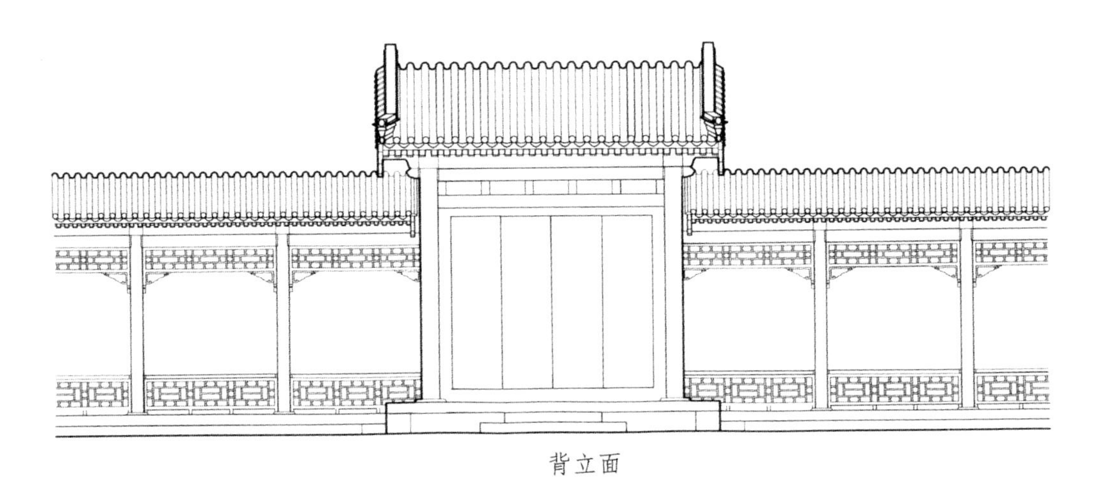

背立面

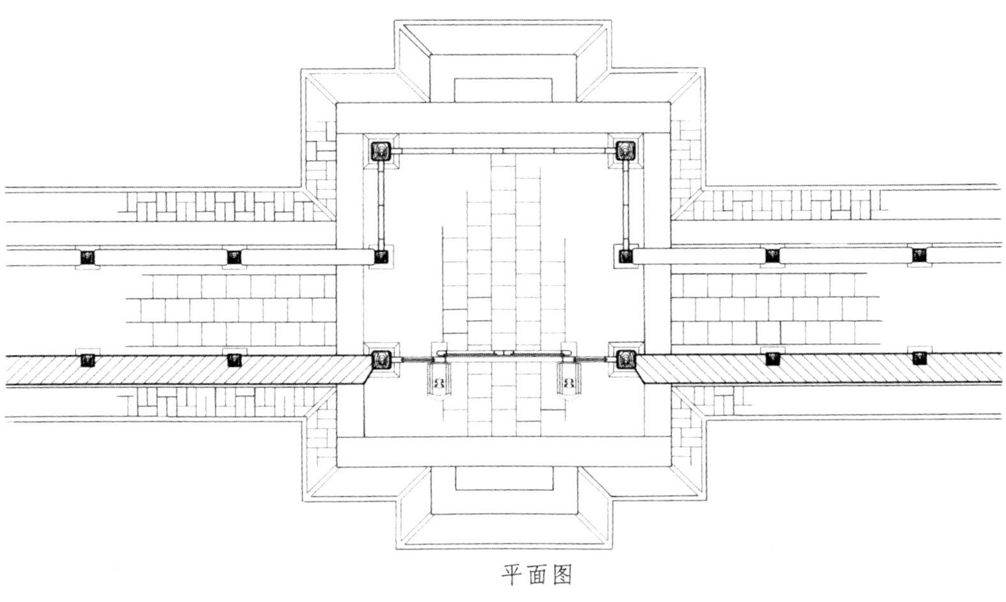

平面图

图5—2—1.1　垂花门
平面、正立面、背立面

垂花门的前檐柱和后檐柱间各安装一樘门扉。前檐柱间安装的是可进行开关的门，由抱框、门框、横槛、余塞、门扇以及抱鼓石、门簪等构件组成，白天开启，夜晚关闭，有防卫功能。后檐柱间安装的是一樘屏门，门扉分作四扇，除去红白喜事或重大活动需要“大开中门”以外，平时都不开启，起着遮挡视线的作用，以免外院的客人或男仆窥见内眷们的活动。

在有抄手游廊的院落中，垂花门的两侧与抄手游廊相连接。抄手游廊的进深较小，一般只有4尺，最大也不超过5尺（约合1.3～1.6m）。抄手游廊外面的柱子与垂花门的外檐柱排列在同一条轴线上，与垂花门两侧的障墙——看面墙也同在一条轴线上，这样，抄手游廊的外侧就被看面墙封砌起来，使之成为仅向内院开敞的游廊。为了装点墙面并使内外院的空间适当沟通，有时在看面墙上设什锦窗（图5–2–1）。

图5—2—1.2　大型宅院的豪华型垂花门

图5—2—1.3　一殿一卷式垂花门（两侧看面墙作砖雕）

图 5—2—1.4　豪华垂花门正面的雕饰和彩画

图 5—2—1.5　一殿一卷式垂花门内侧及与抄手游廊的关系

图 5—2—1.6　一殿一卷式垂花门（两侧看面墙为软心做法，带什锦窗）

图5—2—1.7　双卷棚垂花门（两侧看面墙有什锦窗）

四合院里的垂花门，常见的有两种形式，一种是一殿一卷式，一种是单卷棚式。

一殿一卷式垂花门在四合院住宅中用得最普遍。它是一座造型美观、装饰华丽的建筑，屋顶由一个带正脊的悬山和一个卷棚悬山组合成勾连搭形式。从正面看，是一个带正脊的悬山屋面，从背面看是一个卷棚悬山屋面。一殿一卷式垂花门平面有四根柱，前面两根为前檐柱，后面两根为后檐柱。垂花门有特色的构造是它向外挑出的麻叶抱头梁和其下悬吊的垂莲柱。麻叶抱头梁是垂花门的主梁，后端做成一般梁头形状，搭置在后檐柱的柱头上，前端与前檐柱相交并向外挑出，梁头做成麻叶云头形状。梁头下面各悬吊一根约1m长的垂柱，柱间有檐枋（又称帘笼枋）、罩面枋等构件连接，垂柱端头做仰莲、串珠、二十四气（俗称风摆柳）或莲瓣等雕刻（图5–2–2）。

图5—2—2　一殿一卷式垂花门侧立面、剖面图

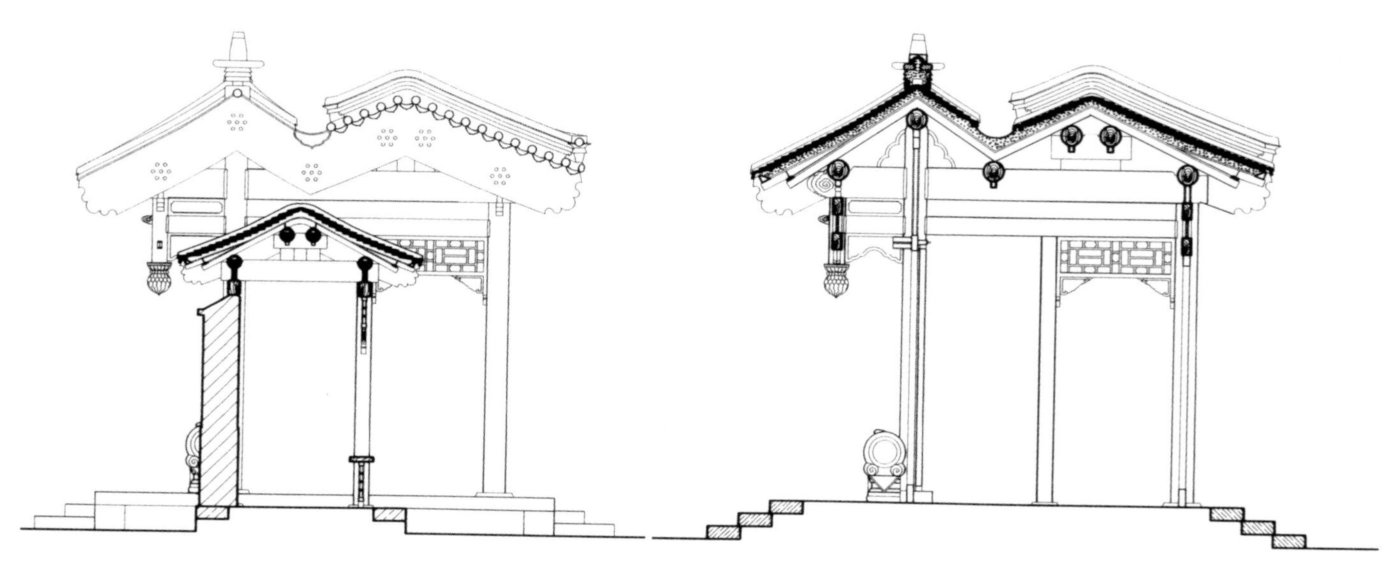

一殿一卷式垂花门的木构架，主要由柱、麻叶抱头梁、月梁、瓜柱、角背、枋、檩等组成，在进深方向使用六根檩，分成4.5步架（卷棚部分的双脊檩间为0.5步架）。其中，麻叶抱头梁自前檐柱向外挑出尺寸为1步架，前后檐柱之间为3.5步架。垂花门每步架间距离一般为2.5尺（约合80cm）。这样，垂花门的实际进深大约是2.8m，加上挑出的0.8m，共3.6m。规模较大的深宅大院，垂花门的进深还可略微加大，面宽也可略为加宽。由这个尺度构成的空间，可供女眷上下轿，遇有红白喜事时也可在此举行某些仪式，有的家庭还把垂花门作为小型堂会的临时舞台。

另外一种单卷棚式垂花门在四合院中也常被采用。这种垂花门的屋面构成比较简单，仅一个卷棚悬山。与它相对应的木构架也做成单檩或双檩卷棚的形式。如果是单檩卷棚，则通进深（含垂步在内）分为4步架，用5根檩，称为五檩卷棚，这种五檩的垂花门也有做成带正脊的例子。如果是双檩卷棚，则通进深分为4.5步架，用6根檩，称为六檩卷棚。但不管何种做法，麻叶抱头梁向外挑出，并悬吊垂莲柱的独具特色的做法不变。单卷棚式垂花门的屋面形式不如一殿一卷的形式丰富，且屋面比较高，不适用于院落偏小的宅院（图5–2–3）。

图5—2—3　单卷棚垂花门剖面图

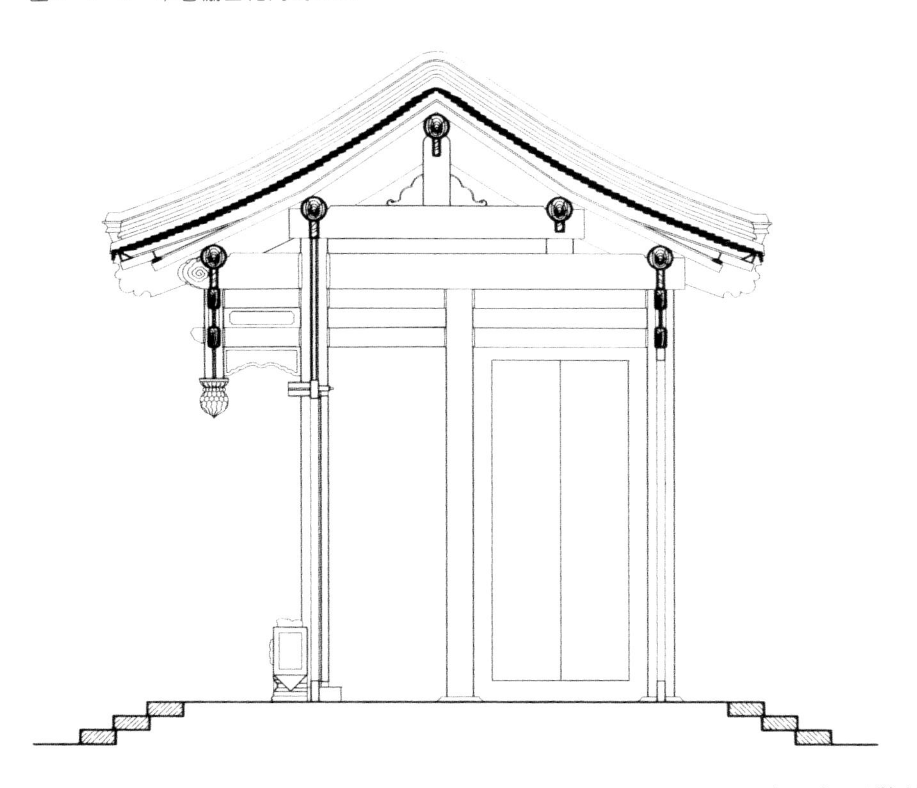

在面宽较小的二进四合院中，由于受尺寸限制，东西厢房一般不做外廊，这样，抄手游廊也就失去了存在的意义。这种没有抄手游廊的二进四合院，在二门处往往建一座随墙门，这座随墙门称为屏门。作为宅院二门的屏门，做法是比较讲究的。它建在突出的台基上，由墙垛、过木、檐口、屋面等部分组成，尺度明显高于两侧的看面墙。屏门的门扉是由四扇镜面板门组成，安装在槛框内。平时，两侧的门扇固定，只开启中间两扇，具有屏蔽和通行双重作用。台明内外安装如意踏跺，以方便出入（图5–2–4）。

但是，也有的房主人对垂花门情有独钟，尽管院内没有抄手游廊，还是要在二门处建垂花门，并且在两侧的看面墙上开什锦窗作为装饰。位于鼓楼东大街的某宅就是一例（图5–2–5）。还有在二门处建独立柱垂花门的，在实际中也不乏其例。

图 5—2—4 用于较小型院落的屏门

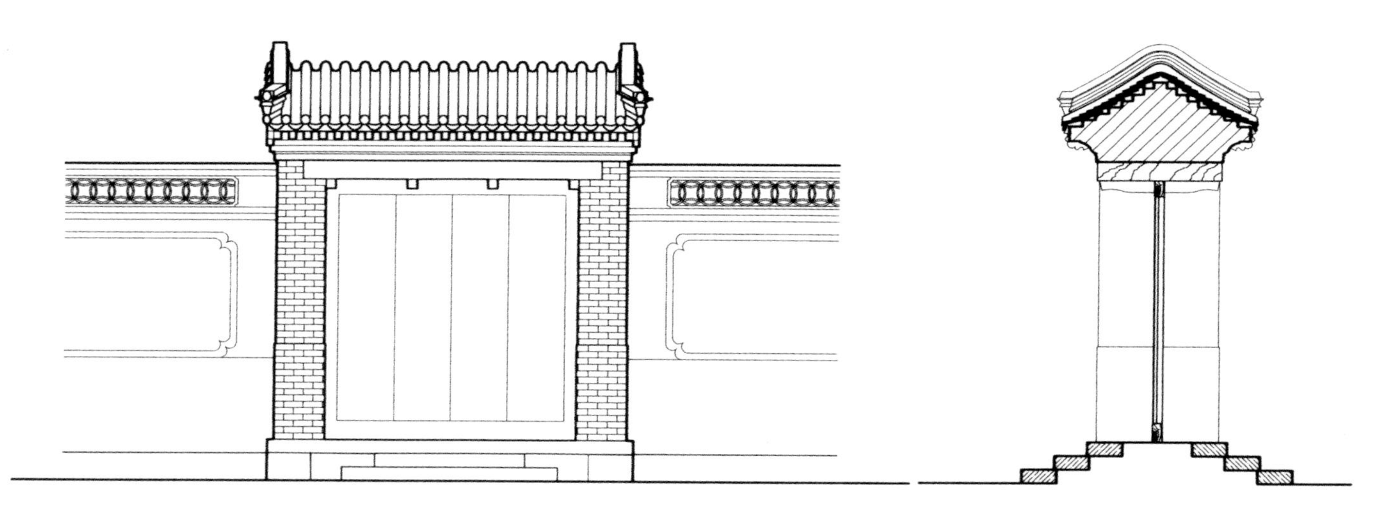

图5—2—4.1 屏门平、立面图

图 5—2—4.2 用于较小型院落的二门——屏门

图 5—2—5 鼓楼东大街某宅垂花门及看面墙

垂花门的一个突出特点是装饰华丽、雕刻精美。它的前檐除垂莲柱上面有雕刻外，在迎面的檐枋和罩面枋之间还有折柱（用于分隔花板的短柱）和透雕花板，檩、枋之间有荷叶墩一类饰件，罩面枋下安装雀替或透雕的花罩。在前檐柱与垂柱之间，还有镶嵌在麻叶抱头梁和麻叶穿插枋之间的透雕花板以及骑马雀替。梁架上的角背、驼峰也常做出雕饰。垂花门的抱鼓石上还有镌刻精细、寓意吉祥的图案花纹。一座垂花门不啻为一件精美的艺术品（参见图5—2—1及图6—1—34）。

垂花门的油饰彩绘也是宅院中最为华丽讲究的，宅主人对垂花门的装点可谓煞费苦心。这一点我们在后面还要谈及。

上面我们已经提到了屏门，这里还要补充一些内容。在四合院中除去二门处可做屏门以外，在外院还有两处常常采用屏门来分隔空间，一处是在宅门的两侧，另一处是前院西侧的第一间或第二间倒座房的位置。

在院落的宽度只能排开五间房的小型四合院中，宅门一般都设在倒座房东端的第一间。进入宅门以后，向左转可以进入院内。人们往往在宅门左侧设一道屏门，使人进入宅门后，不能马上看到宅院，需要通过屏门才能进入宅院。此处如设屏门，在与它对称的位置——倒座房西面第一间的右侧也设置一道屏门，把五开间宽度的前院划分为三个空间，这样划分不仅使平直呆板的外院增加了空间层次，而且也与功能要求相吻合。宅门以内是由门道、影壁围成的空间。它是从院外进入院内的过渡区域，与外院的功能还不尽相同，应当加以区分；倒座房西侧的第一间房一般是库房或厕所，与中间的客厅、书房在功能上大相径庭，也应当加以分隔；中间三间是客厅、书房（包括门房），功能近似，可以放在一个空间。这种以屏门划分空间的手法，在大中型宅院中采用得更为普遍。

如果宅院的宽度是七间房的话，宅门一般占据倒座东端的第二间。做这种安排时，宅门的两侧要各设一道屏门。这种在宅门内侧左右两边均设置屏门的做法，不仅将门道与宅院进行了分隔，而且将门道以东的空间也单独划分出去，使门道前面的过渡空间更具独立性。出于功能要求及对称的考虑，在外院西侧，沿第二间房东墙也设一道屏门，将外院划分成四段不同功能的空间（图5—2—6并参见图1—9）。

图 5—2—6　屏门举例

图 5—2—6.1　门内影壁和两侧屏门

图 5—2—6.2　沟通宅内其他院落的屏门

用于划分空间的屏门，有的采用砌砖墙、中间留门洞安装木质板门的做法；有的采用木结构外框，中间安装板门；还有采用月洞门形式的。采取何种做法，应视院当尺寸大小和环境情况而定。

二、看面墙和抄手游廊

看面墙是分隔内外宅的一道主要隔墙，又是位于垂花门两侧、位置非常显要的墙面。所以，一般房主人都非常重视对看面墙的装饰。我们通过前面的介绍已经了解到，看面墙和抄手游廊是相互关联的，看面墙的构造、做法与抄手游廊有直接关系。由于看面墙与抄手游廊的后檐柱同在一条轴线，所以，看面墙又是抄手游廊的后檐墙，当抄手游廊的后檐做椽子向外挑出（这种做法称为老檐出做法）时，看面墙只能砌至廊子檐枋下皮，墙头做馒头顶。采用这种做法时，看面墙一般砌丝缝墙，墙面没有更多的装饰，廊子里的什锦窗就成了墙面的主要装饰。如果抄手游廊的后檐做封护檐时，看面墙就要将廊子后檐的木构件都包砌在里面，并与廊子的屋面相交接，在檐口做出冰盘檐一类砖檐。采用这种做法时，看面墙的墙面比较完整，有下碱、上身和墙帽（即廊子的屋面部分）。这种完整的墙面做法，与影壁墙做法相类似，因此往往采取影壁墙的做法，在墙的上身部分做出马蹄磉、砖柱子、大枋子、线枋子等框线，在墙心部分摆砌斜方砖（俗称膏药幌子），更为讲究的还可以在中心、四岔做砖雕刻（图5-2-7并参见图5-2-5、图5-2-1.4）。

这里，我们还要对什锦窗的种类、形状、构造、做法进行一些介绍。

什锦窗是一种装饰作用极强的漏窗，多用在园林建筑中，具有美化墙面、沟通空间、借景、框景等作用，在四合院住宅中也常采用。上面提到的看面墙就是适宜放什锦窗的部位。用于民居中的什锦窗，主要有什锦漏窗、什锦灯窗两种形式。什锦漏窗又称单层什锦窗，是指在窗洞内居中安边框，在框内安棂条、仔屉或玻璃屉子。什锦灯窗是在窗洞内安装两层窗框仔屉，屉内安装磨砂玻璃，在玻璃上绘画题字，在两层玻璃之间装灯。每逢佳节吉日，各种形状的窗内灯光齐明，映照两壁诗画，营造出无比美好的气氛，颇有装饰效果。

什锦窗的形状有多种。这些形状均采自各种造型优美的器皿、花卉、蔬果和几何图形，如玉盏、玉壶、笔架、卷书、扇面、套方、玉瓶、银锭、蝙蝠、月洞、双环、石榴、贝叶、寿桃、五方、六方、八角、梅花等等（图 5-2-8）。

什锦窗的艺术魅力不仅来自它的艺术造型，还来自窗套的色彩与装饰。什锦窗的窗套（包括窗口和贴脸）有两种做法：一种是木质的，一种是砖质的。木质窗口用窄板条拼接而成，外面采用木质窗套，表面刷红、绿或黑色油漆。砖制窗套则用砖料砍磨而成，镶在外缘的砖贴脸往往是砖雕艺人大展才华的地方。北京东城区鼓楼东大街某宅的砖雕什锦窗用料之精、雕刻之细，实属罕见，堪称精品（图5-2-9）。

由于什锦窗一般都是安装在园林或宅院内的隔墙上，并常与尺度小巧的游廊配合使用，因此它的尺度都不大，一般在2～3尺之间。就是说，一个什锦窗的形状不管多么不规则，它最长的边不应超过90cm，最短的边可在60cm左右，不宜太大也不宜太小。

由于什锦窗具有借景、框景的功用，所以，它在墙面上的高度，应与人的视线要求相适应。什锦窗中心的位置应与常人眼睛的高度相近。什锦窗的安装应以它自身十字中线的交点（即中心点）为准，而不应以上下边为准。

什锦窗如与游廊相配使用，应每间一樘，居中安置，相邻或相近的窗形应富于变化，不能重复（图5-2-10）。

最后，谈一谈抄手游廊的建筑和构造特点。

抄手游廊是四合院中的附属建筑，由于它的作用主要是沟通各个房间、供人通行或小坐休息，所以体量较小，构造也比较简单。它的构造形式是四檩卷棚，柱高7～8尺（约合2.2～2.4m），进深4～5尺（约合1.3～1.6m），柱头以上承四架梁、月梁，梁上面承檩。抄手游廊的檐枋下面安装倒挂楣子，柱根之间安装坐凳楣子；坐凳可供人坐憩，倒挂楣子具有装饰功能，上面也可以悬挂鸟笼一类玩物供人观赏（参见图5–2–1）。

垂花门、屏门、看面墙和抄手游廊虽然不是四合院的主要建筑，但它们在装点宅院、分隔空间、衬托主要建筑、烘托环境气氛方面有着非常重要的作用，巧妙地运用这些元素对营造优美典雅的居住环境是至关重要的。

图 5—2—7 看面墙举例

图 5—2—7.1 垂花门两侧的看面墙

图 5—2—7.2 倒座房后檐临街的看面墙

图 5—2—8 各种形式什锦窗举例

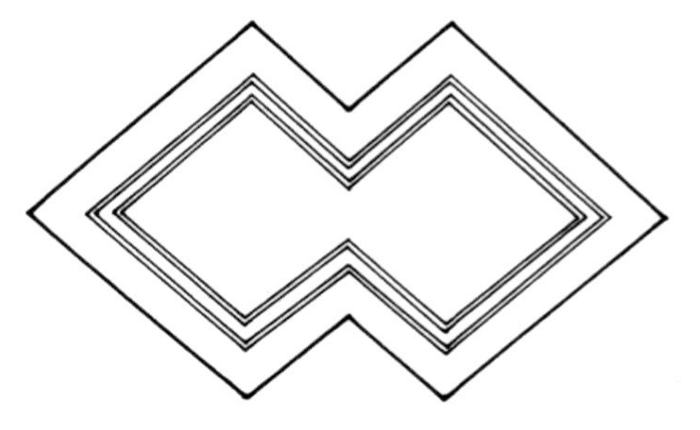

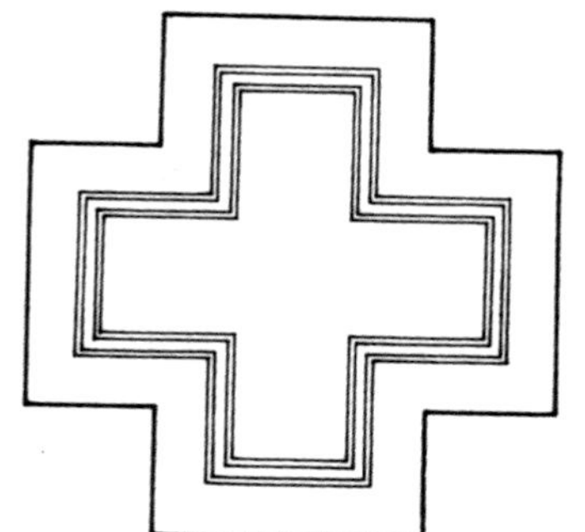

图 5—2—9.2　木质什锦窗

图 5—2—9.3　砖雕什锦窗一例

图 5—2—9　什锦窗的雕刻与色彩

图 5—2—9.1　抄手游廊内的什锦窗

图 5—2—9.4　砖雕什锦窗一例

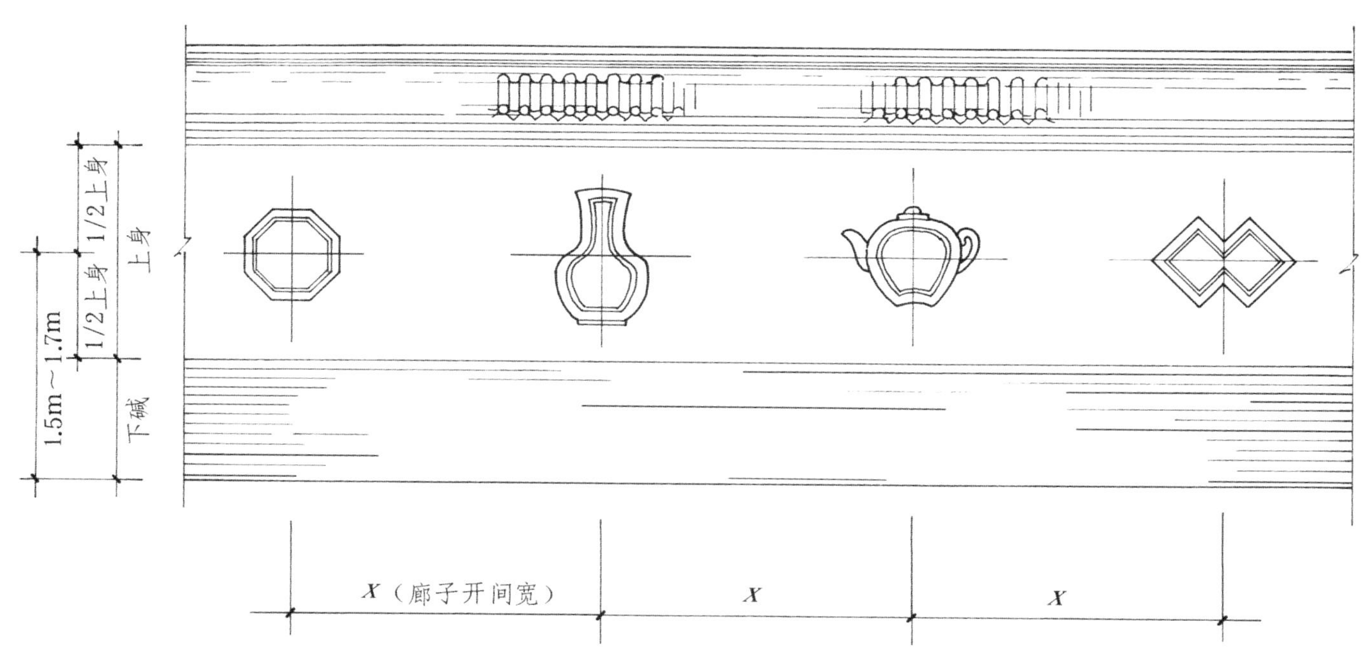

图 5—2—10 什锦窗安装示意图

第三节 主要建筑及其构造

在对宅门、影壁、垂花门、屏门、抄手游廊这些富有特色的建筑物和附属物作了基本了解之后，我们再来探究四合院的主要建筑及其构造。

四合院的主要建筑有正房（含耳房）、厢房（含厢耳房）、倒座房和后罩房（或后罩楼）。在介绍它们的构造之前，首先看一看建筑尺度和院落空间的关系。

一、建筑尺度和院落空间的关系

北京四合院有小型、中型、大型之分，同种类型的院落在尺度上也不尽相同。尽管在这些院落中都有正房、厢房、倒座房，但建在不同空间的同类建筑，它们的体量（包括开间、进深、柱高以及由这些因素构成的面积、体积）是不相同的。同样一幢建筑，如正房，坐落在大空间，则体量较大，坐落在小空间，则体量较小。这种尺度上的差别，首先取决于院落的宽度和深度。如一座小型四合院，占地宽16m（这个宽度在实际中很常见），要在这个宽度内建三正两耳共五间房的话，按一般规律分排，正房明间面宽可为3.3m,次间面宽3m,耳房面宽2.4m;厢房进深为3.5～4m（无外廊），院当宽度可剩7～8m。如果一座大型四合院占地宽25m（这个宽度在实际中亦很常见），要在这个宽度内建三正四耳共七间房的话，按一般规律分排，则正房明间面宽可达3.9～4.2m,次间面宽3.3m,耳房面宽3m;厢房进深如为5.5m（含外廊），院当宽度还余13m左右。通过以上两个例子的比较可以看到，同是主要建筑的正房，建在不同空间的院内，它的开间尺寸会产生明显的差别。这种差别又会影响到柱高、进深和其他部位的尺寸。院落空间不仅对正房的尺度产生影响，对其他房子也会产生同样的影响。院落宽大，房子也随之高大；院落窄小，房子也随之矮小——这是一条基本规律。如果不加分析地将一成不变的建筑尺寸到处套用，就会造成建筑空间比例的失调。当然，上面提到的“高大”和“矮小”都是以满足功能要求为前提的。如果超出了功能要求允许的限度，则要进行根本性的调整。

二、正房、厢房、倒座房、后罩房的一般构架形式

明确了建筑尺度和院落空间的相互关系之后，再分别看一看四合院各主要建筑的一般构架形式。

1. 正房

正房是四合院中最主要的建筑，一般采取七檩前后廊的构架形式，平面的进深方向列4排柱子，它们是前檐柱、前檐金柱、后檐金柱、后檐柱。前檐柱与前檐金柱之间为外廊，门窗装修安装在前檐金柱一缝。后檐墙位于后檐柱一缝。正房的前后金柱上面承五架梁，檐柱和金柱之间施抱头梁和穿插枋。这种七檩前后廊的构架形式，进深可达7m以上（含外廊），适合于大中型四合院的正房。它的明间面宽一般在3.9～4.2m,次间面宽3.3m左右，檐柱高度3.3～3.5m。这个尺度作为民居已经相当宏伟了（图5—3—1）。

图 5—3—1　七檩前后廊式构架及各部位名称示意图

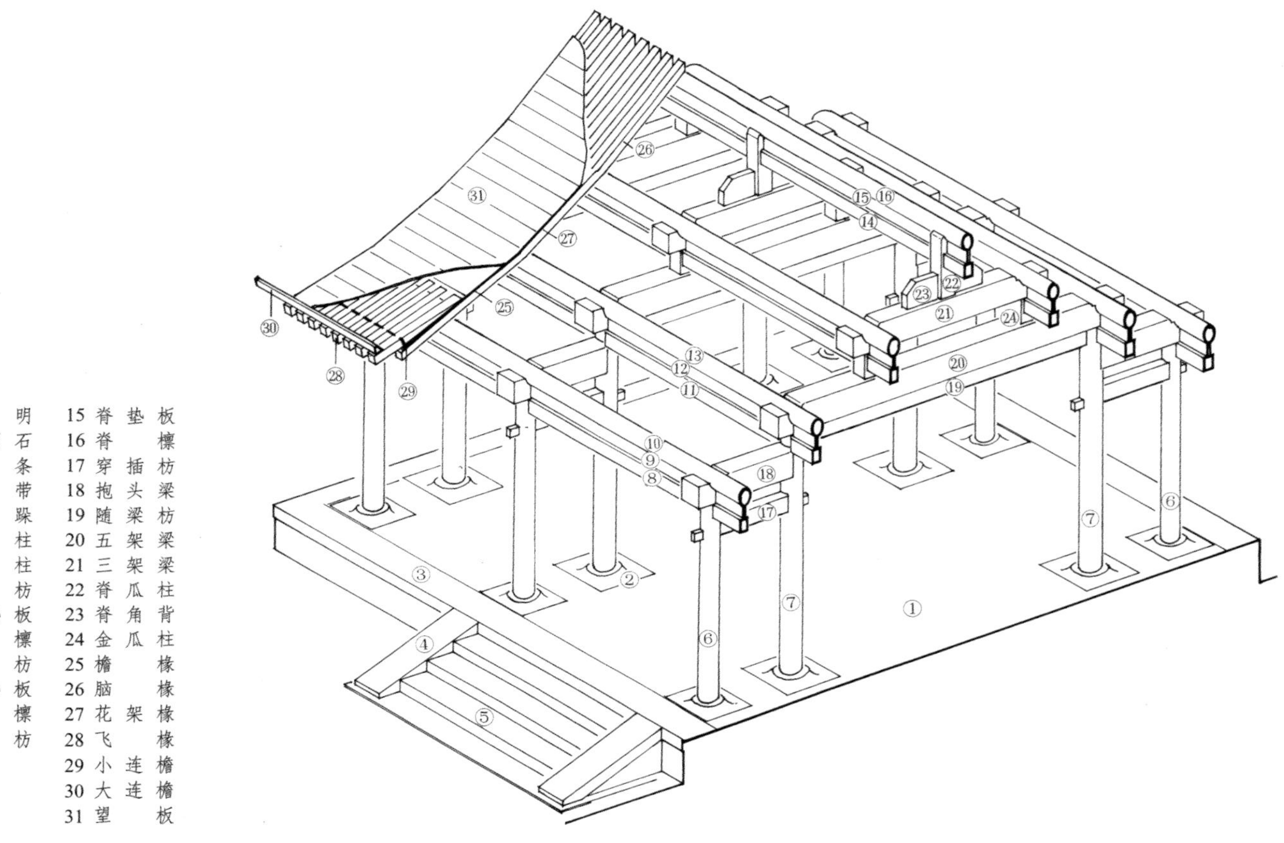

1 台明
2 柱顶石
3 阶条
4 垂带
5 踏跺
6 檐柱
7 金柱
8 檐枋
9 檐垫板
10 檐檩
11 金枋
12 金垫板
13 金檩
14 脊枋
15 脊垫板
16 脊檩
17 穿插枋
18 抱头梁
19 随梁枋
20 五架梁
21 三架梁
22 脊瓜柱
23 脊角背
24 金瓜柱
25 檐椽
26 脑椽
27 花架椽
28 飞椽
29 小连檐
30 大连檐
31 望板

坐落在较小型院落中的正房，则要适当减小尺寸，不仅面宽要酌减，进深也要酌减。减小进深的一般方法是将前后廊式改为前廊后无廊式，即去掉后檐柱及其相关的抱头梁、穿插枋等，平面剩三排柱。作这种改动以后，会出现屋脊不居中，后檐高于前檐的现象，即如匠师们所说：“前廊后无廊，必是撅尾巴房。”这种“撅尾巴房”无论正面、背面、侧面都不甚美观，而且仅仅去掉后廊部分，不降低屋顶高度，也没有完全达到减小尺度的目的。要解决这个问题，需对木构架进行调整。方法是：将原来的后檐金柱降为檐柱，前檐金柱改为钻金柱（直通金檩的柱子称钻金柱），将钻金柱与后檐柱之间的梁改作插梁，使之处在与抱头梁相等的标高。然后通过重新分配步架，令屋脊居中。这样做不仅可以解决房子“撅尾巴”的问题，使立面、侧立面更加美观协调，而且明显降低了屋顶高度，达到了全面减小建筑尺度的目的（图5-3-2）。

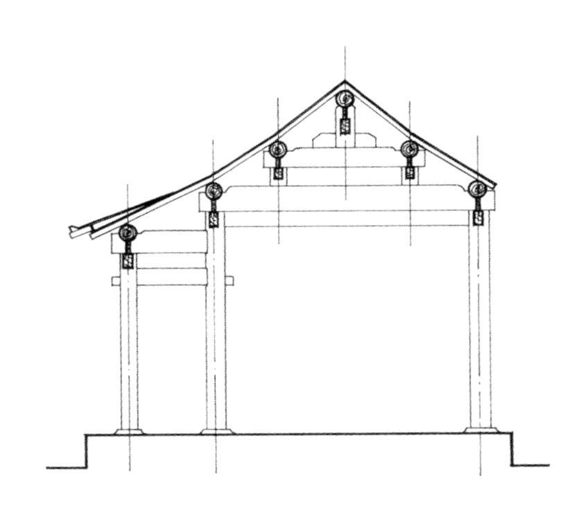

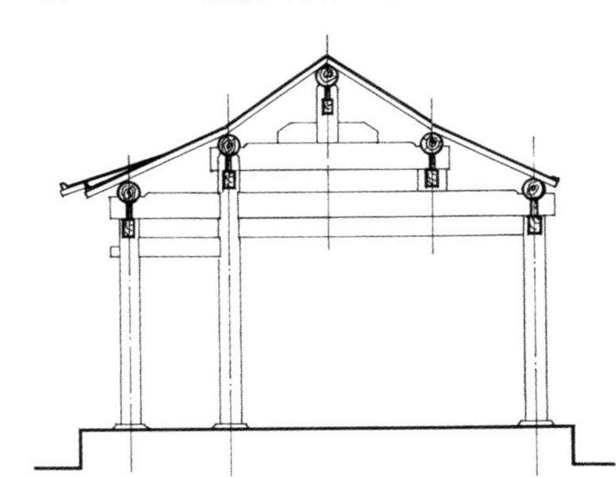

图 5—3—2　前廊后无廊式两种屋架形式比较

2. 厢房

大中型四合院的厢房、前檐一般都带有外廊，其构造采取六檩前出廊的形式。这种构架与上面所举前廊后无廊的形式是一样的，也会出现“撅尾巴”的情况。如果没有特殊需要，也可按上述方法，将其调整为屋脊居中的形式。

在小型四合院中，由于院子宽度不够，厢房无条件做外廊，只能采取五檩无廊的形式，柱上承五架梁，进深方向分作四步架。厢房的前檐有椽子、飞椽向外挑出，后檐因与邻舍房屋相接，不能出檐，要做成封护檐（图5-3—3）。

3. 倒座房

倒座房作为临街建筑，一般都体量不大，多采用五檩构架。倒座房的进深以4～5m为多，而且极少有带外廊的，即便是大中型院落也是如此。这大概是因为它属外宅，在诸房中地位不很重要所致。但位于倒座房东侧的宅门，则比倒座房高大、突出得多。由于宅门的尺度与倒座房不同，它们的木构架互不相连，是相对独立的。宅门的柱子一般要比倒座房的柱子高出半尺到一尺，它的屋架形式也是根据宅门的形式等级而定的。

倒座房的后檐临街，檐子的做法可以灵活多样，既可做椽子向外挑出，也可做封护檐。这一方面要看主人的要求，另一方面也要看院子的档次。一般说来，有官阶的人住的房子出檐做法较多，而一般百姓的房子做封护檐的较多（图5—3—4）。

4. 后罩房（或后罩楼）

后罩房是四合院最后的一排房，一般由家庭中的女眷或未出嫁的女子居住。房间的进深、尺度与倒座房比较相似，多采用五檩构架，而且一般不做外廊。有的后罩房做成二层楼形式，称为后罩楼。这种后罩楼只有大型住宅才可能采用。后罩楼的构架形式并不复杂，柱子一般用通柱，以增强构架的整体性。一层顶部进深方向柱间施承重梁，面宽方向柱间施间枋，形成首层的框架。在承重梁上面，沿面宽方向搭置楞木，楞木上面铺木楼板。楞木间距二尺有余，楼板厚一寸半至二寸，其上再铺方砖作二层室内地面。通柱直通二层，柱头承五架梁及全部屋顶构架（图5—3—5）。

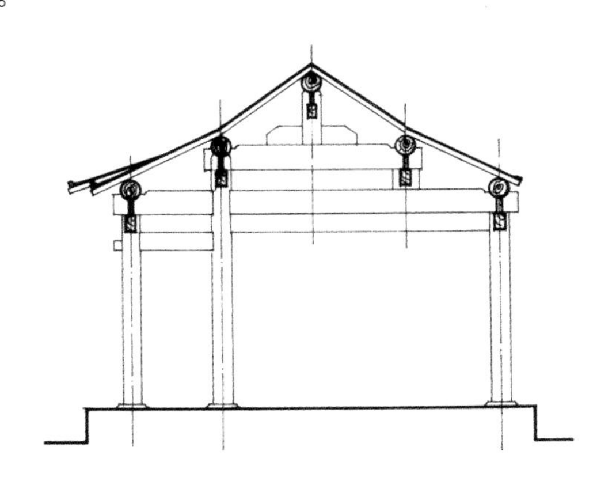

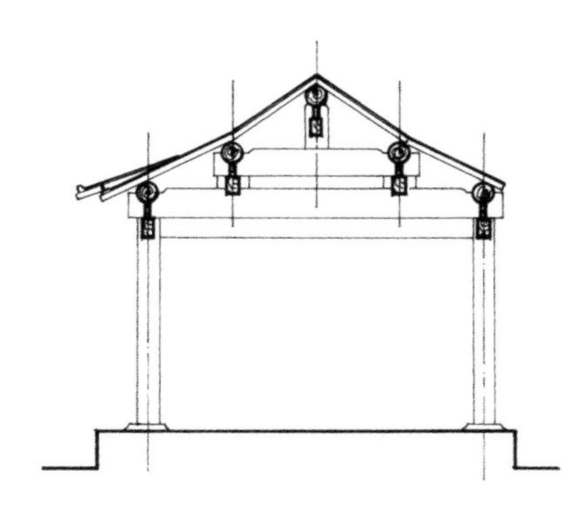

图 5—3—3　厢房的两种常见构架形式

图 5—3—4 倒座房及与大门的关系

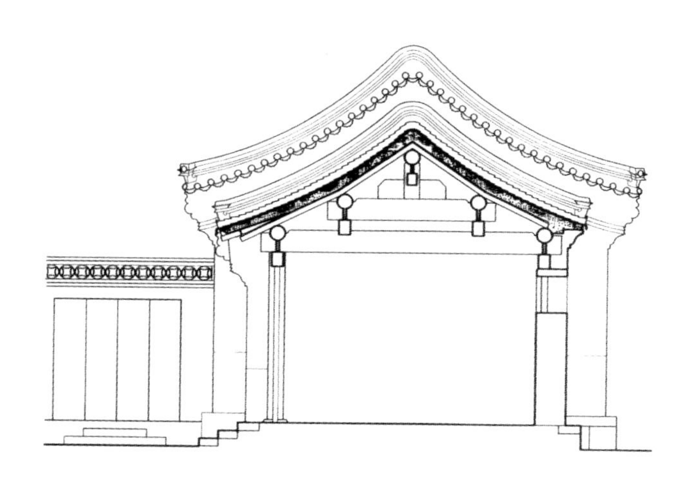

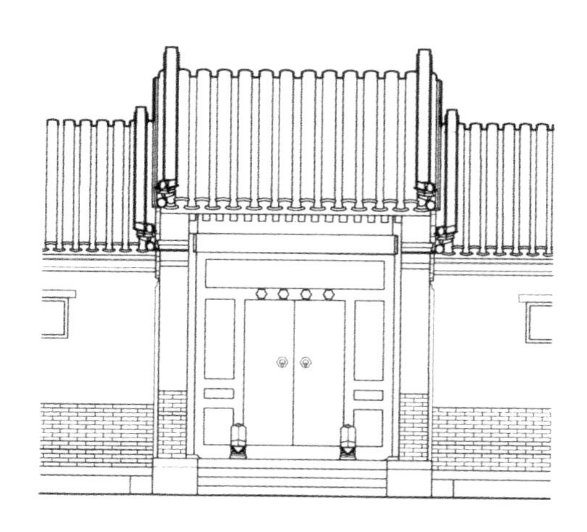

三、墙体的形式及其构造

四合院建筑，除去垂花门是悬山式建筑外，其他都是硬山式建筑。硬山式建筑的墙体有山墙、檐墙、槛墙、廊心墙等。

1. 山墙

硬山式建筑侧面的墙体称为山墙。

山墙将建筑物侧面的木构架包砌起来使之不外露，有保护木结构的作用。

山墙的高度和宽度按建筑物木构架的高宽和进深尺寸而定。墙体的厚度根据建筑体量的大小，按权衡尺度确定。

山墙两端靠近台明边缘的部分称为墀头。墀头在建筑的正立面或背立面，是山墙最显眼的部分，因此格外注重装饰。樨头分为上、中、下三段。下段为下碱，约占墙身高的1/3；中段为上身，约占墙身高的2/3；上段为戗檐盘头，是墀头最精彩的部分，由层层挑出的砖檐和向外倾斜的方砖构成（图5-3-6）。

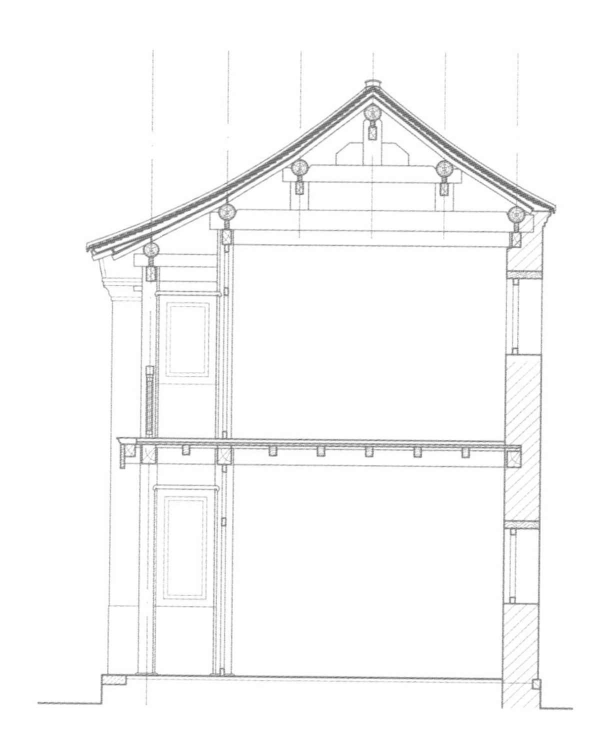

图 5—3—5 后罩楼的一般构造

硬山建筑山墙的侧面也是由上、中、下三部分构成的。下段为下碱，高度约占墙身高的1/3；中段为上身，约占墙身全高的2/3；上部为山尖，是山墙的高起部分，其形状随屋顶曲线而变化。山尖与屋面相交的边缘镶砌的装饰性方砖称为博缝。博缝上面即是屋面。民居中硬山建筑山墙的做法也分为若干档次，最讲究的做法是通体磨砖对缝，做干摆下碱、丝缝上身（含山尖部分）、干摆方砖博缝。稍微经济一些的做法是墙体四角和山尖做干摆丝缝，墙心部分糙砌砖墙，在表面抹灰，即软心做法。软心和墙腿结合部位常采用五出五进的咬槎做法，并用陡砖镶边，既坚固又美观。再一种等级较低的做法是砌淌白墙面。采取这种做法时，砌墙用的砖料不做精细的砍磨，只做简单的加工。砌出的墙体偏于粗糙，但比糙砌砖墙又整齐讲究得多。传统四合院墙体的砌筑一般都是外层一皮好砖，内里用碎砖衬里，墙里皮抹灰。这种做法可以节省大量好砖，使拆下的旧、碎砖得到充分利用。四合院的山墙敢于碎砖垒砌，是因为木结构的建筑是凭木构架承重，墙体只起防寒隔热及围护作用。如果墙体承重的话，是断不能用碎砖垒砌的（图5-3-7）。

图5—3—6 山墙墀头盘头戗檐细部

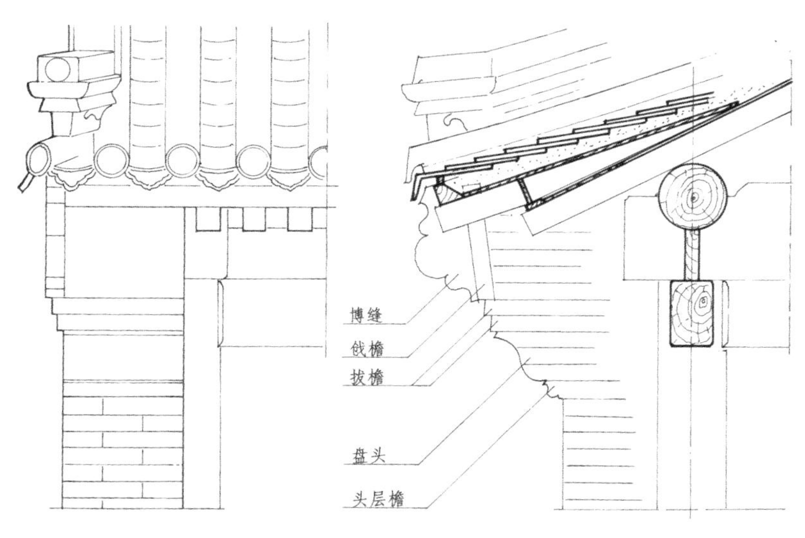

图 5—3—6.1 山墙墀头盘头戗檐细部名称

图5—3—6.2 山墙墀头盘头戗檐细部做法（外侧）

图 5—3—7 山墙、檐墙的几种常见砌法

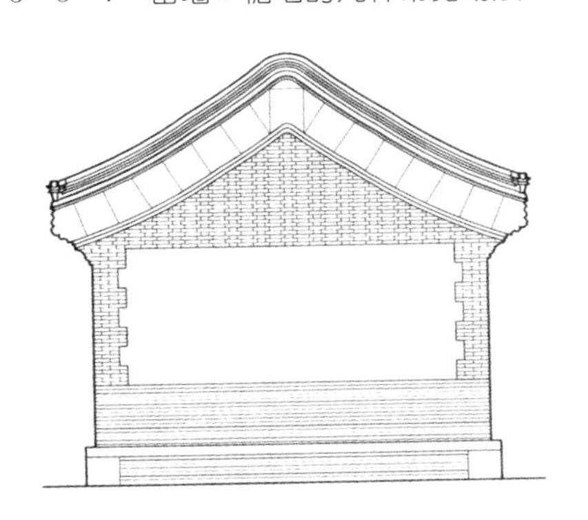

五出五进软心做法（山墙）

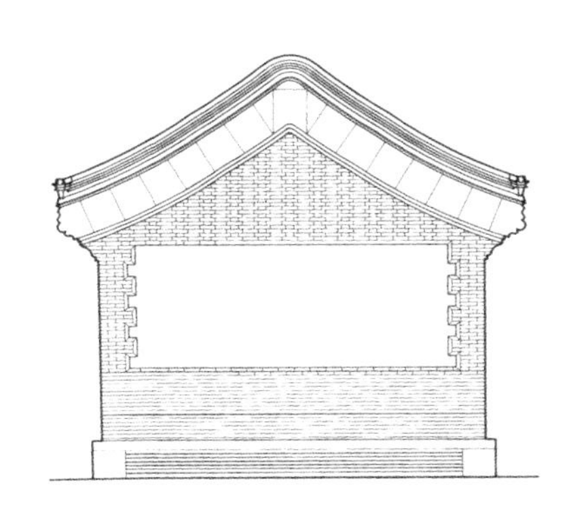

圈三套五软心做法（山墙）

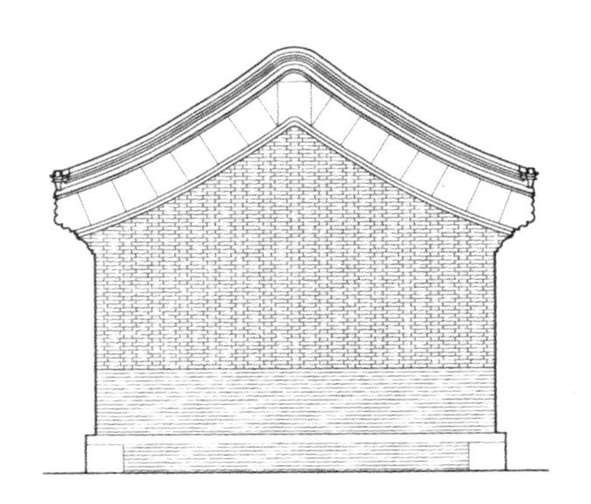

干摆、丝缝做法（山墙）

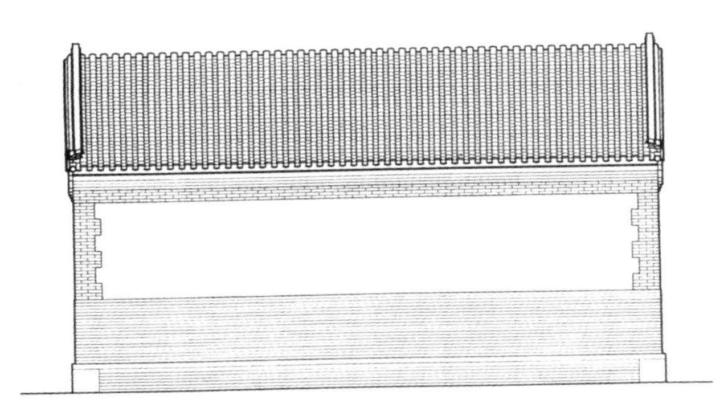

五出五进软心做法（后檐墙）

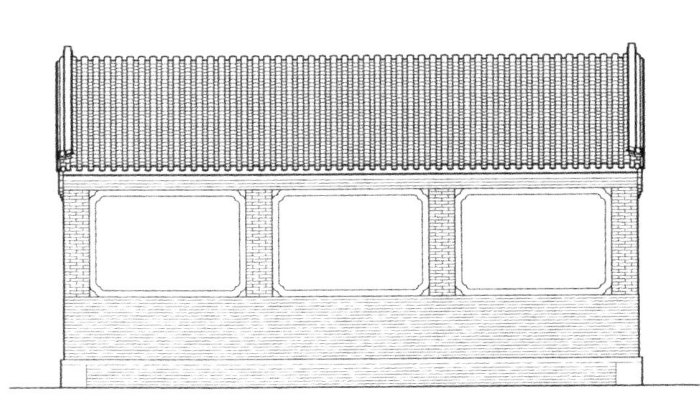

海棠池软心做法（后檐墙）

2. 合抱山墙

两堵独立的山墙贴在一起时，称为“合抱山墙”。合抱山墙是在两幢结构各成体系的建筑并排建造时出现的山墙与山墙之间的构造关系。四合院中采用合抱山墙的地方主要在正房和耳房之间、宅门和倒座房之间以及厢房和厢耳房之间。

组成合抱山墙的两堵山墙必须是互相独立的，两墙之间要留出5～7cm的缝隙。这种做法与现代建筑中的沉降缝十分类似。传统建筑合抱山墙之间的缝隙要做表面处理，一般是沿墙外皮向内退回1～2寸抹出一道凹槽，称为白子。两墙之间缝隙的作用，一方面是为分开两座体积、重量都不同的墙体，以防不均匀沉降，另一方面是为两座山墙合并在一起后让出博缝砖所占的尺寸（图5–3–8）。

由于构成合抱山墙的两座建筑往往是一高一矮、一大一小、一主一次，如正房和耳房、宅门和倒座房，因此，两个山墙的墀头的厚度、尺寸做法都不会相同，要根据设计或传统做法来确定。

3. 檐墙

在习惯于前檐满做木装修的北京传统住宅建筑中，所谓檐墙一般都是指后檐墙。檐墙可用于正房、厢房、耳房、倒座房等各类建筑中。

檐墙的做法常见有两种。一种是当后檐的檐椽、飞檐向外挑出的情况下，椽子下面的檐檩、垫板、檐枋也都要显露出来。在这种情况下，檐墙和屋面檐口之间互无关系，两者不能交在一起，檐墙只能砌至檐枋下皮。做这种处理时，檐墙的墙头可以做成馒头顶、道僧帽或硬顶等形式。

另一种檐墙做法是房屋的后檐椽只搭置在后檐檩上，不向外挑出。这时后檐墙应一直向上砌筑，并与后坡屋面的檐口相交。这种做法叫做封护檐（又称封后檐）。封护檐与屋顶相交时，形成的檐口有各种做法，可根据建筑的等级和宅主人的需求任意选取（图5–3–9）。

图 5—3—8　合抱山墙构造举例

图 5—3—9　后檐墙檐口做法举例

图 5—3—9.1　五层做法

图5—3—9.2 六层做法

图5—3—9.3 五层做法

图5—3—9.4 七层做法

图5—3—9.5 六层做法

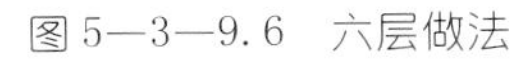

图5—3—9.6 六层做法

图5—3—9.7 三层菱角檐做法

图 5—3—9.8　三层鸡素子檐做法

图 5—3—9.9　三层抽屉檐做法

图 5—3—9.10　七层做法

4. 槛墙

槛墙是房屋前檐窗下的矮墙，高度一般在三尺左右。槛墙的顶面是一块厚为2～3寸的木板，称为榻板。榻板以下为槛墙，以上为木窗。

槛墙虽然构造极为简单，但它也依房屋的等级高低分为很多档次。尤其它处在建筑物正立面非常显眼的位置，是一处可供人们着意装饰的墙面。因此，槛墙的表面也有多种做法，最简单的是砌磨砖对缝的干摆墙面。这种做法简洁大方也不失高雅，因而用得最广。讲究一些的槛墙，在表面做海棠池子。这种做法是在槛墙的外圈砌大枋子、线枋子，墙心内砌斜方砖心或做干摆、丝缝墙面，使槛墙具有线条的装饰美。现存的最讲究的槛墙做法是在大枋子上和海棠池内再加砖雕。具体做法是，在约15cm宽的大枋子内划分出若干长方形或多样形的小池子，在池子内雕刻以花草为主要内容的图案；槛墙的墙心部分，在中心和四角做砖雕，题材与外圈小池子雕刻题材相一致。也有仅仅在海棠池子内做雕刻，周围仅做素面枋子的。如何做法须由房主人或设计人员定（图5—3—10）。

图 5—3—10 槛墙的各种做法举例

图 5—3—10.1 周边圈海棠池子、墙心贴方砖

图 5—3—10.2 墙心内刻卍字不到头图案

图 5—3—10.3 池子大及墙心内满做雕刻

5. 廊心墙和廊门筒子

廊心墙是建筑物外廊两侧的窄墙，位于山面的檐柱和金柱之间，宽度为檐、金柱之间的净宽度外加两侧包金尺寸，高度自廊内地面至穿插枋下皮。廊心墙立面高度分为两段，下段为下碱，占全高的1/3或1/3弱；上段为上身，占全高的2/3或2/3强。下碱做法简单，一般为干摆墙面。上身做法种类较多，最常见的是沿墙外圈做15cm左右宽的大枋子，在大枋子里圈做约6cm宽的线枋子，线枋子以内为墙心，通常做干摆斜方砖心，俗称“膏药幌子”。廊心墙的墙心宽度应为方砖对角线尺寸的1份、1.5份或2份，墙心高度应为宽度的2倍或2.5倍，具体尺寸须视廊心墙的整体尺寸而定。廊心墙还有一种软心做法，是在墙心内抹白灰，表面或素白，或绘山水花鸟画。这种做法在园林中较多，但民宅中也可见到。廊心墙的最上面有一个突出出来、形似画轴的构件，称为小脊子，宽度5～6cm，截面呈半圆形，系用条砖制成。在廊心墙上方的穿插枋和抱头梁之间有一段窄长的墙面，其长度同廊心墙之宽，高同檐枋之高，称为“穿插当子”。其上也有装饰，一般是在灰色的砖面上刻出如意一类阴纹图案。在穿插当子的上方，由抱头梁上皮和檐椽下皮圈出的三角形墙面称为“象眼”。象眼一般是抹白灰面层罩青灰。在灰未干时，用镂刀按画谱镂画出花纹图案，称为“镂活”，也颇具装饰性。这种在象眼处抹灰镂画的做法称为“软做法”，与软做法相对应的硬做法是按墙体的做法砌砖墙。

廊心墙也有做得非常讲究的，那就是在墙心部分做雕刻，或仅做四岔，或做中心四岔，甚至还有在大枋子内卡小池子，在小池子内刻花卉图案的，格外华丽讲究（图5–3–11）。

当带外廊的房子与抄手游廊相连通时，往往要在廊心墙处开辟门洞，这时，廊心墙就变成了廊门筒子。

这种在廊心部分开辟的、供人行走的门筒子，其宽度和高度都要合乎吉门尺寸（一般宽在90cm，高在190cm左右）。门口上方至穿插枋下皮之间的空余部分称为门头。门头也要做装饰。一般做法是在四周做大枋子，沿大枋子里圈做线枋子，线枋子以内为墙心。墙心内砌干摆墙面，墙面或做素活，或在其上做雕刻。门头墙心部分雕刻内容较多的是由两字组成的题额，如“蕴秀”“扬芬”“竹幽”“兰媚”“含珠”“集萃”等等，以蕴含丰富的词汇把人带入幽雅深邃的意境之中（图5–3–12）。

廊门筒子有砖制和木制两种做法。砖制门筒子用砖料摆砌，需预先加工出砖料，雕刻也应事先做好，然后按顺序摆砌成活。木制门筒子，是在墙内预砌木筋（或木砖），然后在外面包筒子板，板表面做麻灰地仗，刷深灰色或黑色油漆。

图 5—3—11　廊心墙的不同做法举例

图 5—3—11.1　素作廊心墙

图 5—3—11.2 墙心内刻卍字不到头图案

图 5—3—11.3 墙心及大边满做雕刻

图 5—3—12 廊门筒子及其门头雕刻装饰举例

图 5—3—12.1 素作廊门筒子及门头

图 5—3—12.2　廊门筒子上门头做花草雕刻

图 5—3—12.3　做油饰彩绘的廊门筒子及门头

图 5—3—12.4　雕有吉祥图案的廊门筒子及门头

四、屋面的种类及其构造

四合院建筑的屋面，有合瓦屋面和筒瓦屋面两种。合瓦屋面多用于中小型民居，筒瓦屋面则多见于大型豪宅。

中国传统建筑的屋面擅长用脊饰进行装点，屋面与脊饰相配，又可以形成多种不同形式。

筒瓦屋面如加正脊，脊饰多采用大屋脊形式。这种大屋脊是按大式做法砌筑的屋脊，脊饰构成自下而上依次为当沟、瓦条（二层）、混砖、陡板、混砖、帽子。在屋面两端排山部分，还要做垂脊与正脊配套交圈。做这种大屋脊时，脊端要安吻兽，垂脊下部要安垂兽和跑兽。跑兽的数量根据建筑物的体量、出檐和瓦号来确定，一般以五只为多。这种采用筒瓦屋面施大屋脊安吻兽的房屋，决非一般官品人家所能采用，通常只能用于王府或相当于王府的建筑。

比上述做法等级稍低的是筒瓦屋面配小式脊。小式脊的构成主要有当沟、瓦条（二层）、混砖、帽子。它是在大式屋脊的构造中减掉了一层混砖和一层陡板，因而脊的高度要低得多。采用筒瓦屋面的房子通常不做正脊，而是将屋脊做成圆弧形状，前后两坡瓦面从脊部漫卷而过，这种做法称为过陇脊。屋面两头采用箍头脊并附以排山勾滴，称为“铃铛排山”（图5－3－13）。

图 5—3—13 筒瓦屋面及其脊饰

图 5—3—13.1
铃铛排山箍头脊

图 5—3—13.2
清水脊蝎子尾

图 5—3—13.3　筒瓦屋面披水排山箍头脊

中小型住宅常采用的合瓦屋面，配以脊饰之后也形成几种不同形式。

一种是合瓦屋面带正脊做法。这种正脊是小式建筑常用的清水脊，其构造自下而上依次为当沟、瓦条、混砖、楣子。正脊的两端是最精彩之处：楣子端头向斜上方（30°～45°）翘起，称为蝎子尾；混砖端头加宽加大并施雕刻，称为花草盘子（这种花草盘子有两种常见做法：水平放置在蝎子尾之下的称为"平草"；陡立在蝎子尾两侧的称为"跨草"）；当沟尽端采用雕有花草纹的型砖，称为盘子和圭角。圭角下方向外侧延伸的部分抹成圆背形状，称为小脊子（其长度相当于两陇瓦宽）。由于小脊子的高度已大大低于正脊的高度，迫使边上与小脊子相交的瓦陇（通常是一陇合瓦，一陇筒瓦）只能压低高度，变成低坡陇。这种做法，旨在突出端头脊饰。这种带正脊的屋面，其两侧不再做箍头脊，仅在梢陇（最外边的一道用筒瓦瓮就的瓦陇）外侧加披水砖檐，称为"披水梢陇"（图5–3–14）。

另一种是合瓦屋面无正脊的做法。这种做法相似于上面提到的筒瓦过陇脊做法。但由于合瓦与筒瓦的构造不同，合瓦屋面做成的过陇脊呈马鞍形，称为"鞍子脊"。屋面两侧做排山脊（亦称箍头脊）。排山脊外面，如做筒瓦勾滴，称为铃铛排山；如做披水檐，称为披水排山。二者比较，铃铛排山较为复杂讲究，是一种较为高级的做法（图5–3–15）。

民间做法中，还有仰瓦灰梗屋面、干槎瓦屋面、棋盘心屋面、灰背顶屋面等多种简易节约的屋面做法。仰瓦灰梗屋面是屋面只铺底瓦，不铺盖瓦，在底瓦瓦陇接缝处（俗称蛐蜒当子）用麻刀灰塑起4~5cm的灰梗以防漏雨。这种做法比起合瓦屋面可节省约1/2的瓦件。干槎瓦屋面是利用板瓦自身的弧度和斜度，在瓮结时令其互相遮压，不用灰梗遮挡依旧达到防漏效果。这种做法一方面要求瓦件造型规整，另一方面要求瓦工师傅有高超的手艺，是一种很有特色的民间做法。棋盘心屋面是将合瓦屋面的当心一部分做成灰背或石板瓦，以节省部分瓦件。一般是在每一间的中部偏下部分做灰顶或石板瓦顶，其余脊部和分间部分依然瓮合瓦，称为"麦穗"。这种屋面既节约材料又颇具艺术特色，是民间常用的屋面做法。以上三种简易的屋面都不加脊饰，简单、朴素、经济，应当加以继承（图5–3–16）。

灰背屋面是表面不铺瓦的屋面，凭灰背形成的密实的面层抵防雨水的渗透。这种做法，或用于屋顶的局部（如棋盘心屋面的屋心部分），或用于灰平台屋面以及盝顶、天沟等处。传统工艺中凭青灰、白灰加麻刀赶轧密实作为防水层的做法，有非常好的防水效果，值得借鉴。

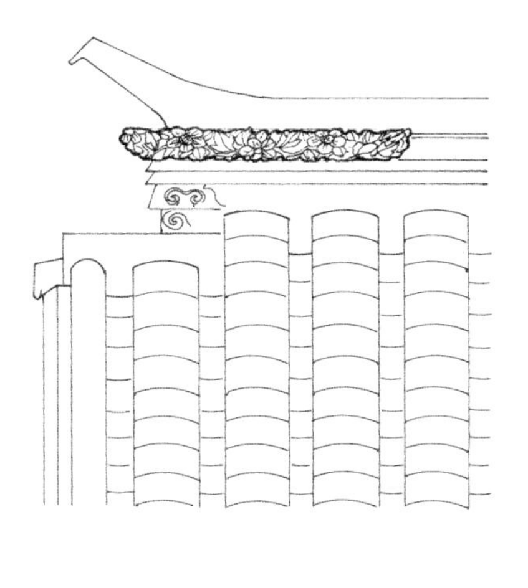

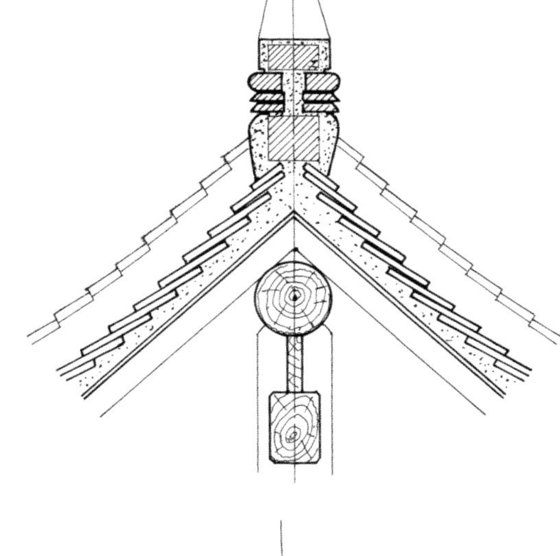

图 5—3—14 合瓦屋面清水脊细部构造

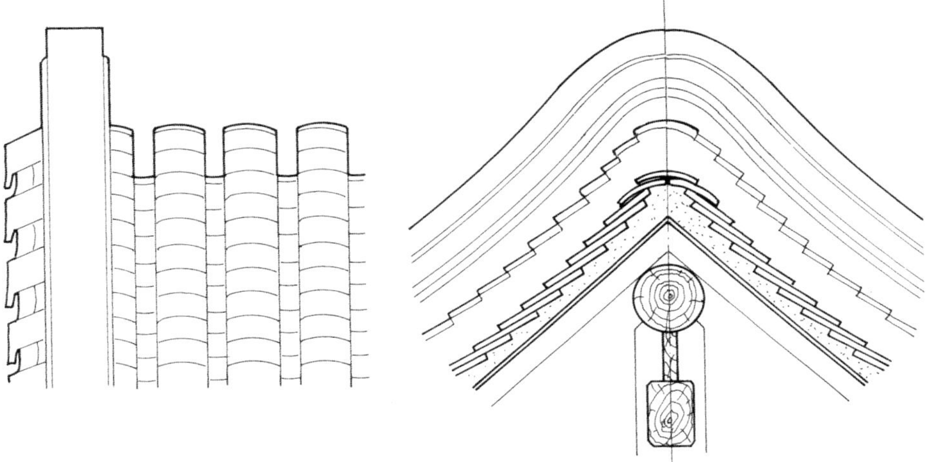

图 5—3—15 合瓦屋面鞍子脊细部构造

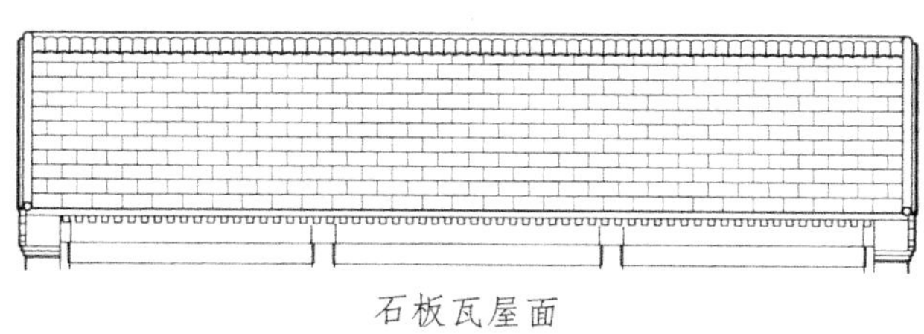

石板瓦屋面

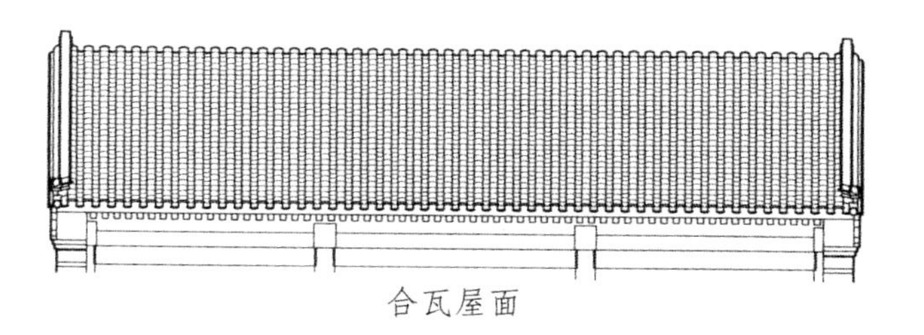

合瓦屋面

仰瓦灰梗屋面

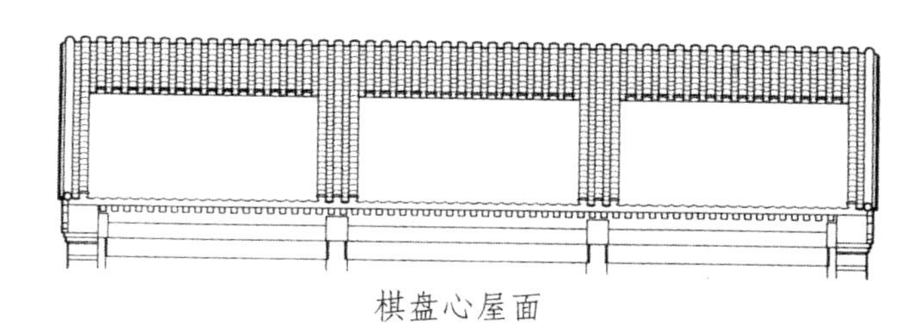

棋盘心屋面

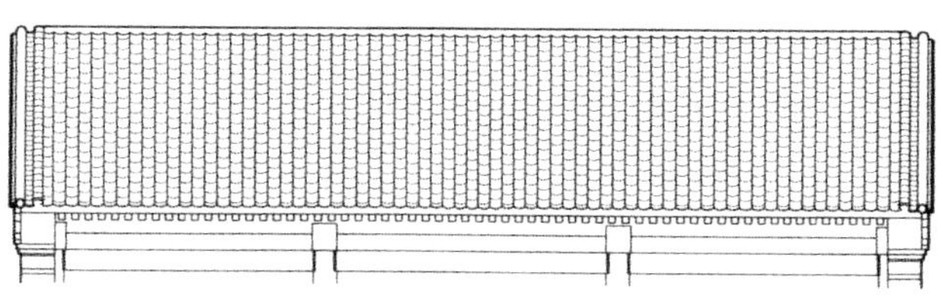

干槎瓦屋面

图 5—3—16 干槎瓦、仰瓦灰梗和棋盘心屋面举例

五、台基与地面

中国传统建筑是建在台基之上的。在一进或二进的四合院中，正房、宅门和垂花门的台基在同一个水平高度上，倒座房、东西厢房、游廊的台基比正房要低一层台阶（约13～15cm）。这种高低的差异，是为了突出正房和宅门、垂花门。

如果是三进院落的四合院，第三进院落的室外水平高度要比前边的院落高出13～15cm（即一层台阶的高度），后院的厢房或正房，也要随室外地面的长高而提高自身台基的高度。后院各房的台基提高以后，其正房台基比前院正房要高出一阶（13～15cm），厢房的台基也比前院厢房的台基高出一阶，处在与前院正房台基等高的位置。如果后院只有后罩房的话，后罩房的台基可与前院正房的台基相平。

由于北京四合院大多为坐北朝南院落，院内的雨水要从东南方向排出院外，故院内的地平至少要高出院外1~2步台阶，以保证雨水排放的通畅。这样，我们就得到了一个室外地平和台基标高以等差级数增减的关系（图5–3–17）。

室外地面的标高变化，保证了雨水的排放；而台基标高的变化，则确保了台基对于地面的相对高度，具有防潮、保护木柱根部不受雨水侵蚀等功能。

四合院各房间的台基中，厢房台基（倒座房、游廊同）高出室外地面2~4步，正房（宅门、垂花门同）台基高出室外地面3~5步。台基的边沿安装阶条石，台基转角部分安装埋头石，台基下部安装土衬石，柱根下部安装柱顶石，房间的台阶安装踏跺石，两侧安装垂带石。这些石活，对保护和加固台基、衬垫柱根、方便出入、美化基座都有非常重要的作用（图5–3–18）。

图 5—3—17　四合院室外地平和台基标高变化示意图

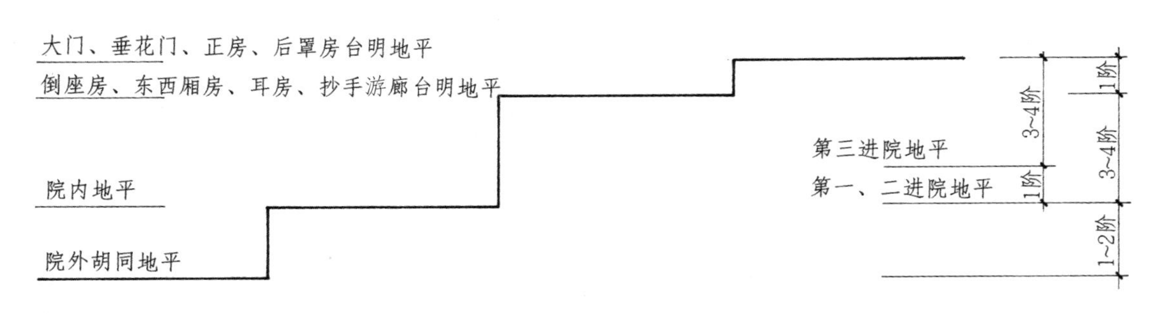

图 5—3—18　台基及基石活部位名称

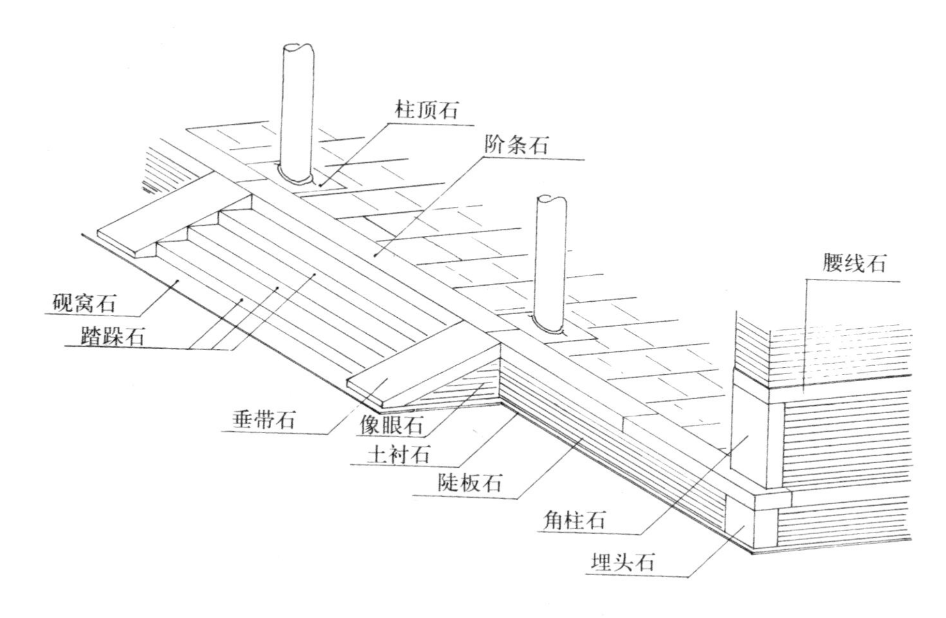

四合院的地面，有室内地面、廊内地面以及庭院甬路、散水等部分。传统地面做法多用砖料铺墁，但依材料、做法不同，可分为若干等级，差别很大。最高等级的“金砖墁地”多用于宫殿坛庙建筑，极少见于民宅。民居中地面做法较讲究者为细墁地面。墁这种地面时，要求对砖料进行砍磨加工，使砖料规格统一、尺寸精确、棱角完整、表面光平。铺墁后还须用桐油浸泡，称为“钻生”。用这种材料和工艺墁出的地面平整、细致、整洁、美观、坚固耐用。

稍逊于细墁地面的，称为淌白地面。这种地面只要求砖料砍磨四个小面，称“干过肋”，不要求砍磨大面，墁出的地面稍显粗糙。

最为简易的地面做法为糙墁，砖料不进行砍磨加工，砖块之间缝隙较大。

糙墁地面只能用于室外，或做甬路，或做散水；细墁地面用于讲究的室内；淌白地面则多见于一般民宅的室内地面做法。

由于传统建筑的地面是由规格不同的砖块铺墁，因此，就产生了多种多样的砖块排列形式。这种排列颇具艺术性，也是值得借鉴的（图5–3–19）。

图 5—3—19 地面的砖块排列形式

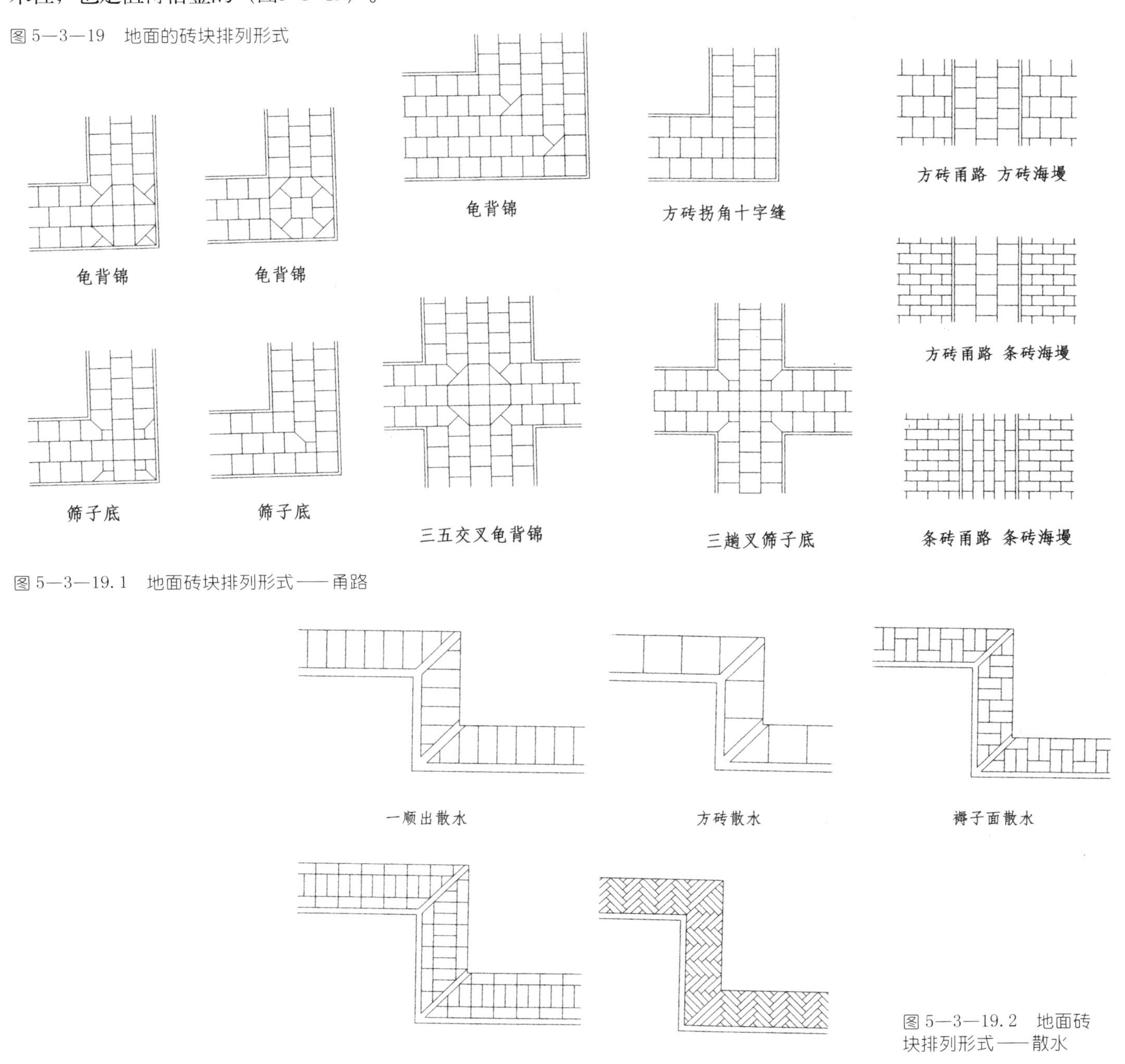

图 5—3—19.1 地面砖块排列形式——甬路

图 5—3—19.2 地面砖块排列形式——散水

第四节　外檐装修

中国传统建筑将木质的门、窗、户、牖、帘架、隔断、楣子、花罩、天花、吊顶等等，统统称为“装修”。装修依其所处位置不同，又分为外檐装修和内檐装修。位于室外或用来分隔室内外空间的装修称为外檐装修，用于室内的装修称为内檐装修。

北京四合院的外檐装修主要有街门、屏门、隔扇、帘架、风门、支摘窗、倒挂楣子、坐凳楣子、牖窗、什锦窗等，现分述如下。

一、街门

这里所说的街门，是指安装在宅门柱间或墙洞内的木质门框、门扇及其附属部分。街门可以分为两类：一类是用于广亮门、金柱门和蛮子门的街门，另一类是用于如意门、西洋门和随墙门的街门。

1. 用于广亮门、金柱门、蛮子门的街门

用于广亮门、金柱门、蛮子门的街门尽管安装的位置不同（用于广亮门时，安装在中柱间；用于金柱门时，安装在金柱间；用于蛮子门时，安装在檐柱间），但是它的构造大同小异，都是由槛框和门扇及其附件组成的。这种门的槛框部分由抱框、门框、上槛、中槛、下槛、走马板、余塞板、余塞腰枋、连楹（又称门龙）、门簪以及抱鼓石等构件组成。

槛框是大门外框的骨干构件，其水平者称为槛，垂直者称为框。横槛之中，附在脊枋（或金枋或檐枋）之下的为上槛，附在地面之上的为下槛，两抱框分别附在两柱子侧面，这四根构件将大门的槛框与木构架结合在一起。中槛位于上槛和下槛之间，它是确定门口高度的水平构件，中槛的下皮与下槛的上皮之间的距离，即门口的高；门框是确定门口宽度的垂直构件。它们位于两抱框内侧，上端交于中槛，下端交于下槛，两门框之间的净距离即为门口的宽。门口的宽与高都应当合乎门光尺的吉尺寸。门口以外的部分用木板封堵起来，上方的木板称为走马板，两侧的木板称为余塞板。卡在下槛下面的两块抱鼓石是大门槛框的重要附件，抱鼓石临街一面尺度较大，雕成圆鼓形或方墩形作为象征和装饰；院内一面为扁方形，尺度较小，其上嵌有铸铁海窝，为安装门扇之用，称为门枕。两门枕石海窝之间的距离由门扇的宽度决定。在中槛内侧，附有一根通长的横木，与中槛成⊢形相交，两端交于柱上，称为连楹。连楹上面与门枕石海窝对应的位置挖有两个洞眼，是用来安插大门上端的门轴的。连楹与中槛凭四只门簪锁合。门簪头部呈六角形（亦有圆形和方形的，但较少见），其径寸略小于中槛之宽。门簪头长约为直径的1.2～1.5倍，后尾为一长榫，穿透中槛和连楹，并穿出楹外。透榫与连楹间凭木梢锁合（图5–4–1）。

图 5—4—1 用于广亮门、金柱门、蛮子门的街门的槛框及门扇构造图

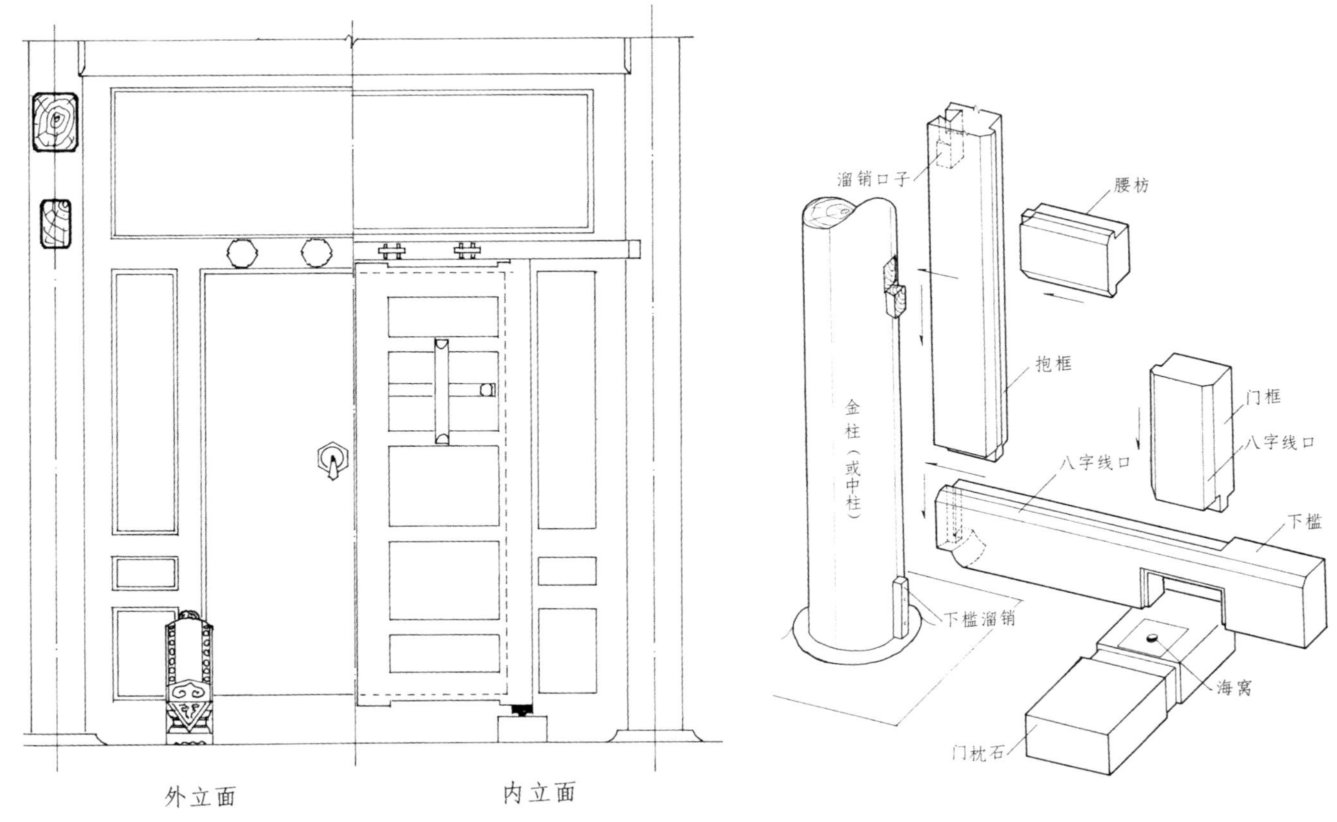

图 5—4—1.1 外立面与内立面

图 5—4—1.2 槛框构造做法之一

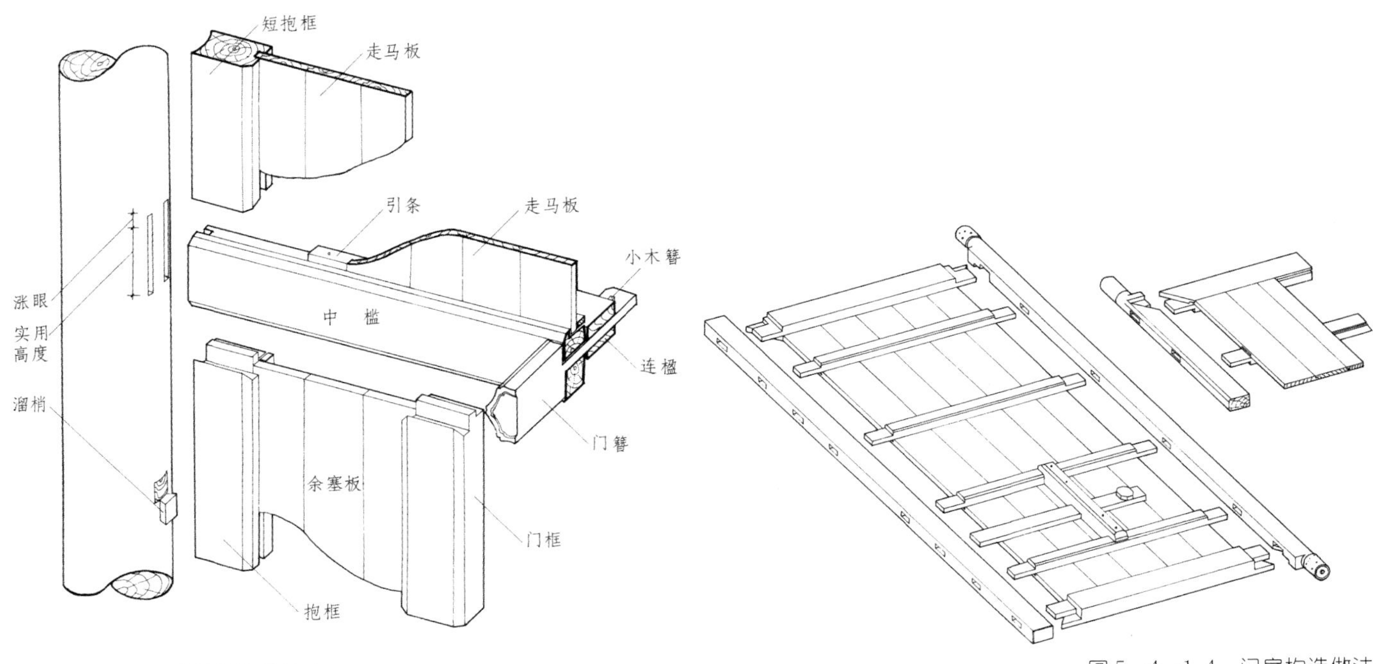

图 5—4—1.3 槛框构造做法之二

图 5—4—1.4 门扇构造做法

街门的门扇为对开的棋盘门。棋盘门又称攒边门，是由门边、门轴、抹头和门心板等构成的木质大门，其正面表面光平，背面显露出门边、抹头、穿带等构件，并附有门插关。门扇内侧的一根边梃略长于门扇，做上下碰头，外侧一根做上下门轴。门轴截面为圆形，外面安铁制套筒，上端插入连楹的碗口之内，下端装踩钉置于门枕石的铸铁海窝之内，上下两点形成转动枢纽以利开启关闭。用于门轴转动部位的铁件，在清工部《工程做法则例》中称之为"寿山""福海"，上部的称为"寿山"，下部的称为"福海"（图5-4-2）。

街门外面安装有铙钹形状的铜制饰件，是扣门用的响器，称"门钹"。门钹还有装饰作用。门扇下角附有保护门板的铁皮包叶，剪成如意云状，用小泡头钉钉于门板表面，称壶瓶叶子（图5-4-3）。

2. 用于如意门、随墙门、西洋门的街门

用于如意门、随墙门、西洋门的街门，有一个共同点，就是门口开在门墙之上，洞口窄小，四周没有大面积的槛框余塞，因此构造较为简单。这种门的门口仅有上下左右四框，分别称为上槛、下槛和门框。槛框的里口为吉尺寸。门扇为对开扇，下有门枕，上有连楹。由于门口窄小，锁合上槛和连楹的门簪只用两只。门扇做法与广亮门的街门类似，唯体量较小。

民宅中还有诸如栅门、大车门之类的临街大门，尽管不是正式宅门，但仍有保留和参考价值（参见图5-1-9）。

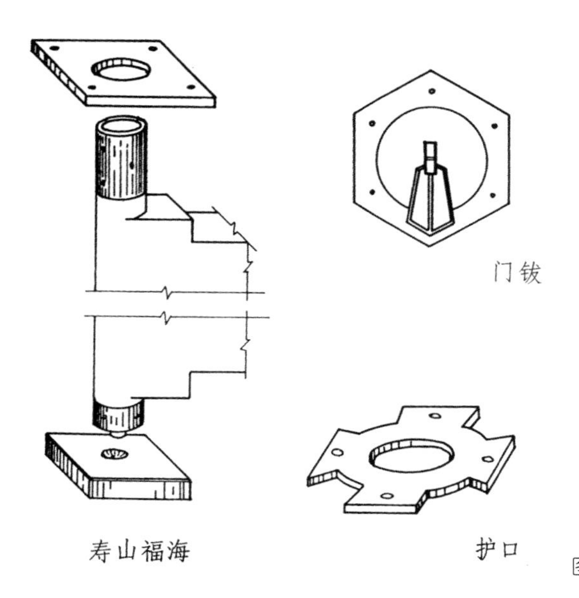

图 5—4—2　街门转轴及铁件图

图 5—4—3　门钹和门包叶

图 5—4—3.1　门钹

二、屏门

图 5—4—3.2 门包叶—壶瓶叶子

用于四合院的屏门主要有两种：一种是分隔院内空间的屏门；一种是安装在垂花门后檐柱间的屏门。

1. 用于分隔院内空间的屏门

屏门，顾名思义，是起屏蔽遮挡作用之门。它仅起屏风作用，无防卫功能。

关于屏门分隔院内空间的功用，在本章第二节已经提到。它通常是用在第一进院内宅门两侧和院子另一端与之对称的位置，将宅门、客厅（兼书房）、厕所或库房等几部分功能不同的空间分隔开来。

四合院中用以分隔空间的屏门及其隔墙，有纯木质的，也有做砖隔墙安木质屏门的。这是因为空间的划分不是永久不变的。遇有变动时，木质屏门及隔断拆改起来比较方便。在固定不变的部位则多用砖墙来作为隔墙（参见图5–2–6.1）。

2. 用于垂花门后檐柱间的屏门

用于垂花门后檐柱间的屏门，是专门用来屏蔽外宅视线的，非遇特殊事件终年都不开启。由于屏门只起屏挡作用，不具防卫功能，因此，制作门用的木板也仅有1.5寸厚（约4.5～5cm），用于固定屏门的金属构件也极为纤细小巧。它们分别是鹅项、碰铁、屈戌、海窝及木制的楹子等（图5–4–4）。

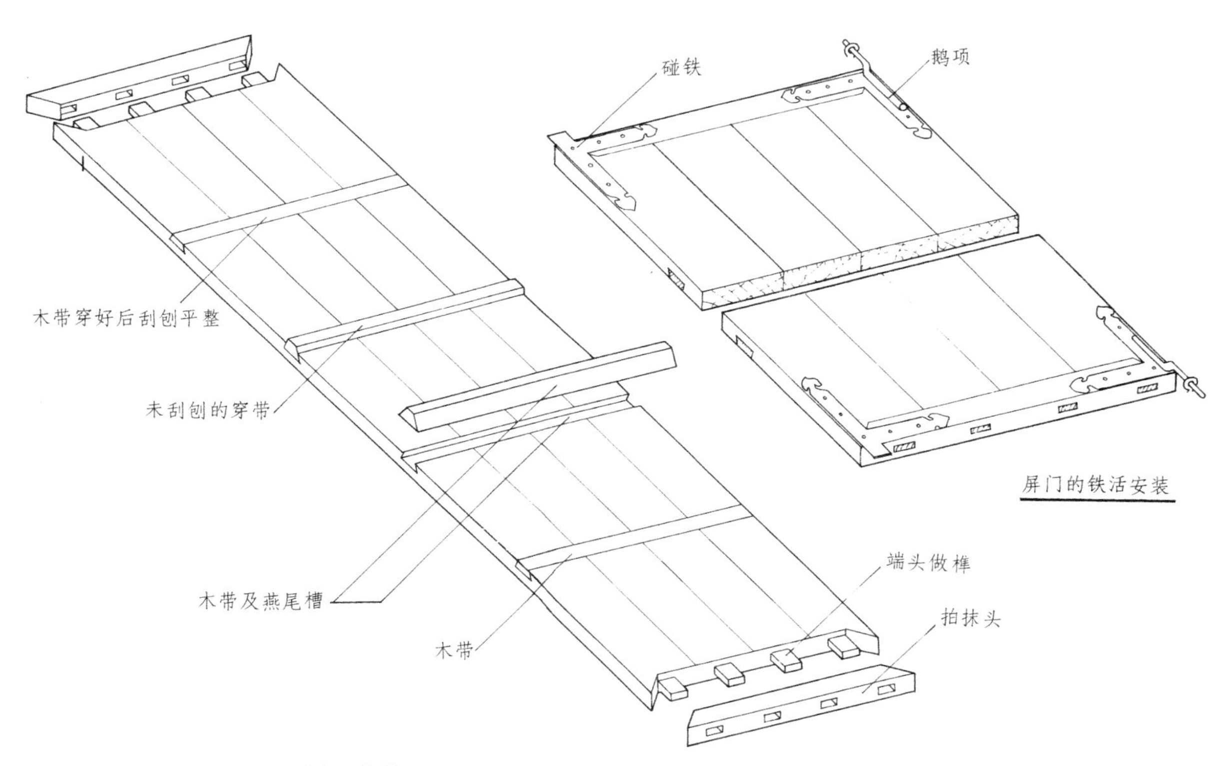

图 5—4—4 木质屏门及其金属构件

三、隔扇门、支摘窗、帘架及风门

1. 隔扇门

隔扇门在宋代称“格子门”，是安装在建筑物的金柱或檐柱间用以分隔室内外空间的木装修门。四合院民居的隔扇门安装在正房或厢房的明间，一樘四扇。隔扇门的外围有槛框与柱、枋等大木构件连接固定。槛框的构成与隔扇所在位置有直接关系。隔扇安装于金柱间时，在它的上方要设横陂窗，这主要是为了限制隔扇的高度。带横陂装修的槛框构成为上槛、中槛、下槛以及两侧的抱框。下槛与中槛之间的净尺寸是隔扇的高，这个高度要合乎吉门尺寸的要求。在一般情况下，确定中槛的位置时，要参考檐柱的高度，如果檐柱高度适中的话，可将中槛下皮定在檐枋下皮再下返一个上槛宽的位置。中槛的位置确定以后，横陂的高度也就随之确定了。

隔扇安装于檐柱间时可不设横陂窗，如果因檐柱过于高大造成隔扇尺度失调时，可将隔扇限制在适当高度，空余的部分做横陂窗。不过在民居中这样的情况极为罕见。

横陂窗是固定窗，不开启，每间房有3～5扇，扇与扇之间用短间框隔开（图5—4—5）。

2. 支摘窗

建筑的明间安装隔扇，次间安装支摘窗，二者不仅相邻，在尺度上也有内在的联系。

支摘窗是设在槛墙之上的，窗的高度决定于槛墙的高度，而槛墙的高度又与隔扇的尺度有关。传统装修中有“四六分隔扇”之说，指将隔扇全高定作10份，以隔扇中绦环下抹头下皮为界，上部占6份，下部占4份。这条分界线，就是槛墙榻板上皮的位置。现在，我们以图5—4—6为参照，试计算一下槛墙的高度。假定该建筑檐柱高为300cm，檐枋高为30cm，下槛高为24cm，上槛（假设檐枋之下有上槛）高为16cm。根据以上尺寸可算出隔扇净高230cm，以四六分之，下部高为92cm，加下槛之高24cm，得槛墙高为116cm。如果觉得这个尺寸偏高，还可适当进行调整。槛墙的尺度确定之后，用中槛下皮至地面的尺寸减去这段高度，所得即为支摘窗尺寸（参见图5—4—5）。

支摘窗是北京传统民居用得最多的一种窗式。其构成是，沿次间两侧安抱框，在圈定的范围内，居中安间框，将窗分为左右两部分，每部分又分为上下两段；窗扇均设内外两层，上面的窗能支起，下面外侧的护窗能摘下，支摘窗也因此而得名。

图5—4—5 隔扇、横陂、支摘窗立面图

3. 帘架和风门

前文提到，明间隔扇门一樘四扇，平时两侧的扇固定不开启，仅开启中间；两扇。尽管如此，作为居住用房，两扇隔扇形成的门洞也显得太大而不合用，不仅开启不便，对冬日保暖、夏日防热也很不利。为限定门口 尺寸及为挂门帘方便，人们在开启的两扇隔扇门外侧又贴附一层装修，称为帘架。

帘架，顾名思义，为悬挂门帘的架子。帘架的宽度相当于两扇隔扇宽，高度至中槛下皮，它外圈的框架由两根边梃和两根抹头组成，称为帘架大框。两抹头之间为帘架横陂，帘架的里口（上至中抹头底皮，下至下槛上皮，左右至边梃里皮）应符合吉门尺寸。这种仅有大框的帘架，适合于坛庙、寺院等公共建筑，用于民居时，门口仍显太大，不便使用，须在此范围内另辟一适合于居住用的门口，于是便产生了适合于家居用的帘架和风门。这种帘架，是按吉门口尺寸，在帘架大框内居中留出门洞，在门洞两侧和上方的空间分别安装余塞和楣子。这个由余塞（俗称帘架腿子）和楣子围成的小门口就是适于家居的门口。门口内装单扇门，称为风门。风门一般向外开启。风门内侧，冬天可挂棉门帘，夏天可挂竹门帘，以便保温通风。在风门下面贴附在下槛之外的门槛称为哑巴槛。帘架下端有固定边梃的木构件，其上雕有荷叶纹饰，称为荷叶墩；固定边梃上端的是刻有荷花图案的木构件，称为荷花栓斗。这两件木质构件既有装饰功能又有吉祥寓意（图5—4—6）。

四、牖窗、什锦窗

中国古代把开在墙面上的窗户称为牖窗，牖窗与满装修的窗是不同的。前文谈到的什锦窗，即是典型的牖窗。北京四合院中，除什锦窗以外的牖窗一般用在建筑物的山墙或楼房的山墙和后檐墙上，体量、形状都须根据功能要求和环境情况而定。牖窗体量要大于什锦窗，形状以六角、八角、圆形等规则几何图形为主。因牖窗是用于居住建筑之上的正规窗户，故不能似什锦窗那样活泼随意、变幻多端。

关于什锦窗，前文已有较详细的叙述，此处不再重复。

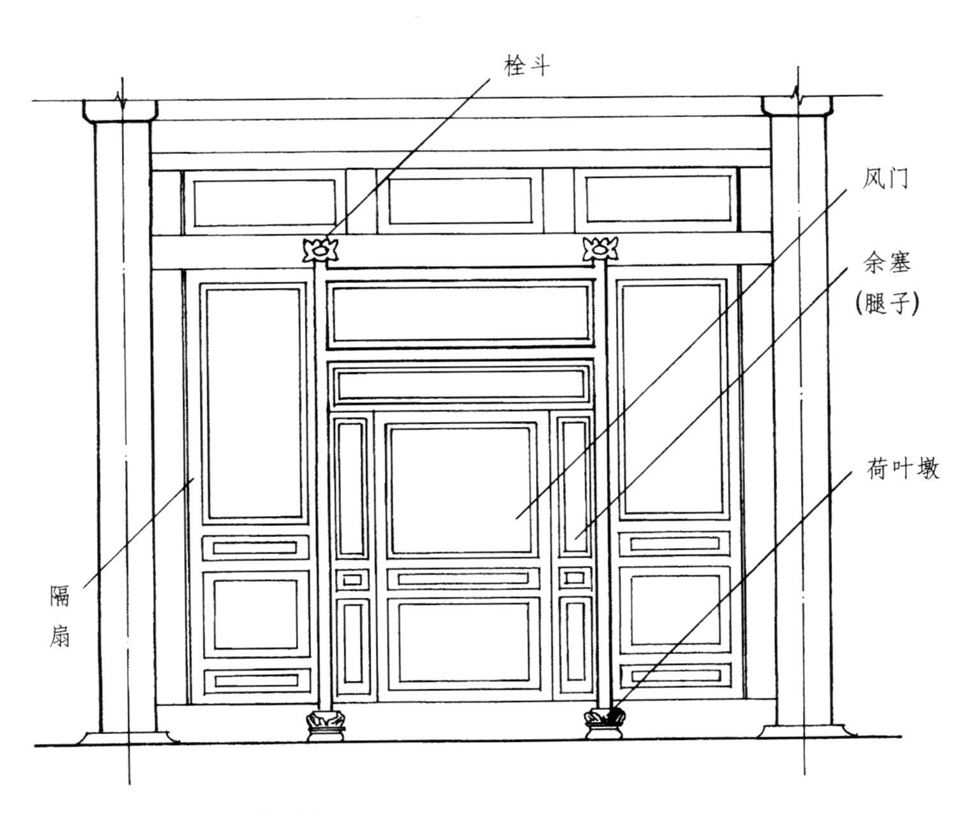

图5—4—6 用于民居的帘架

五、倒挂楣子和坐凳楣子

倒挂楣子是安装在房屋外廊檐枋之下的木装修，抄手游廊的檐枋之下也常安装倒挂楣子。倒挂楣子有棂条楣子和雕花楣子两种形式。倒挂楣子的长为柱间净尺寸，宽（高）1～1.5尺，由边框、棂条及花牙子组成。雕花楣子则由边框和花心组成（图5－4－7）。坐凳楣子是安装在房屋外廊檐柱之间可供人坐息的木装修，抄手游廊的柱间也安装坐凳楣子。坐凳楣子由楣子和坐凳面构成，其高度在1.5～2尺之间，以适合人坐息为度。

六、外檐装修的棂条花格

不论是隔扇帘架还是支摘窗、楣子、风门，其窗心部分都由棂条花格组成。排列规律有序的棂条图案不仅给人以美的视觉感受，还以它蕴含的丰富寓意给人以美的心理感受。

四合院常用的棂条花格有步步锦、灯笼框、灯笼锦、龟背锦、盘长、冰裂纹及由这些基本图形组合演变出的各种图案。

步步锦是由长短不同的横竖棂条按规律组合排列的一种窗格图案，在四合院民居中应用很广。步步锦窗格棂条之间有工字、卧蚕或短棂条（俗称矮佬）连接支撑，按序排列出各种不同形式。步步锦窗格受人赏识，不仅在于它的图形优美，富有韵律，“步步锦绣，前程似锦”的美好寓意更令人对它喜爱有加。

灯笼框是将古代人夜间照明用的灯笼形状加以提炼抽象形成的棂条图案。灯笼框窗格棂条排列疏密相间，棂条间巧妙地用透雕的团花、卡子花进行连接，既有构造作用，又极富装饰功能。灯笼框中间形成的较大空白不仅利于采光，而且用作室内装修时还可以在其间作画题诗，尽情装点，使之充满文化气息。而灯笼的图形，又向人们隐示着“前途光明”的吉祥寓意。

龟背锦是以正八角形为基本图案组成的窗格形式。古人以神龟为长寿吉祥之物，以采自龟甲纹的八边形图案做窗格棂条，不仅给人以美的感受，还向人们传达着“延年益寿”的吉祥信息。

盘长是用封闭的线条回环缠绕形成的图形，采自印度古代的传统图案。盘长是佛家八种宝器的一种，它的寓意是“回环贯彻，一切通明”，因而是人们喜用的图案之一。

冰裂纹这种将冰面炸裂产生的自然纹理，经过概括提炼和艺术加工，作为窗格图案用于装修之中，给予人们的不仅有美的感觉，更有回归自然、融于自然的怡悦感受。

四合院外檐装修采用的窗格棂条形式多样、内容丰富、寓意深刻、美轮美奂，是高雅的艺术结晶。

外檐装修的棂条截面及其棂间空当，均有较固定的尺寸比例。以步步锦窗格为例，棂条看面宽为6分（合19mm），进深为8分（合25mm），表面做成指甲圆（又称泥鳅背）形状。棂条间的空当相当于3～3.5倍横条看面之宽，约合6～7cm,称为“一空三棂”，如果过 疏或过密，都会使比例失调，失掉传统装修原有的风格（图5—4—8）。

图5—4—7　倒挂楣子和坐凳楣子举例

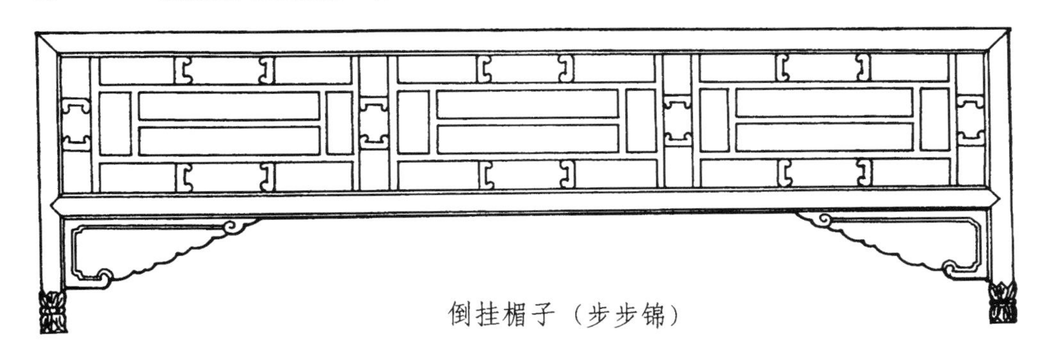

倒挂楣子（步步锦）

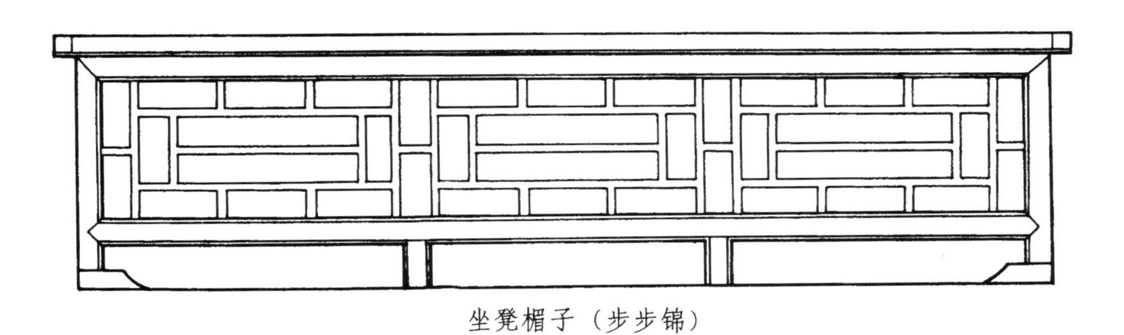

坐凳楣子（步步锦）

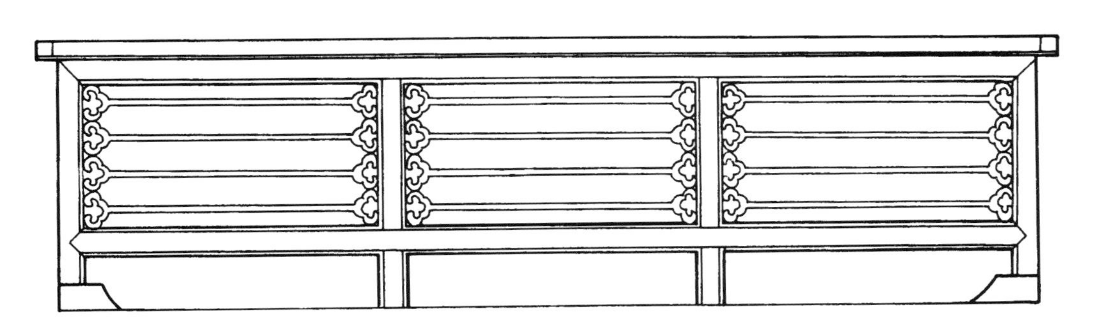

坐凳楣子（金线如意）

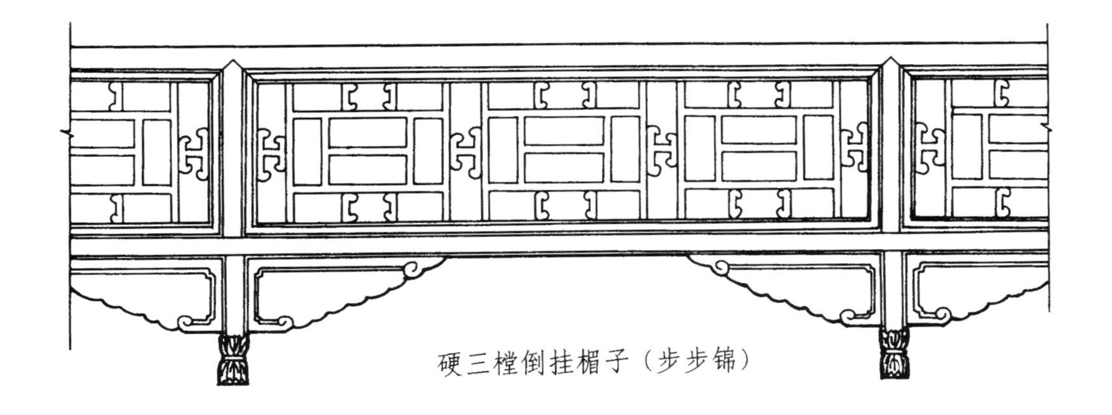

硬三樘倒挂楣子（步步锦）

图5—4—8　各种棂条花格举例

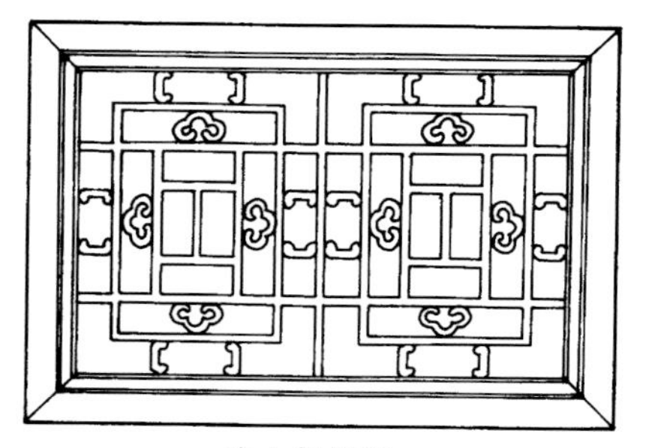
套方灯笼锦

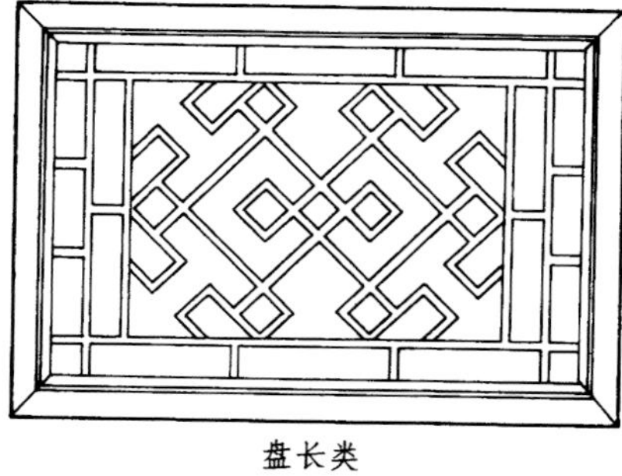
盘长类

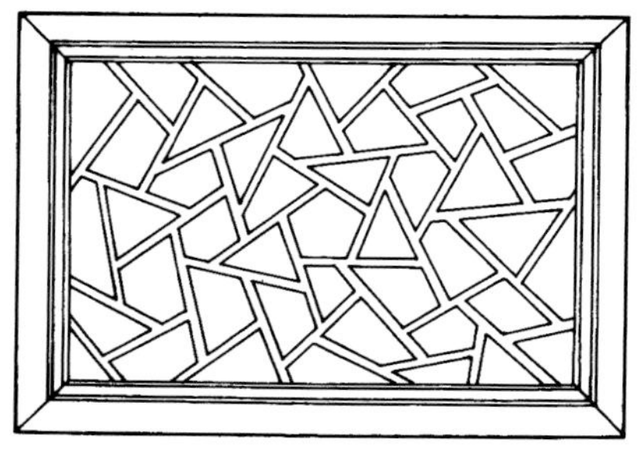
冰裂纹

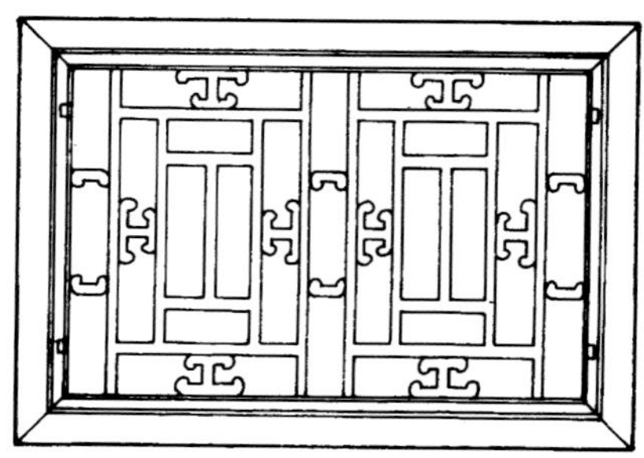
工字卧蚕步步锦

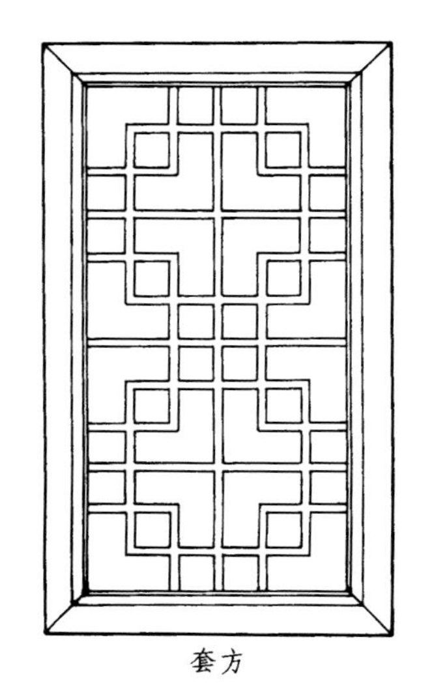
套方

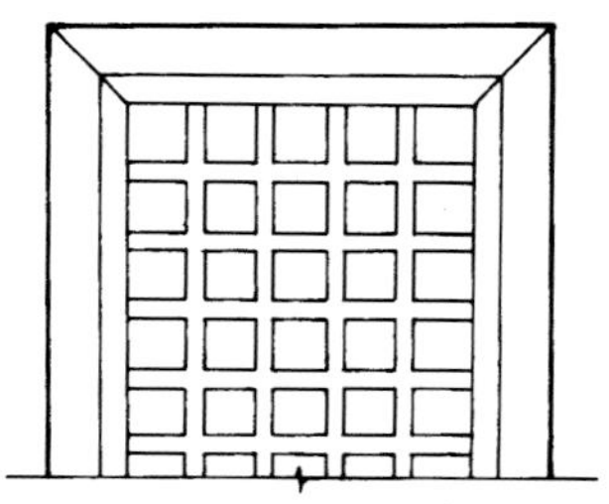
正搭正交方眼槅扇

套方灯笼锦

正搭正交万字窗

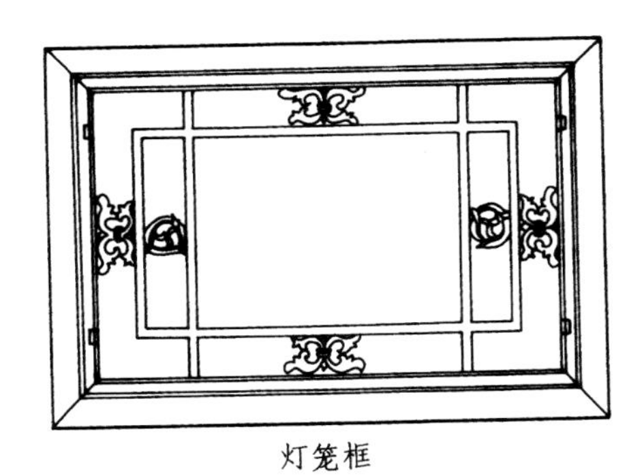

灯笼框

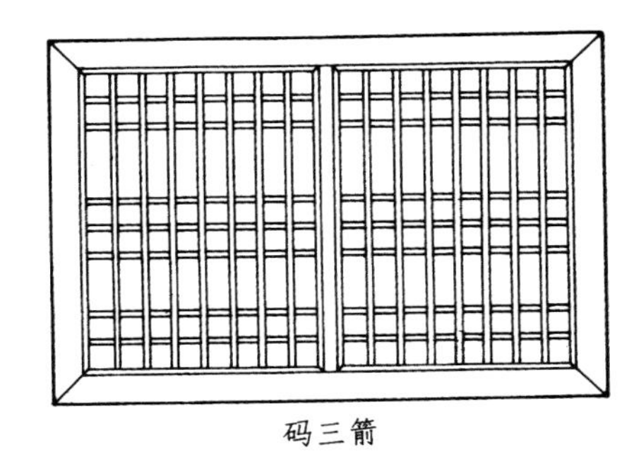

码三箭

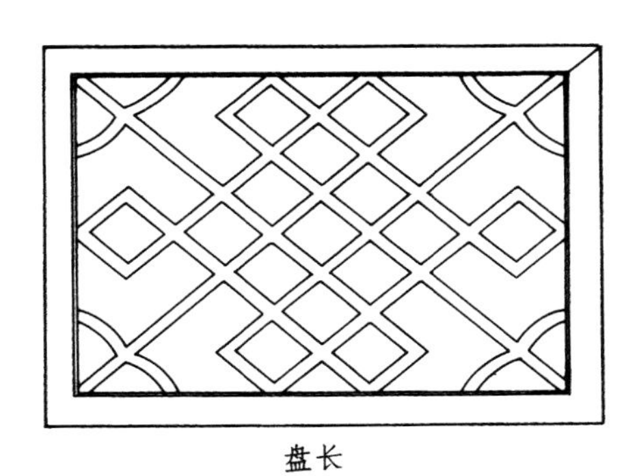

盘长

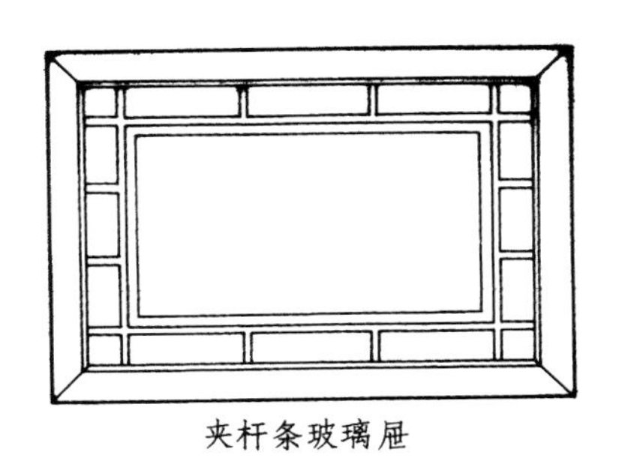

夹杆条玻璃屉

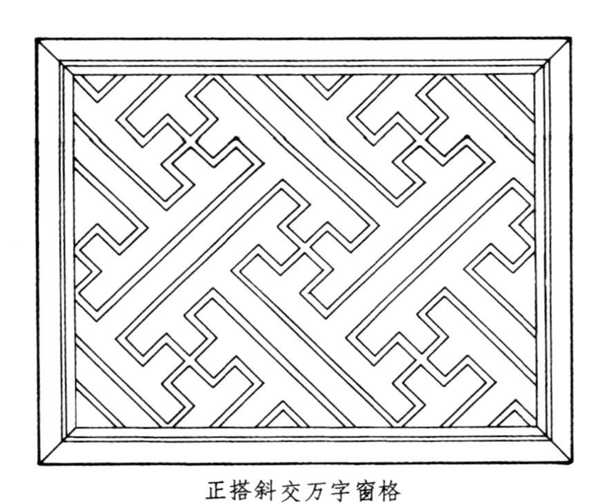

正搭斜交万字窗格

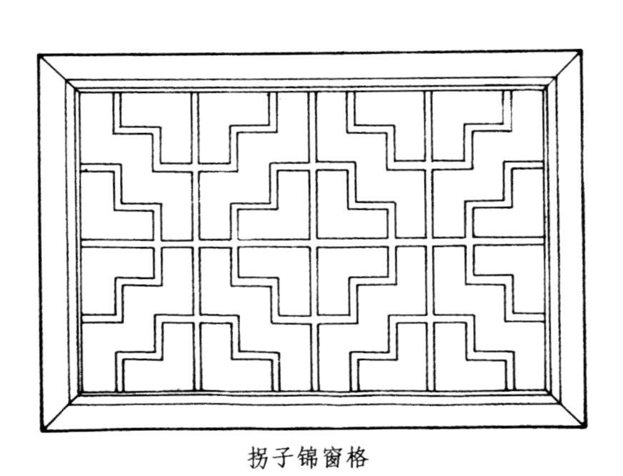

拐子锦窗格

龟背锦

第五节　内檐装修

四合院住宅中常见的内檐装修主要有：碧纱橱、几腿罩、落地罩、落地花罩、栏杆罩、床罩、圆光罩、八角罩、博古架、板壁、墙裙、天花顶棚等。现分述如下。

一、碧纱橱

碧纱橱是用于室内的隔扇，常用在进深方向柱间，由外框和隔扇组成。外框主要有上槛、中槛、下槛、抱框、短抱框，其构造与外檐隔扇边框相类同。组成碧纱橱的隔扇数量，由房间进深的大小决定，一般有6扇、8扇、10扇、12扇不等，且多为偶数，每扇宽度40～50cm。碧纱橱的功能主要用以分隔空间，中间（或一侧）要留门供人出入，在门口还要挂帘子，所以也要安装帘架。用于室内的帘架与外檐帘架略同，只是由于在室内，不考虑防寒或通风问题，故不再安装腿子、风门，只有帘架大框即可。室内碧纱橱隔扇之高一般应与外檐隔扇高相同，以使内外交圈。如外檐隔扇有横陂时，内檐碧纱橱也应有横陂。民居当中碧纱橱隔扇多采用灯笼框棂条图案。隔扇花心（包括仔屉、棂条、团花、卡子花在内）部分做成两层，一层为固定的，一层为可以取下来的，中间夹以绢纱，称为两面夹纱做法，碧纱橱即以此得名。绢纱或为单一素色，或在灯笼框中心空白处绘画题诗，十分雅致。中槛以上的横陂棂条，也随隔扇棂条纹样，统一而和谐。碧纱橱是可以移动和改装的，每扇隔扇、横陂与大框间都有暗梢固定，需要拆改移动，重新组合室内空间时，只需将隔扇摘下，重新组装即可（图5－5－1）。

图5—5—1　碧纱橱（本图引自《紫禁城宫殿建筑装饰内檐装修图典》）

图5—5—1．1
碧纱橱正立面图

图5—5—1．2 碧纱橱举例

二、几腿罩、落地罩、落地花罩、栏杆罩、床罩

几腿罩、落地罩、落地花罩、栏杆罩、床罩均属花罩类，在功能和构造上有共通之处。一般多用在进深方向。它们是用来划分室内空间的，但又与碧纱橱的功能不同，它既有分隔作用又有沟通作用。

1. 几腿罩

几腿罩是花罩中最简单也是最基本的一种，其他罩都是由它发展而来。几腿罩由两根横槛（上槛、中槛）和两根抱框组成，两横槛之间是横陂，分成五当或七当。空当内安装棂条花格横陂窗。中槛与抱框交角处各安花牙子一块。从立面看，这种罩很像一个八仙桌或一个茶儿，两侧的抱框，恰似几腿，这也许就是几腿罩名称的由来（图5–5–2）。

2. 落地罩

如果在几腿罩两侧，贴抱框各附一扇隔扇，则变成了落地罩。“落地”系指罩的两侧有隔扇或雕刻饰件延伸至地面。落地罩的落地隔扇，上端做暗梢与中槛下皮固定，下端并非直接落地，而是落在一个须弥墩（用木头做成的类似须弥座形状的构件）上面。须弥墩与隔扇下抹头之间也有暗梢固定。隔扇与中槛交角处，各安装花牙子一件作为装饰（图5–5–3）。

图5—5—2 几腿罩

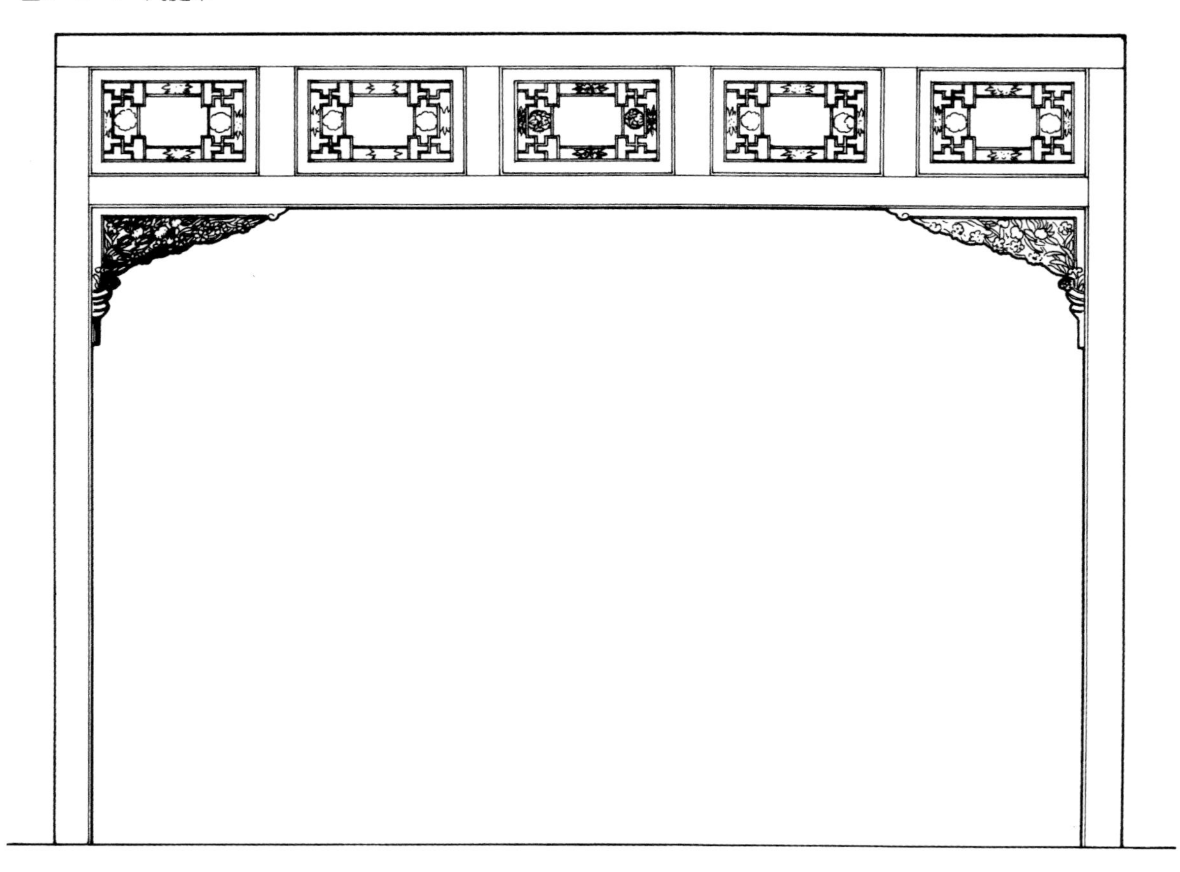

图5—5—3 落地罩

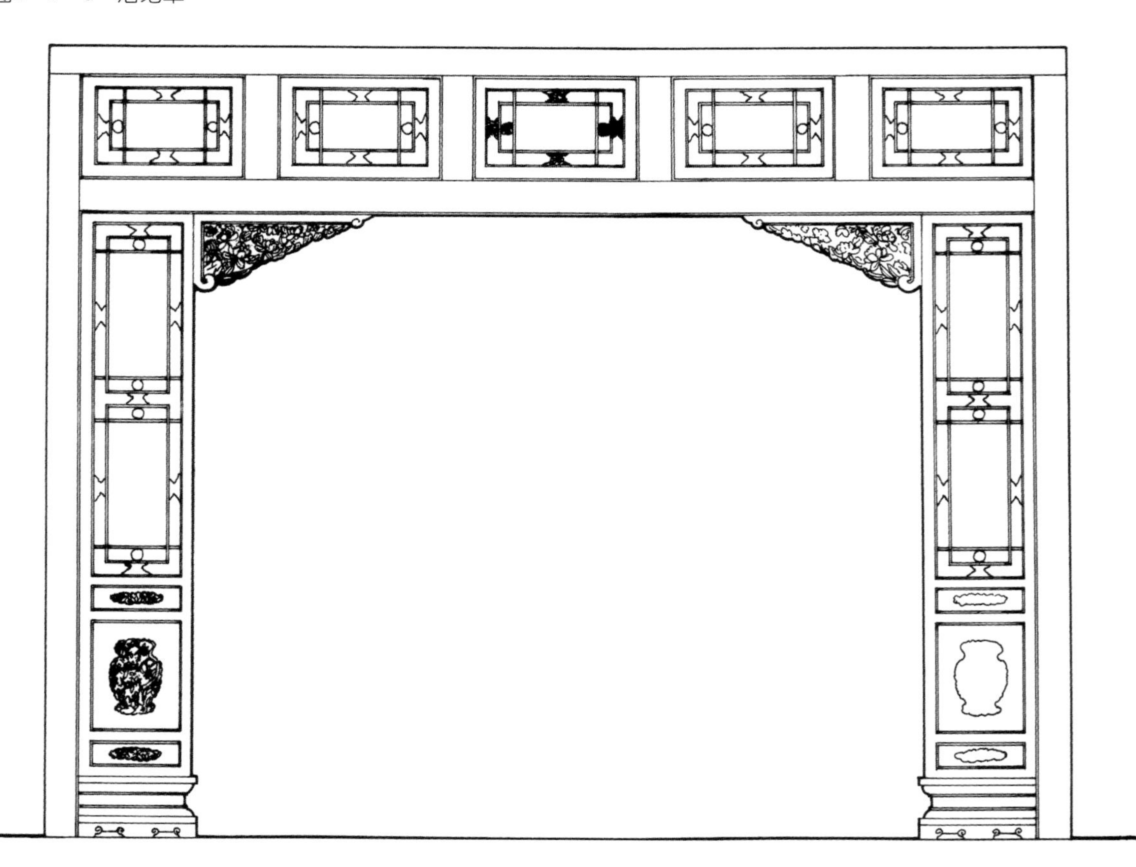

3. 落地花罩

落地花罩是花罩中十分华丽的一种，其构造是沿几腿罩的中槛下面通长安装透雕花罩。花罩两端沿抱框向下延伸直达地面，形成“冂”形三面雕饰。

这种镶在大框内侧的巨型花雕，由边框、透雕大花罩和须弥墩等部分组成。透雕的大花罩分为三块。上面横向是一整块，凭边框上的暗梢固定于中槛下皮；两侧的两块下端落在须弥墩上面；上端与横向的花雕两端相接。接缝处不仅要使纹饰雕刻顺畅自然，内里还要有榫卯暗梢互相连接固定。两侧的花罩与抱框也须有木梢固定。这种华丽无比的巨型雕饰，多取“岁寒三友”“玉棠富贵”“鹤鹿同春”等内容吉祥、图案优美的题材，雕镌精细，做工考究，堪称艺术佳品（图5—5—4）。

4. 栏杆罩

栏杆罩，顾名思义，是带有栏杆的花罩。栏杆罩的构成，是在几腿内侧加两根立框，将空间分成中间宽两边窄的三段，在两侧的抱框和立框之间加装栏杆。栏杆高在三尺左右，一般采取寻杖栏杆形式，由寻杖扶手、净瓶及抹头、牙子、木雕花板等组成。上部均安装透雕的花罩，形成上有花罩、两边有栏杆、中间供通行的格局（图5—5—5）。

栏杆罩两侧的栏杆，是凭溜梢安装固定的，随时可以取下来。上边花罩的构造类似于室外檐枋下面的雕花楣子，由边框、花心组成，花罩的两个边框下端做成花篮状，并留有插梢眼，各凭一根带有装饰的插销固定于大框之上。花罩的上边，也有暗梢与中槛固定。需要移动时，只要拨下插销即可将花罩摘下。

5. 床罩（炕罩）

床罩是安装在床榻前面的花罩。床罩的形式与一般落地罩相同，多用于面宽方向沿床榻外侧安装。罩内侧挂幔帐。白天将幔帐挂起，夜间睡觉时放下。如果室内空间高，床罩上面还要加顶盖（图5—5—6）。

图5—5—4 落地花罩

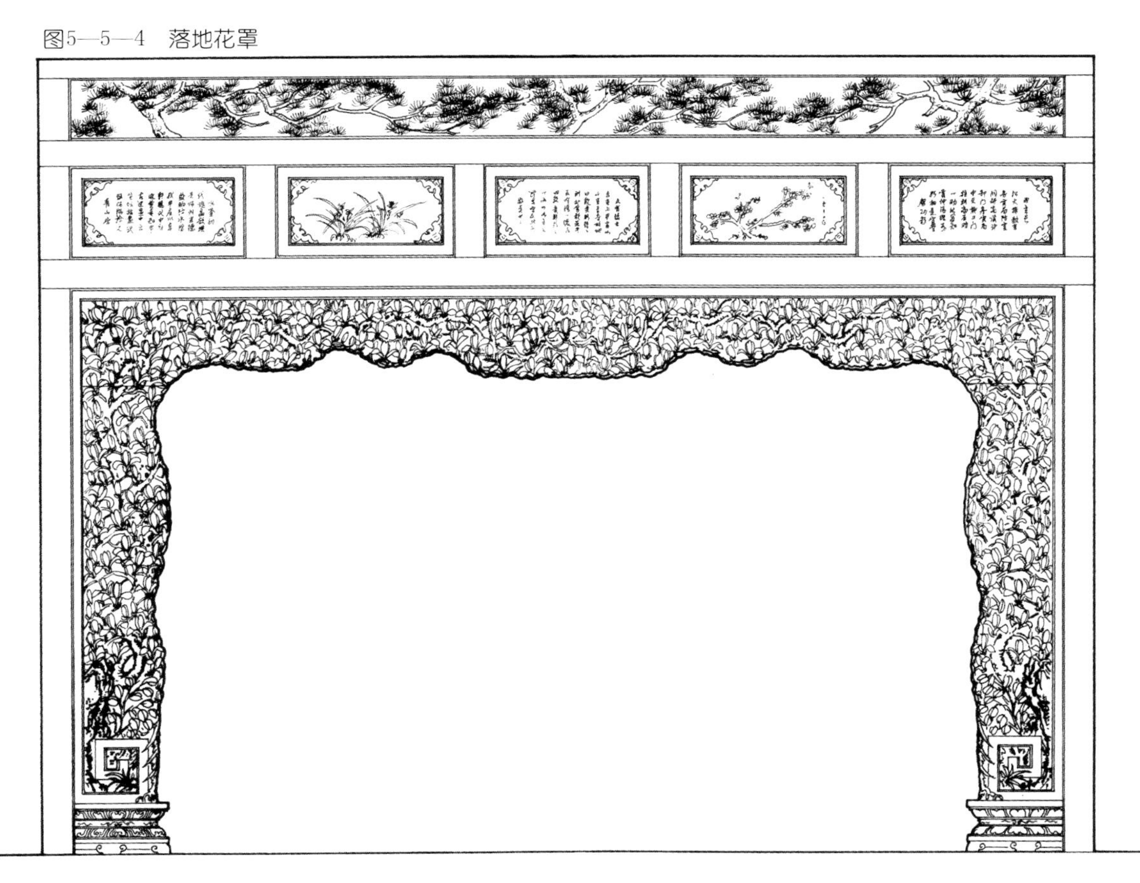

图5—5—4．1 落地花罩正立面

图5—5—4.2　落地花罩举例（本图引自《紫禁城宫殿建筑装饰内檐装修图典》）

图5—5—5　栏杆罩

图5—5—5.1　栏杆罩平立面图

图5—5—5．2 栏杆罩举例（本图引自《紫禁城宫殿建筑装饰内檐装修图典》）

图5—5—6 床罩

三、圆光罩、八角罩

圆光罩、八角罩是与上述各种花罩形式不同的另一种花罩。这种罩是在几腿罩的基础上，居中留圆形或八角形洞口，洞口的上部和两侧均满做棂条花格。常采用的花格有卍字不到头图案、冰裂纹图案等。圆光罩和八角罩的作用主要用于分隔空间，同时也是很好的室内装饰（图5—5—7）。

四、博古架

博古架又称多宝格，是一种兼有装修和家具双重功用的室内木装修，多用于进深方向柱间，具有分隔空间、陈列古董玩器等多种功用。博古架也有设在面宽方向或贴墙摆设的。如何用法，须根据空间要求或主人意旨而定。

博古架具有一定厚度，一般为1～1.5尺，以便摆放器物。它由上、中、下三部分组成，下部为板柜，是用来储存玩器或书籍的，中部是博古架的主体，由不规则的架格组成，格子的大小，要依所陈列的器物的形状、尺寸而定；博古架的上部可做朝天栏杆一类装饰。博古架如用来分隔空间，中部还要留出供人通过的洞口。这种极为讲究奢华的室内装修陈设，只有家财万贯的豪门富户或酷爱古董的收藏家才有，一般人家是不具备的（图5—5—8）。

图5—5—7　圆光罩和八角罩

图5—5—7．1　八角罩

图5—5—7．2 圆光罩

图5—5—8 博古架

五、板壁

板壁是用于分隔室内空间的板墙，多用于进深方向柱间。板壁的构造，是在柱间立槛框，在框间满装木板，木板表面刨光，或糊纸，或做油饰彩绘，或做雕刻绘画。板壁也可做成碧纱橱形式，分成若干扇，每扇立面分为上下两段。下段似碧纱橱那样做裙板绦环板，上段满装木板。板上或刻诗词古训，或刻古人名画，十分高雅别致。

六、室内裱糊

裱糊是室内装修的重要手段。室内裱糊主要有顶棚、墙面、木装修的内侧及门窗裱糊等。

1. 顶棚裱糊

顶棚的作用在于限制室内高度并保温、防尘。由于传统四合院住宅房屋多为大屋顶，室内很高，所以，一般室内都加顶棚。室内顶棚有三种基本造型。

（1）三锭　三锭又称一平二切，其基本形状是一个平面、两个斜面，形成⌒形，这是一种等级较高的做法。它的优点是既保温防尘，又保持了一定的空间高度，适用于室内空间较大的房屋。

（2）卷棚顶　卷棚顶顶棚形状为拱券状曲面，利于保持空间高度，其优点是坚固耐久。

（3）平顶　平顶顶棚为平面，吊于梁下皮，为一般做法，室内空间相对较低（图5–5–9）。

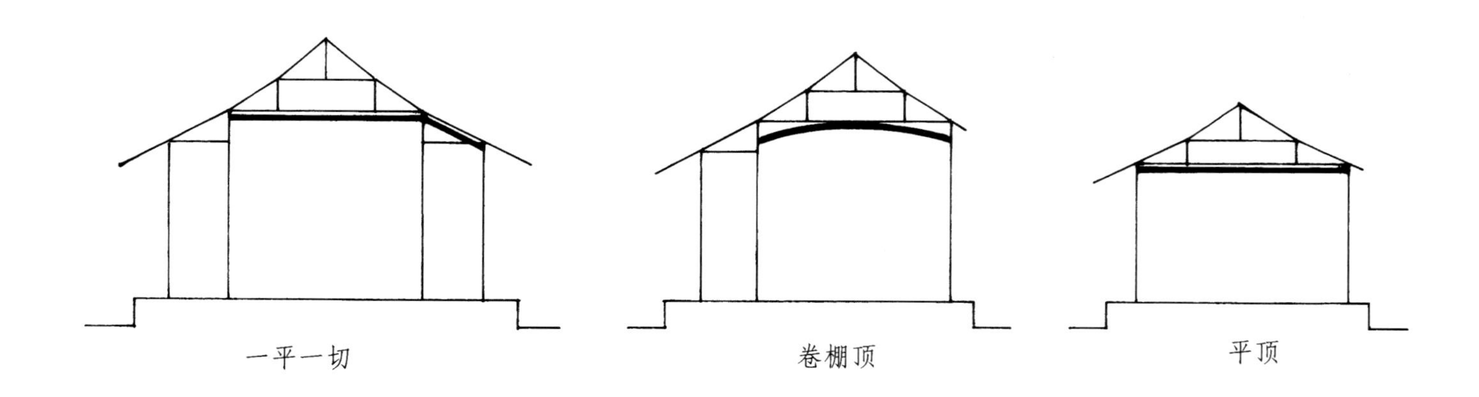

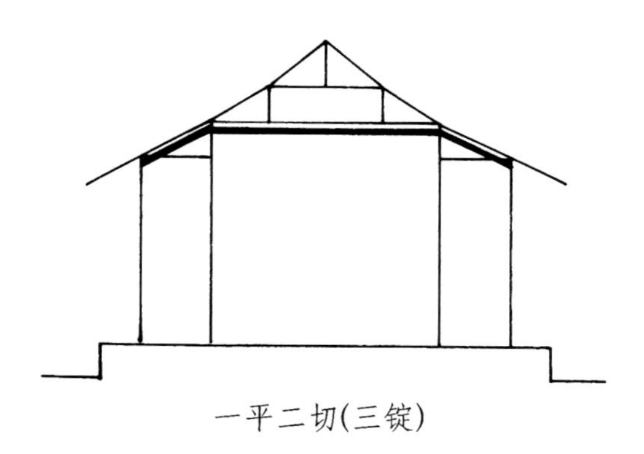

图5—5—9　室内顶棚的几种常见形式

传统室内顶棚是在表层糊纸，里面的架子称顶棚架子。顶棚架子常见有三种。一种是木架子，采用截面1寸左右的小木方做成类似豆腐块窗格形状的木质方格子，称为“木顶隔”，又称 “白堂篦子”。这是一种高档次的做法，多在高级住宅中运用（图5—5—10）。另外一种为秫秸架子，系用高粱杆绑扎而成，造价不高但轻便实用，多用于一般民宅。还有一种苇帘顶棚，亦用于一般民宅。

顶棚表层糊纸有不同做法，高档者有“纸附布”做法，以布打底；一般做法是以纸打底。所用纸张多为“高丽纸”“成文纸”。顶棚面层的用纸比较讲究，早期是用生宣纸，这种材料正好与墙面上的书画挂屏、博古器皿等室内陈设相配，共同营造出一种文雅大气的书卷氛围。晚期纸张品种增多，出现了“银花纸”“大白纸”，但效果远逊于生宣纸。

2. 墙面裱糊

传统民居室内墙面主要有木板墙面和抹灰墙面两种。墙面糊纸也分为底层和表层，所用材料与顶棚大体一致。

3. 装修裱糊

装修裱糊主要有门窗裱糊和槛框、榻板等部位裱糊。在玻璃普及之前，传统民居的门窗棂条内侧是以糊纸为主，一般多用高丽纸，有“横糊窗户竖糊门”的说法，即糊窗的纸横纹使用，糊门的纸竖纹使用。槛框、榻板介于墙面和门窗之间，其色调、质感应与墙面、门窗相一致，故也常糊纸。从室内看，墙面、门窗及槛框的洁白无瑕，与室内的文玩字画、家具陈设协调搭配，造成一种洁净明亮、高雅脱俗、书香浓郁的文化气息。

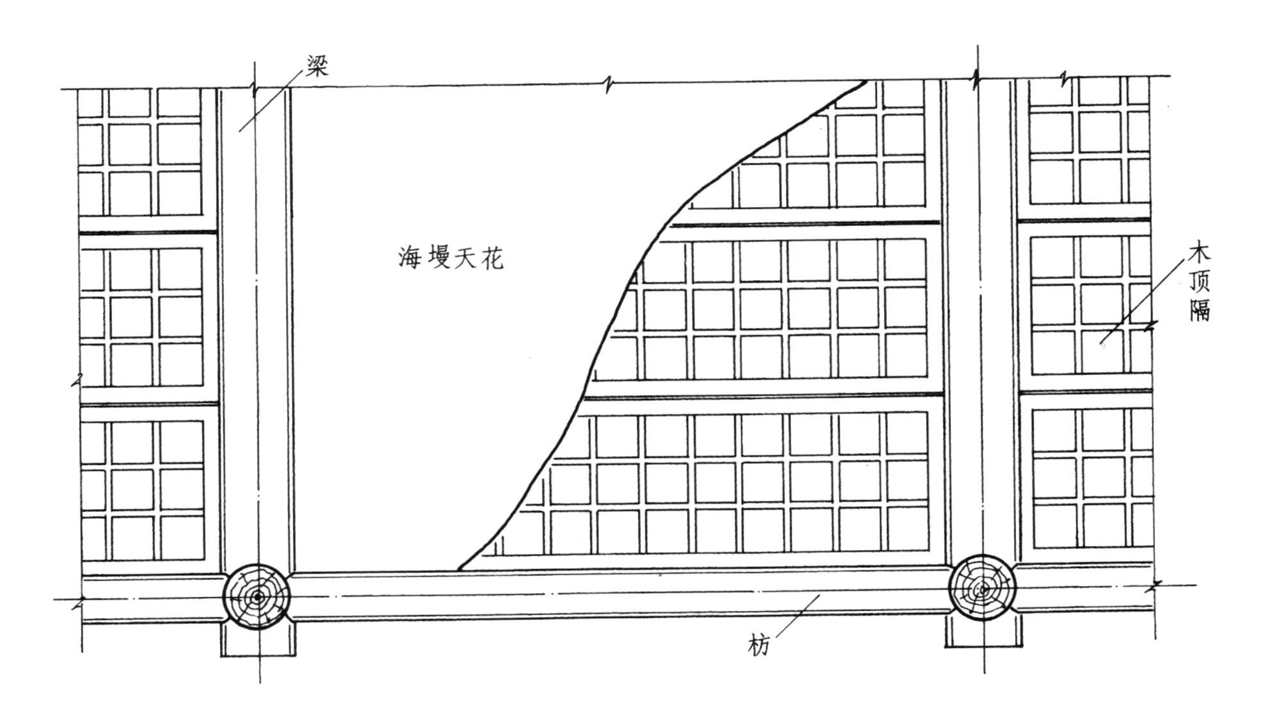

图5—5—10 木顶隔构造举例

七、内檐装修的特点和艺术特色

北京四合院的内檐装修是十分讲究和极具艺术特色的，它与外檐装修相比，有许多不同之处。

1. 用料讲究，做工精细

外檐装修由于经常风吹日晒，一般都用质地松软、不易变形、耐糟朽的材料，大多使用松木；而室内装修则不然，它不受风吹日晒，不易变形褪色，且兼有陈设作用，因而用料十分讲究。碧纱橱花罩等多用楠木、楸木，讲究的甚至用花梨、红木、鸡翅木等。有些截面较大的构件，如抱框、横槛，不宜使用红木、花梨时，则以松木为心，外面用红木包镶，称为贴皮子，也十分考究。由于用材讲究，做工也格外精细，无论线条肩角，割对拼粘，均严丝合缝，平整密实，其雕刻纹饰，更极尽功力，工细精致，无与伦比。

2. 截面纤巧，做法考究

内檐装修虽然也有横槛抱框、边梃仔屉，但截面一般较外檐装修要小，而且里口棱角部分大多要做出凹角线，颇具装饰性。用做碧纱橱的内檐隔扇，无论体量或用料，都较外檐隔扇精巧得多。内檐碧纱橱的边梃抹头，看面最大不过5～6cm，有的仅4cm。内檐装修的棂条，其看面仅有四分（合公制13mm），进深只有六分（合公制18～20mm），称为四六分条。这比用于外檐装修的六八分条要纤巧得多。内檐装修的做法也是极为考究的。仍以碧纱橱为例，那种两面夹纱的仔屉棂条，两层仔屉原是做在一起的，成活以后再用极薄的小锯从中一破为二，这样做是为了保证两层仔屉棂条重叠以后无任何错位现象。

3. 色彩素雅，凝重脱俗

四合院建筑尽管外檐用红柱绿窗、青绿彩画，但室内绝少施用色彩，内檐装修一律木材本色——或为楠木色，或为红木花梨色。表层一般都做本色烫蜡，更显凝重脱俗。这种由楠木、花梨木构成的本色调，与室内的硬木家具极为协调一致，是绝好的色彩搭配。

4. 内涵丰富，感染力强

室内装修所采用的窗格形式、棂条花纹及雕饰纹样、图案搭配，都具有深刻的思想内涵和极强的艺术感染力，置身其间，有如置于艺术的殿堂之内，令人目不暇接。关于装修中雕刻的内容，下文还要列专题介绍，在此暂不赘述。

四合院的内檐装修，是实用性与艺术性结合得最完美的部分，是传统居住文化体现最丰厚的部分，也是展示古代能工巧匠聪明才智最充分的部分，是需要我们认真加以发掘、研究和继承的。

第六章
四合院建筑的装饰和室内家具陈设

第一节 砖雕、石雕和木雕

在北京四合院的装饰艺术中,砖雕、石雕、木雕艺术占有相当重的分量。古代的建筑设计师和能工巧匠在砖、石、木构件上通过精雕细刻的艺术手段去描绘生活、抒发情感、寄托理想，创造出不朽的艺术佳作，取得了高超的艺术成就，为中国的传统居住文化长卷增添了绚丽的色彩。

一、北京四合院的砖雕艺术

1. 砖雕的装饰部位

北京四合院的砖雕，在墙面、屋脊等醒目部位均有表现，各个部位的砖雕都有不同形式和内容。

（1）宅院门头砖雕　北京四合院的砖雕首先应用于宅门上，在广亮大门、金柱大门、蛮子门、如意门、墙垣式门等多种不同形式的宅门上，都有精美的砖雕点缀。广亮大门最突出的部分——墀头上端，往往做突出醒目的砖雕。这个部位的砖雕一般由戗檐、垫花和博缝头组成。金柱大门所采取的砖雕形式与广亮大门相似，也是着重装点墀头部位。但也有在檐柱与金柱之间的廊心墙部位做砖雕装饰的，属更为讲究的做法。蛮子门与金柱大门大致相同，着重在墀头部位做戗檐雕刻（图6—1—1）。

图6—1—1　戗檐垫花砖雕一组

图6—1—1. 1　竹鸟、花卉、松鸟

图6—1—1．2 兰草、菊、竹石

图6—1—1．3 松鼠葡萄、喜鹊登梅、花鸟

图6—1—1．4 （戗檐）太师少师、（垫花）牡丹花篮、博古

图6—1—1．5 垫花、戗檐垫花

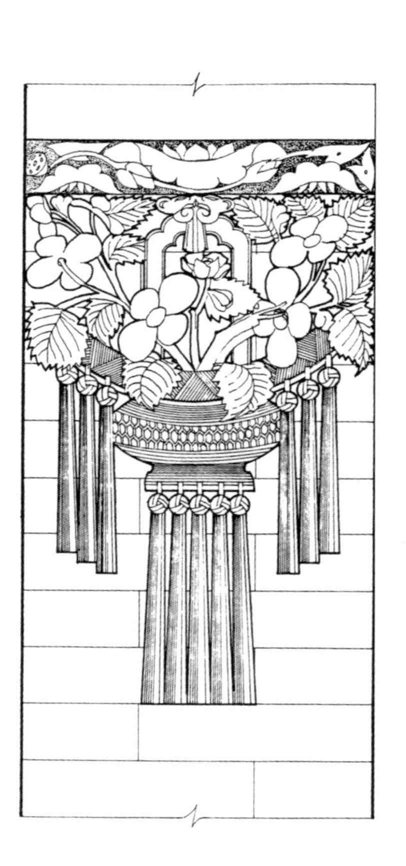

图6—1—1.6 戗檐砖雕（子孙万代）

图6—1—1.7 戗檐砖雕（松鼠葡萄）

图6—1—1.8 戗檐、垫花砖雕（牡丹、福寿）

图6—1—1. 9 戗檐砖雕(花草)

图6—1—1. 10 戗檐砖雕(草龙、团寿字)

图6—1—1. 11 戗檐砖雕(居家欢乐)

图6—1—1．12 戗檐、垫花砖雕（菊花、牡丹）

图6—1—1．13 戗檐砖雕（狮子绣球）

图6—1—1．14 戗檐砖雕（荷花茨菇）

图6—1—1．15 戗檐（太师富贵）、垫花（子孙万代、福寿眼前）

图6—1—2 北京某宅如意门门楣砖雕

图6—1—2.1 某宅如意门砖雕

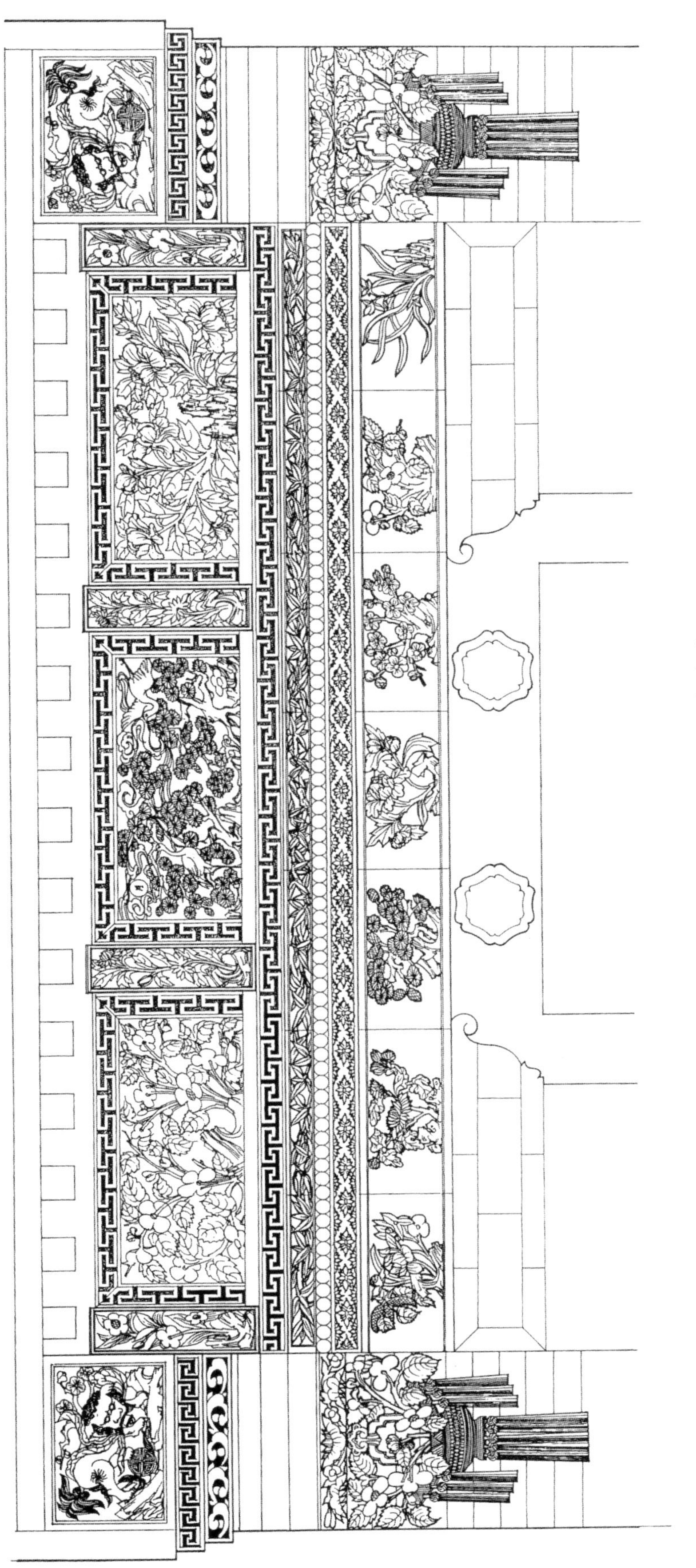

图6—1—2.2　某宅如意门砖雕

值得一提的是如意门的砖雕装饰，它是北京四合院宅门装饰的代表。如意门砖雕除两侧墀头上的戗檐、垫花、博缝外，主要是门楣雕刻。门楣雕刻一般是在门洞上方安装砖挂落，在挂落上方出冰盘檐若干层，冰盘檐上方安装栏板、望柱。讲究的如意门，其挂落、冰盘檐、栏板、望柱上均布满砖雕，极为华丽美观，令人百看不厌（图6－1－2）。

如意门门楣砖雕也有采用其他形式的。如宣武区牛街某宅，在门楣挂落板上面摆砌出须弥座形式，由下枋、覆莲、串珠、束腰、连珠、仰莲、上枋等部分组成，在须弥座上面置栏板柱子，其上全部做砖雕。这种实例，在北京其他地区也不鲜见，是如意门雕刻的另一种形式（图6－1－3）。除此之外，还有在门楣部分用一大块富贵牡丹花板（外加线枋子、大框）来代替冰盘檐、栏板、望柱的，令人有一种新鲜感。由此可以看出，民间的砖雕装饰在形式上并非一成不变（图6－1－4）。

图6—1—2．3　门楣砖雕（局部）

图6—1—2．4　门楣砖雕（局部）

图6—1—2．5　门楣砖雕（局部）

图6—1—2. 6 门楣砖雕（局部）

图6—1—3 牛街某宅如意门砖雕

图6—1—4 东四八条某宅如意门砖雕

最简朴的墙垣式门，也不乏砖雕装饰。这种门一般为硬山形式，两侧与院墙相接。在门楣上方，有挂落板，其上为冰盘檐，正脊做清水脊蝎子尾。这种小门楼大多为素活，但其中也有采用砖雕者，雕刻多用于挂落板、头层檐及砖椽头等处，正脊蝎子尾下方还置平草或跨草盘子（图6－1－5）。

西洋门的砖雕装饰主要用在门楣之上的砖匾上，其余部分则以枭、混等线角为主（图6－1－6）。

图6—1—5　小门楼砖雕

图6—1—5．1　随墙门上的砖雕

图6—1—5．2　小门楼雕上的砖雕

图6—1—5．3　小门楼上的砖雕

图6—1—6　西洋门上的砖雕

图6—1—7　影壁心砖雕

图6—1—7.1　影壁心砖雕（中心花为荷花）

(2) *影壁砖雕*　影壁上面的砖雕是很丰富的，尤其大门内的影壁，更是重点装饰部位。这种影壁的下碱，有直方形的，也有的做成须弥座形式。成须弥座形式时，可在上下枭、束腰等部位加以雕饰。影壁的上身多做成模拟柱、枋的砖框，称为柱子、马蹄磉、枋子，枋的侧面箍头部分做成三岔头或霸王拳头形状，内饰花纹；迎面之箍头做成宝瓶形状，称为“耳子”。在柱枋内侧，附有一圈线枋子，使外框富有层次感。影壁心多用方砖斜砌而成，有素做的，但更多的是在中心和四角加中心花和岔角花，非常富有装饰性。影壁的檐口和墙帽部分也有很华丽的雕刻，一般多在第一层砖檐、连珠混等处做雕饰，讲究的影壁，还要做砖椽子，椽头部分也做雕刻。影壁墙帽部分如有正脊时，还要在正脊两端做平草或跨草雕饰（图6－1－7）。

图6—1—7.2　影壁心砖雕（中心花为松鼠葡萄）

图6—1—8 廊心墙的雕刻

图6—1—8．1 礼士胡同某宅廊心墙雕刻

用于大门外侧的影壁，其雕刻部分略为简单，但也有将这种影壁做得很讲究的。不论内外影壁，在其中心部分，还往往雕出砖匾形状，匾内题刻“如意”“平安”“吉祥”“鸿禧”“迪吉”等吉辞。

（3）廊心墙、看面墙砖雕　廊心墙指房屋外廊两侧的墙面和金柱大门外廊两侧的墙面。廊心墙虽面积不大，但因其位置重要，常作为重点加以装饰。金柱大门内的廊心墙砖雕，还可以和墀头上的砖雕共同构成大门的装饰。

廊心墙砖雕（图6—1—8）一般做在上身墙心部分，在中心刻中心花，四角刻岔角花。更为讲究的做法是将外圈的大枋子也做出雕刻。北京礼士胡同某宅的大门外廊的廊心墙雕刻就非常精美，很具代表性（图6－1－8．1）。

图6—1—8．2
廊心墙字雕

图6—1—8．4
廊心墙砖雕一例（凤栖牡丹）

图6—1—8．3廊心中心花（凤栖牡丹）

图6—1—8．5　廊心墙砖雕一例（富贵如意）

看面墙，是指四合院中除影壁之外，其他为人所注目的墙面，如垂花门两侧的墙面、倒座房的后檐墙，以及其他显著位置的墙面等。因这些墙面的位置重要，常常被人们作为重要部位加以装饰。

垂花门两侧的墙面上，如没有什锦漏窗，则需要加以装饰。主要装饰手段为：或做素面墙心（这种墙心又分为硬心和软心两种：硬心者贴方砖膏药幌子；软心者抹灰，做白墙面）；或在墙心内加砖雕装饰，做法略同影壁（图6－1－9）。

位于街面上的看面墙，做雕饰的较为鲜见。北京礼士胡同某宅的倒座房后檐墙临街一面，在墙心内做中心、四岔花卉雕刻，每间一组，接续排列，十分雅致（图6－1－10）。

图6—1—9　垂花门两侧的看面墙砖雕

图6—1—10　礼士胡同某宅倒座房沿街的看面墙砖雕

（4）廊门筒上门头板砖雕　四合院正房、厢房外廊两侧的廊门筒子上方为门头板，由八字枋子、线枋子和墙心组成，门头板处的砖雕多做在墙心和八字枋上，在横枋、立枋和搭脑上画出池子，内刻花卉或其他图案，墙心部分的雕刻则按匾心处理。在上面题字。北京礼士胡同某宅的门头板砖雕极其精美丰富，匾心处雕字有“竹幽”“兰媚”“隐玉”“含珠”“蕴秀”“伫月”等佳词，蕴含丰富，意境深远，令人玩味无穷。有些稍低矮的房子，廊门筒上方门头板尺度较小时，匾心内则不刻字，留做空白，也颇有余韵。做门头板雕刻时，往往将上面的穿插枋当子一并刻出，题材与门头板处相同，以求上下协调一致（图6－1－11）。还有在门头板处刻汉纹瓦当和吉祥图案的，也颇具特色。门头板部分，如不做雕刻，即为素活，由八字枋、线枋子及板心等组成，与门筒子融为一体。

（5）槛墙雕刻　讲究的四合院，在槛墙上面也做雕刻。槛墙砖雕的布局一般是在外圈横枋及立枋子上圈出小的海棠池子，在池子内做雕刻；里圈的线枋较窄，一般不做雕刻；在面积较大的墙心内，则做中心四岔雕刻图案。雕刻题材多为花卉，这种题材构图灵活，适合于各种大小不同的画面（图6－1－12）。北京西四北某宅槛墙在海棠池子内满雕卍字不到头图案。这是民居中极为鲜见的做法，估计为帝制灭亡以后所做。槛墙上面仅圈海棠池（仅做出内外圈枋子和墙心）而不做雕刻的，称为素做，是一种简易做法，还有仅在墙心部分雕刻中花岔角、外框不做雕刻的。槛墙雕刻形式多样，不一而足（槛墙雕刻的其他做法参见图5－3－10）。

图6—1—11　廊门筒子上的门头板砖雕

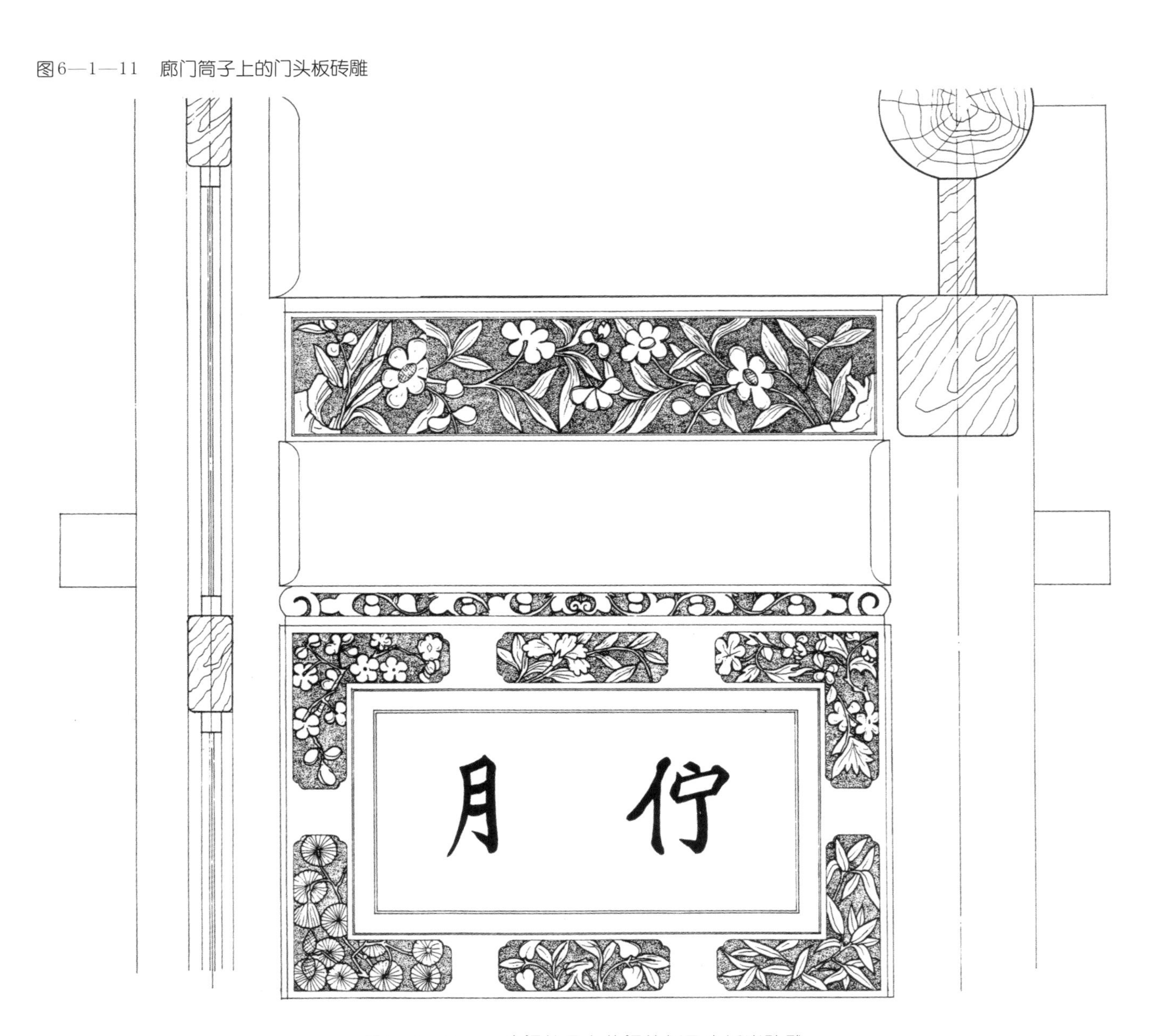

图6—1—11．1　廊门筒子上的门头板及穿插当砖雕

图6—1—11．3　门头板砖雕（竹幽）

图6—1—11．2　门头板砖雕（吉祥图案）

图6—1—11．4　门头板砖雕（兰媚）

图6—1—11．5 门头板砖雕（撷秀）

图6—1—11．6 门头板砖雕（扬芬）

图6—1—11．7 门头板雕刻一例

图6—1—12 北京礼士胡同某宅槛墙砖雕

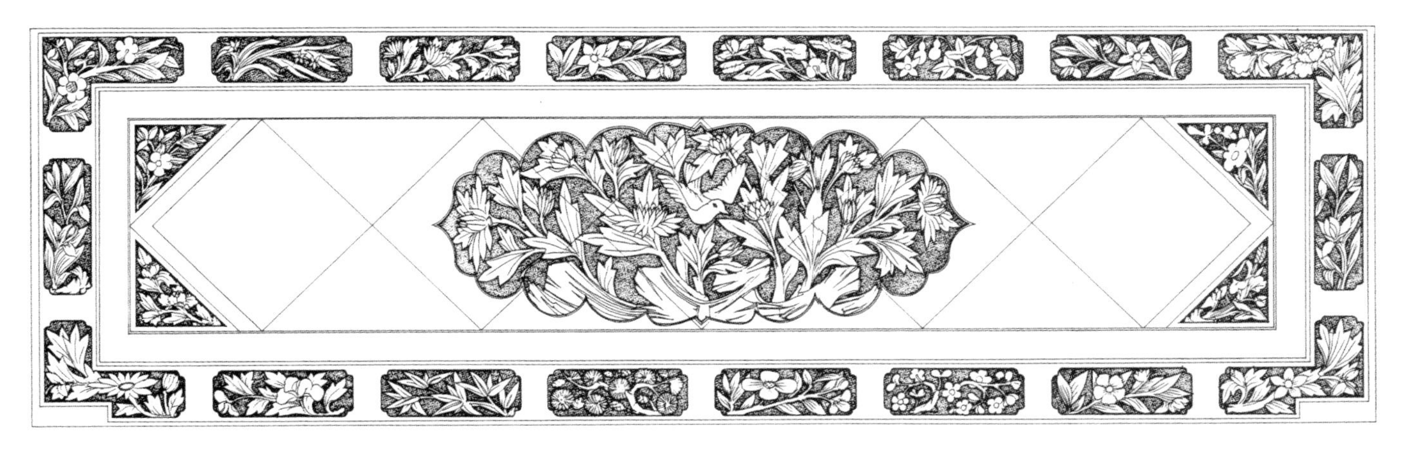

图6—1—12.1 礼士胡同某宅槛墙砖雕

图6—1—12.2 槛墙砖雕局部一

图6—1—12.3 槛墙砖雕局部二

（6）什锦窗砖雕 什锦窗是采用“月洞”“八方”“六方”“卷书”“扇面”“玉盏”“玉壶”“宝瓶”“海棠”“蝠磬”“桃”“柿”“方胜”“双环”等各种图案做成的，用以装点墙面的小型牖窗。这种窗多见于园林建筑中，其形式丰富多样，颇具艺术特色和趣味性。在四合院中，多用于垂花门两侧的看面墙上，其功能除了装饰墙面，同时也有沟通内外宅的作用。

什锦窗的砖雕主要用在窗外侧的砖质贴脸上。贴脸宽度一般在四寸左右，在这宽度极其有限的范围内，需依据图形变化，圈出若干个形态不一的池子，在池子内做雕刻，或按外框形状将窗套贴脸分为若干部分，在其中设计图案，进行雕刻。北京东城区礼士胡同某宅以及东城区鼓楼东大街某宅的什锦窗砖雕，从设计到雕刻，都相当秀美别致，堪称精品（图6－1－13）。

图6—1—13 什锦窗砖雕刻举例

图6—1—13．1　什锦窗砖雕刻举例（墨线图）

图6—1—13．2
什锦窗砖雕刻举例（实物照片）

（7）砖雕栏杆　栏杆有两种：一种用于台基外围，功能是遮拦行人，以保安全，同时又有装饰作用；一种用于平台房屋顶四周，既有安全作用，又有装饰功能。

四合院的砖雕栏杆多用于平顶房屋面外围，属朝天栏杆一类。栏杆的形式多种多样，有寻杖栏杆形式，也有花栏杆形式。北京西四丁字街某宅砖雕栏板即采取寻杖栏杆形式，由地袱、望柱、栏板等部分组成。栏板下半部分为卍字不到头图案，上部为牡丹净瓶图案（图6－1－14）。

（8）屋脊上的砖雕　屋脊是古建筑屋顶的重要装饰部分，北京民居四合院多为小式建筑，屋脊以清水脊居多。有正脊的屋面在清水脊两端要做鸥尾（蝎子尾）作为装饰。在鸥尾下方，还要安装花草盘子。其中，平砌在混砖一层的为平草，陡跨在脊两侧的为跨草。这部分砖雕的雕刻内容多为松、竹、梅及花草一类。

施鞍子脊或过垄脊的屋面不做正脊，屋面两边做箍头脊（又称垂脊）压在瓦面边垄之上，脊两端下垂至屋檐部分，端头按45°角向外侧扭出，称为趔角。趔角部分的楣子下面有一块平盘式构件，称为趔角盘子。这块盘子上也常雕些花草图案，是脊端檐头的重要装饰（图6－1－15）。

图6—1—14　朝天栏杆砖雕举例

图6—1—15　正脊和垂脊上的砖雕

图6—1—15．1
正脊上的跨草砖雕

图6—1—15．2　垂脊端头的平草盘子及圭脚一例

图6—1—15．3　垂脊端头的平草盘子及圭脚一例

四合院的其他部分，也有做砖雕装饰的，如用做墙内柱子通风防腐的砖雕透风、用做雨水排放的明沟沟眼、用在花砖墙墙帽上的砖雕镶嵌以及瓦头、滴水上面的装饰等等（图6－1－16），充分展示出砖雕艺术在北京四合院中应用之广泛。

图 6—16—1　透风及墙帽砖雕镶嵌

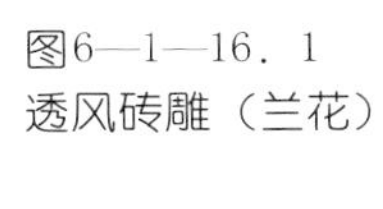

图6—1—16．1
透风砖雕（兰花）

图6—1—16．2
透风砖雕（兰花）

图6—1—16．3
透风砖雕（栀花）

图6—1—16．4
透风砖雕（灵芝）

图6—1—16.5 墙头砖雕及花瓦墙帽

图6—1—16.6 透风砖雕——牡丹花

2. 砖雕题材和纹饰特点

北京四合院的砖雕，题材极为丰富，可归纳为以下几类。

（1）自然花草类 这类以自然界的花草为主要题材，经常入画的种类有牡丹、菊花、松、竹、梅、兰、荷花、栀花、水仙、灵芝、海棠、葫芦、石榴、柿子、茨茹、蒲棒以及江西腊、大丽花等。

用自然界的花草为内容作雕刻装饰题材，并赋予一定的内涵，是中国传统装饰惯用的表现手法，如牡丹象征富贵，菊花象征高雅，松柏象征长寿，竹子象征傲骨，兰花象征清雅，荷花象征高洁，葫芦、石榴、葡萄寓意多子等等。松、竹、梅又称岁寒三友，象征清高，为文人雅士所欣赏。

自然花草类题材，在砖雕中应用非常广泛，无论戗檐、垫花、廊心、影壁、窗套、花盘子等均可采用。这种题材贴近自然，接近生活，易被人接受，且形态优美，易于构图，富于表现力，并可以与其他种类题材交叉使用（图6－1－17）。

（2）吉祥图案类 中国人民向往美好生活，热爱和平，追求富裕，自强不息，这些美好理想和优秀品德经常用吉祥图案形式表现在建筑上，在市俗建筑民居四合院中尤其突出。例如：以如意、柿子、卍字组成“万事如意”；以牡丹、白头翁组成“富贵白头”；以灵芝、水仙、竹子、寿桃组成“灵仙祝寿”；以松、竹、梅组成

图6—1—17 自然花草类砖雕举例

图6—1—17.1 墙心砖雕——菊花

图6—1—17.2 戗檐垫花雕刻——花草

图6—1—17．3 门楣砖雕 ——花草

图6—1—17．4 影壁岔角花雕刻——花草

图6—1—17．5 门楣砖挂落板上的花草池子

图6—1—17. 6 影壁岔角花雕刻——花草

图6—1—17. 7 门楣砖雕——花草

“岁寒三友”；以大象、宝瓶组成“太平有象”；以菊花、麻雀组成“居家欢乐”；以葫芦及藤蔓组成“子孙万代”；以蝙蝠、石榴组成“多子多福”；以花瓶、月季花组成“四季平安”；以鹌鹑、菊花、枫叶组成“安居乐业”；以牡丹、海棠组成“富贵满堂”；以松树、仙鹤组成“松鹤延年”；以松树、仙鹤、梅花鹿组成“鹤鹿同春”；以梅花、喜鹊组成“喜上眉梢”；以马、猴、松树、蜜蜂、印绶组成“封侯挂印”；猴子骑在马上，上有蜜蜂飞舞称为“马上封侯”；以公鸡和鸡冠花组成“官上加官”；以桂圆、荔枝、核桃组成“连中三元”；以蝙蝠、寿字、绶带组成“福寿绵长”；以寿字、蝙蝠组成“五福捧寿”；以寿字加卍字不到头图案组成“万福万寿”等等。这些吉祥图案是采用象形、会意、比拟、谐音等手法，将自然界和人类社会中的一些事物组合起来，表达一种吉祥内容和美好愿望。尽管有些内容组合难免牵强生硬，但通过构图技巧却可以形成一幅美丽和谐的画面，令人百看不厌，玩味无穷。这是中华民族传统装饰艺术的独到之处（图6—1—18）。

图6—1—18 吉祥图案类砖雕举例

图6—1—18．1 垫花雕刻—— 多子、福寿在眼前

图6—1—18．2 博缝头雕刻—— 万事如意

图6—1—18．3 戗檐雕刻—— 喜鹊登梅

图6—1—18．4　瓦头——圆寿字

图6—1—18．5　透风雕刻——子孙万代

图6—1—18．6　门楣雕刻——松鹤延年

(3) 博古图案类 这类砖雕题材，多采自夏商周以来各朝代的青铜器皿、宝鼎、酒具、炉、瓶等形象，加上书案、博古架的图形及文房四宝、画轴拂尘、花草纹饰组成画面的基本内容。由古玩摆饰为基本内容的画面，构图典雅，书卷气浓郁，给人以高雅文化的享受。这种题材的雕刻，多用于文人墨客的居所，用在栏板、戗檐等显赫部位，是砖雕题材中高雅脱俗的上乘之品（图6−1−19）。

(4) 锦纹图案类 以锦纹装饰建筑，在我国有悠久历史。在著名的宋代《营造法式》一书中，即载有各种锦纹图案，如连环纹、密环纹、簟纹、方环纹、罗地龟纹、香印纹等。这些锦纹，最初用在彩画中，后来逐步向相关方面发展，在石雕、木雕、砖雕中均有体现，这就是有名的“宋锦”。明清以来，宋锦的纹饰又有变化。清代砖雕上出现的锦纹，主要有回纹、汉纹、万不断（卍 字不到头）、扯不断、拐子锦、丁字锦以及菊花锦、海棠锦、龟背锦、如意纹、寿字、福字等。

在大幅砖雕图案中，锦纹图案多安排在边框、线角处，作为花边来处理，以衬托主题。这些锦纹构图严谨，极富韵律感，令人赏心悦目，具有很好的装饰效果。而寿字、福字、卍字一类有思想内容的锦纹，还可以作为主题内容置于整幅作品中，起到点题的作用（图6−1−20）。

图6—1—19 博古图案类砖雕举例

图6—1—19．1 门楣栏板雕刻——博古一

图6—1—19．1 门楣栏板雕刻——博古二

图6—1—19．3　戗檐雕刻——博古

图6—1—20　锦纹图案类砖雕举例

图6—1—20．1　挂落板上面卍字锦、挑檐周围的丁字锦图案

图6—1—20．2
门头板外框的丁字锦图案

图6—1—20．3 门楣冰盘檐上面的丁字锦和菊花锦图案

图6—1—20．4 门楣上的扯不断和回纹图案

图6—1—20．5　门楣冰盘檐上面的扯不断和竹叶纹图案

图6—1—20．6　戗檐中心的圆寿字及戗檐四周的丁字锦图案

（5）蕃草图案类　蕃草，是图案化了的一种花草纹饰，基本图形是一反一正向前弯曲伸展的草弯，为二方连续图形。蕃草纹饰多用于窄长的部位，如线枋子、砖拔檐、冰盘檐下面的头层檐等处，它与锦纹类边框装饰形成线条一软一硬、一方一圆、一曲一直的对比效果，惟妙惟肖（图6—1—21）。

北京四合院砖雕的蕃草类纹饰，还有以竹叶纹、兰花纹和栀花纹作连续图案的，多用于砖檐或混砖上面，也有极好的装饰效果（图6—1—22）。

蕃草图案类花纹除去作砖雕的边框线角以外，也有独立组成画面的。如礼士胡同某宅后花园影壁就采取蕃草纹作为中心花，蕃草图案分布均衡，翻卷有致，构图秀美，脉络清晰，给人以美的感受（图6—1—23）。

图6—1—21　蕃草图案应用举例

图6—1—21．1　蕃草与锦纹搭配使用之一例——上为锦纹下为蕃草

图6—1—21．2　某宅门楣挂落板及挑檐——连珠下及挑檐心内均为蕃草图案

图6—1—21．3　戗檐下两层拔檐刻蕃草

图6—1—21．4
戗檐下两层拔
檐——锦纹和蕃草

图6—1—22　竹叶纹、菊花纹等图案举例

图6—1—22．1　连珠以上三层均为菊花纹图案

图6—1—22．2　枭混砖下面为竹叶纹图案

图6—1—22．3
某宅倒座房后檐墙冰盘檐下的竹叶纹图案

图6—1—22．4
栏板地栿菊花纹，匾圈为扯不断纹

图6—1—23 以蕃草纹组成的影壁中心花

图6—1—23．1
以蕃草纹组成的影壁中心花一例

图6—1—23．2 以蕃草纹组成的影壁中心花图案

(6) 宗教法器类 这类题材多用于宗教建筑中，但由于民间的信教者很多，有些民宅也不乏采用这类图案。如为佛教则多用轮（法轮）、螺（法螺）、伞（宝伞）、盖（白盖）、花（莲花）、罐（宝罐）、鱼（金鱼）、长（盘长）作图案的主要内容。其中，法轮意为“大法圆转，万世不息”；法螺意为“具菩萨意，妙音吉祥”；宝伞意为“偏复三千，净一切业”，白盖意为“张弛如如，曲复众生”；莲花意为“出五浊世，无所染者”；宝罐意为“福智圆满，具定无漏”；金鱼意为“坚固活泼，解脱坏劫”；盘长意为“回环贯彻，一切通明”。如为道教者，则采用“扇子”（汉钟离所执法器）、“渔鼓”（张果老所执法器）、“花篮”（韩湘子所执法器）、“葫芦”（铁拐李所执法器）、“阴阳板”（曹国舅所执法器）、“宝剑”（吕洞宾所执法器）、“笛子”（蓝采和所执法器）、“荷花”（何仙姑所执法器），隐喻道教的八仙人，称为“暗八仙”（图6—1—24）。

图6—1—24 宗教法器类砖雕举例

图6—1—24．1 挑檐心内的暗八仙雕刻，从左至右依次为：笛子（蓝采和）、花篮（韩湘子）、葫芦（铁拐李）、荷花（何仙姑）

图6—1—24．2 挑檐内的暗八仙雕刻，从左至右依次为：阴阳板（曹国舅）、渔鼓（张果老）、宝剑（吕洞宾）、扇子（汉钟离）

除以上佛教、道教法器纹饰外，宝相花也是使用得最为广泛的纹饰。宝相花为西番莲（大丽花）的图案化，常用于佛教建筑中，民间也偶有采用。

佛八宝与暗八仙这两种法器纹饰，民间用暗八仙者较多，它有表示神仙来临之意，象征吉祥；而佛八宝因宗教气氛太浓，多用于庙宇建筑的雕刻，极少在民居中出现。

(7) 人物故事类　中国是历史文化十分悠久的国家，各朝各代都有很多著名人物、著名事件、历史故事、演义小说、文学著作、寓言典故以及流传数千年的儒家思想及由此而产生的宗法制度、纲常伦理及形象化了的《三字经》《名贤集》《二十四孝》等教化民众的材料。历史名著《红楼梦》《西游记》《三国演义》《封神演义》《西厢记》等，给人们留下了无数脍炙人口的生动故事。历代的民间艺人、能工巧匠把这些历史故事、著名人物、著名事件、伦理故事绘成图画，绘制在彩画中，雕刻在砖石上，既具有教育意义，又有装饰作用。

在我国，历史人物故事是砖、木、石雕刻的传统题材，尤其是在我国南方如安徽、江苏、浙江、福建、四川、湖南一带颇为盛行，而且题材广泛，镌工细腻。北京四合院中也有一些人物故事雕刻，但比起其他题材来则少得多，且多见于彩画当中。这反映出各地文化传统之差别。在北京四合院的砖雕中，“福”“禄”“寿”三星题材反而常见，表达了北京人对福、寿、官禄的追求和渴望。

(8) 龙凤纹类　在封建社会中，龙、凤的形象几乎成了帝王后妃的象征而为皇家所独有，无论彩画还是雕刻，只要在建筑的装饰上出现龙凤花纹，必然就是皇家建筑。但是，随着时间的推移，特别是封建王朝被推翻之后，龙凤的图案，尤其是经过加工提炼而形成的夔龙、夔凤的图案，也逐渐流行于公共建筑和民间建筑中。无论彩画还是砖石雕刻，都不乏图案化了的龙凤纹饰。它们与回纹、卷草相结合，形成硬线条和软线条的夔龙、夔凤形象，颇有装饰味道。龙凤是人们理想中的吉祥物，这些吉祥物的形象流入民间是必然的。比如民间婚庆喜事，除披挂大红之外，还常有龙、凤的图案，以象征幸福、吉祥、蒸蒸日上。尽管如此，在民居中采用龙凤图案，也仅限于夔龙、夔凤，写实的龙凤形象是很少用于砖石雕刻和彩画的。

以上八种砖雕题材，仅仅是笔者在收集了大量北京四合院雕刻纹饰的基础上人为划分的。无论哪种雕刻题材用于砖雕，其构图均有一个共同特点，即画面图案分布均匀，无大面积的空白，而且越是花纹布局紧密、枝叶重叠交错，越是上乘之作。花草雕刻，一般都是粗枝大叶，主干由一块或一组山石生出，按一定规律盘旋缠绕，以均匀地布满空间为准则。当然，分组分段的卡子、团花等图案并不受以上规律限制。

砖雕刻的构图古拙质朴，民间艺术味道极浓，有些画面在安排时只求总体效果，并不深究其中的比例关系。如“松鹤延年”，以松树和仙鹤为内容，为求画面均衡，仙鹤的大小竟与松叶相差不多。又如“马上封侯”中的马，形体很小，与自然中的松叶、猴子极不成比例，但画面的总体效果很好，既突出了主题，又将画面图案化、美观化了。应该说，这是中国民间艺人在艺术上的一个创造。

构图均衡，无明显的紧密稀疏，形成了北京四合院砖雕的特有韵味——富贵、华丽、高雅。它与我国其他地区，特别是南方民居的砖雕有着不同的风格。

3. 砖雕工艺技术

砖雕有两种做法，一为雕砖，二为雕泥。雕砖，是在已经烧好的砖料上，按设计的图谱进行放样雕刻。雕泥是在泥坯脱水干燥到一定程度时进行雕刻，然后将雕好的成品放入窑内烧结。这两种做法各有优点。

我们通常所指的砖雕，是在烧好的砖料上面雕刻的工艺手段。砖雕的工序大致有以下步骤。

(1) 放样　根据设计图纸或传统砖雕谱子，放出足尺图样。放样时要注意砖雕风格——画面疏密匀称，构图均衡优美，枝、蔓、花头缠绕遮挡合理，层次分明。

(2) 过画　将所放实样，过画到备好的砖料上。如果图案占据几块砖时，要先将砖块拼对严实，然后再依样摊画。由于砖雕层次较多，有些部分在雕刻过程中可能会将线条剔掉。所以，第一次摊样可以画大轮廓，待将大轮廓雕出以后再画其细部，边雕边画，不要一次画全，以免做无用之功。

(3) 耕　用最小的细錾子沿画出的线条刻画一遍，在砖面上留下清晰刻痕，以免在雕刻时将笔道擦掉而无据可依。

(4) 打窟窿（又名钉坑）　用小錾子将图案以外的空隙部分剔掉，露出图案大形，为进一步细刻打基础。依

做法不同，打窟窿又分两种情况。做透雕时，窟窿需打透眼；做落地雕时，只需将落地部分剔刻至需要的深度（一般20～30mm）。

(5) 镳　在打窟窿的基础上用錾子将图案以外的部分刻去。这道工序是对钉坑的细加工。通过这道工序还要将花叶间的遮挡层次关系表现出来。

(6) 齐口　用錾子沿花饰图案的侧面、上面细致地剔凿。至此，图案层次已基本成形。

(7) 捅道　用细錾子雕刻叶筋，这道工序相当于木雕中用“溜沟”刻叶脉。经过这道工序后，花纹已基本成形。

(8) 磨　用磨头将图案内外粗糙不平之处磨平。

(9) 上药　以白灰（七成）、砖面（三成）加适量青灰，再加水调成与砖相同颜色的糊状物，涂抹在砂眼、崩棱掉角等残缺之处。

(10) 打点　待“药”干后，将周围磨净，用砖面水将图案揉擦干净，令其露出清水活的质感。

至此，砖雕全过程即告完成。

砖雕工具主要是宽窄、薄厚、长短各不相同的錾子，一般宽0.3～1.5cm的各备一种即可。此外，还有起地的翘錾子、刻圆的圆錾子（正反圆）、挠子、煞刀、平尺、勾尺、矩尺、制子、活尺、套筒、锉刀、磨头（砂轮）等（图6–1–25）。

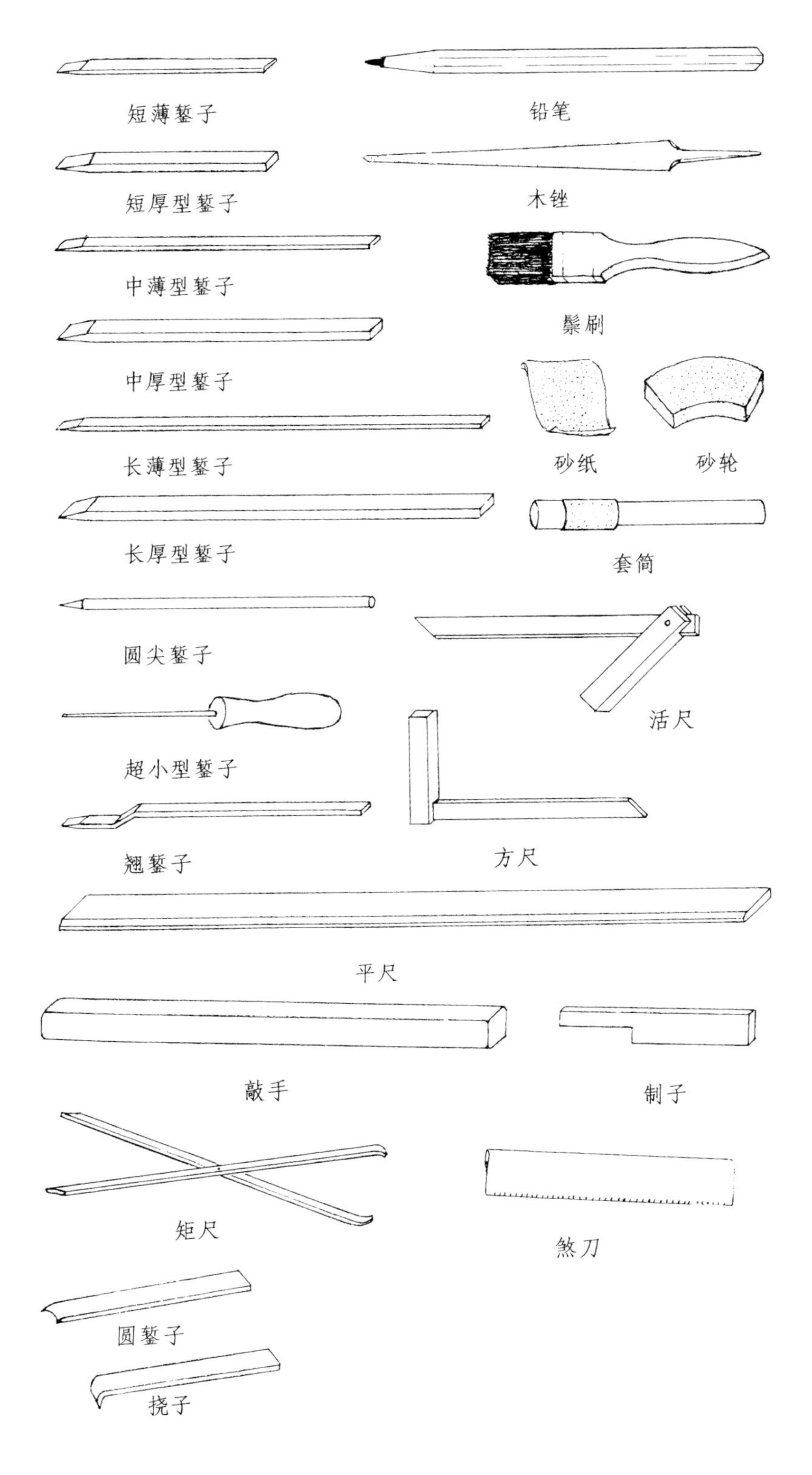

砖雕作品要有艺术性，要有砖雕的性格，依题材不同，当粗则粗，该细则细。粗乃粗犷，并非粗糙，要有力度，见刀功，不能混沌一片；细乃细腻，犹如工笔画，要将内容表现得细致入微，醒目传神。尤其是硬线角的图案，如卍字不到头、回纹、汉纹、扯不断、硬卡子纹等，必须见棱见角，处处跟线；软线条的图案如花卉、蕃草等，则应柔顺自然，线条流畅，不能生硬牵强。

创作砖雕作品，需有深厚的功底和长期全面的艺术修养，决非旦夕之功所能就。北京作为古都，保存许多优秀的砖雕作品。尽管近百年来经历了多次大的劫难，但仍有不少传世佳作被人们保存下来，留传至今。

二、北京四合院的石雕艺术

石雕在中国传统建筑中应用很广，其历史比砖雕要悠久得多。从目前能见到的史料看，它始于殷商，兴于秦汉，成熟于唐宋，明清时期继续发展。历代创作出的石雕佳作，是中国传统建筑艺术的重要组成部分。

石雕主要用于宫殿、坛庙、寺院、陵寝及纪念性建筑，用于普通民居的并不多，甚至远不如砖雕用得广泛，这主要是因为民居中采用石料的部位较少，且做法都比较朴素。尽管石雕在四合院中应用不多，但其艺术价值却是不容忽视的。

1. 石雕的使用部位和雕刻题材

(1) 抱鼓石和幞头鼓子　抱鼓石是用于宅门门口两侧的石构件。抱鼓石是一个笼统的名称，如果细分，应将刻有圆形鼓子的石构件称为圆鼓子，将方形的抱鼓石称为方鼓子，又称幞头鼓子。这两种鼓子石都与门枕石连做在一起。以宅门的门槛为界，外侧是雕有花饰的方圆鼓子石，内侧是用以安置门扇的门枕石。

圆鼓子石一般分为上下两部分，上部为圆形鼓子部分，大约占全高的2/3，系由大圆鼓和两个小圆鼓组成。大鼓呈鼓形，两边有鼓钉，鼓面有金边，中心为花饰。小鼓是大鼓下面的荷叶向两侧翻卷而形成的腰鼓部分。圆鼓子石的下部为须弥座，由上枋、上枭、束腰、下枭、下枋、圭脚组成。须弥座的三个立面有垂下的包袱角，其上做锦纹雕刻。门枕石与须弥座相连，中间有门槛槽相隔（图6—1—26）。

圆鼓子多用于大中型宅院的宅门，其高2～2.8尺（合64～90cm），宽（看面）7～9寸（合22～30cm），进深2～2.5尺（合64～80cm，含门枕石在内）。确定其具体尺寸要看宅门等级的高低和尺寸的宽窄。

圆鼓子上面的狮子，有趴狮、卧狮和蹲狮等不同做法。趴狮的狮身基本含在圆鼓中，前面只有狮子头略略扬起，基本不占立面高度；卧狮是将俯卧的狮子形象刻在鼓子上，其高度约占全高的1/10；蹲狮（又称站狮）前腿站立，后腿伏卧，头部扬起，所占高度为鼓子石全高的1/4～1/3。

圆鼓子两侧鼓心图案以转角莲最为常见，除此之外常用的图案还有麒麟卧松、犀牛望月、松鹤延年、太师少师、牡丹花、荷花、宝相花、狮子滚绣球等等。圆鼓子的正面，一般雕刻如意草、宝相花、荷花、五世同居等图案。

方鼓子略小于圆鼓子，多用于体量较小的宅门（如小型如意门、随墙门等），其高1.5～2尺，进深与高大致相同，厚（看面）5～7寸，由幞头和须弥座两部分组成。幞头上刻有卧狮，其高度（幞头加卧狮）约占全高的2/3，须弥座约占1/3。幞头的金边多做阴纹或阳纹线刻，图案以回纹、丁字锦纹为主。方鼓子侧面及正面的雕刻内容由于不受圆的形状限制，画面安排起来稍显灵活，雕刻内容可有回纹、汉纹、四季花草，也可安排吉祥图案，如松鹤延年、鹤鹿同春、松竹梅等（图6—1—27）。

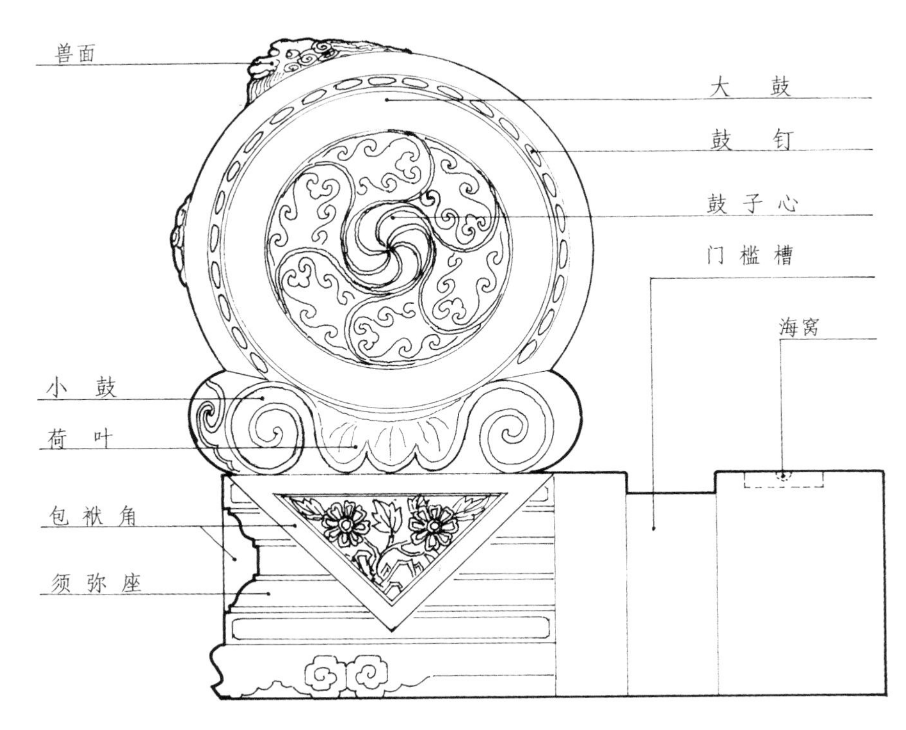

图6—1—26 常见圆鼓子石的部位名称

图6—1—27 各种抱鼓石举例

图6—1—27.1 抱鼓石举例（一）

图6—1—27．2 抱鼓石举例（二）

图6—1—27．3 抱鼓石举例（三）

图6—1—27．4　抱鼓石举例（四）

图6—1—27．5　抱鼓石举例（五）

图6—1—27．6　抱鼓石举例（六）

图6—1—27.7 抱鼓石举例（七）

图6—1—27.9 抱鼓石举例（九）

图6—1—27.8 抱鼓石举例（八）

图6—1—27．10
抱鼓石顶部纹样

图6—1—27．11　抱鼓石举例（十）

图6—1—27．12　抱鼓石举例（十一）

图6—1—27．13　抱鼓石举例（十二）

图6—1—27．14　抱鼓石圆鼓子细部

图6—1—27．15　抱鼓石举例（十三）

图6—1—27．17　抱鼓石正面（一）

图6—1—27．16　抱鼓石举例（十四）

图6—1—27. 18　抱鼓石正面（二）

图6—1—27. 19　幞头鼓子

图6—1—27．20　抱鼓石举例（十五）

图6—1—28　带雕刻的角柱石

（2）滚墩石　滚墩石，有些地方又称其为抱鼓石。这并不是名称上的混淆，而是二者有共同之处。滚墩石是两面对称的抱鼓石，它用于独立柱垂花门或木影壁根部，起稳定垂花门或影壁的作用，同时又富有装饰效果。滚墩石高 2.4 ～ 3 尺（合 80 ～100cm），宽（看面）1.5 尺（合 50cm 左右），进深 4.5 ～ 6 尺（合 150 ～200cm），中间有安插柱子的“海眼”。海眼是透眼，柱子穿过海眼直达基础。

滚墩石的雕刻内容、纹饰与抱鼓石大致相同。立面由大圆鼓子、小圆鼓子、须弥座或直方形座构成，大圆鼓子顶面刻“趴狮”。圆鼓子心常采用的图案有转角莲、太狮少狮、犀牛望月，正面多刻如意草、宝相花等。

（3）挑檐石、角柱石　讲究的四合院，墀头上的挑檐采有石构件，称为“挑檐石”。挑檐石端头的形状，是由半混、炉口、枭三部分组成的曲线，表面一般不做雕刻。墀头下碱部分的角柱石，一般也不做雕刻，但也有在这个部位着意装饰的。北京东城罗鼓巷秦老胡同某宅宅门的角柱石刻满锦纹，是很有特色的一例（图 6–1–28）。

（4）泰山石　泰山石又称“石敢当”，是宅子避邪用的镇物，用于宅院外墙正对街口的墙面上或房角正对街口处，用来压镇街口过强的“衢气”以及其他对宅院有冲犯的邪气。泰山石一般高 3 尺左右，宽 6 ～ 7 寸，上端刻成虎头形状。虎头下面刻有“泰山石敢当”字样，镶砌在墙面或专门为其建造的影壁上；也有不经细致加工，在方正石或随形石上面直接刻字的（图 6–1–29）。

图6—1—29 石敢当举例

图6—1—29．1 镶在照壁上面的石敢当

图6—1—29．2 镶在宅院墙面上的石敢当

（5）陈设墩和绣墩 讲究的宅院中，要摆放盆景、奇石等饰物供人观赏。放置奇石、盆景的石墩台称陈设墩。这种墩台多用汉白玉或清白石雕刻而成，表面遍饰各种花纹图案，颇具观赏性。与陈设墩相类似的，还有置于庭院中供人小坐休息的石桌和石墩。这种石墩称为绣墩，一般刻成鼓形，表面刻出各种花纹和吉祥图案，具有浓厚的传统文化气息（图6–1–30）。

（6）其他石雕构件 除去以上几种，用于四合院民居的石雕构件还有宅门内的闩架石、闩眼石，这是用来安插或放置门闩用的石构件。闩眼石是砌在门后的侧墙内，做插门闩之用。有的宅门在抱框背面安铁环来安插门闩，也具有同等作用。另外还有用在大门外的拴马石和上马石（参见本章第一节）、用于明沟沟眼的沟门石、将雨水排入地沟用的沟漏石等等。这些石构件都很简单，且不具观赏性，但人们依旧将透空部分用钱币、如意等图案加以装饰，使之达到寓意吉祥、图案隽美和实用的高度统一（图6–1–31）。

图6—1—30．1 陈设墩（一）

图6—1—30．3 绣墩

图6—1—30．2 陈设墩（二）

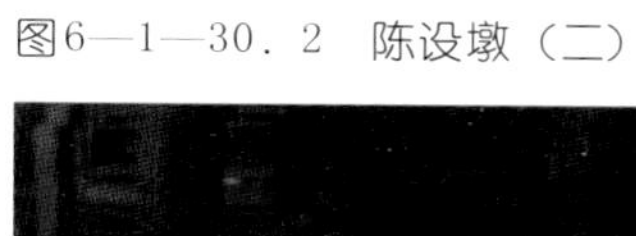

图6—1—30 陈设墩和绣墩

图6—1—31 其他石构件举例

图6—1—31．1 排雨水的沟眼石

图6—1—31．2　放置门闩用的闩架石

图6—1—31．3　将雨水排入地沟用的沟眼石

2. 石雕工艺技术简介

四合院的石雕刻从雕刻技法上主要涉及平雕、浮雕和圆雕三种。

平雕是石雕中最简单的一种。它的基本程序是：①将图案或谱子画在经过初步加工的石料上；②用錾子和锤子沿图案凿出浅沟，称做“穿”；③将图案以外的部分（地子）落下去，并用“扁子”将地子扁光；④将图案的边缘修整好。这种平雕，在石匠中称为“平活”，它多用来雕刻卍字、回纹、丁字锦、鼓钉或其他装饰图案。

浮雕是石雕中用得较多的一种雕刻手法。它的基本程序是：①画样、起谱子，先将设计图案或传统图案画在较厚的纸（如牛皮纸）上，然后沿花纹扎出密排的针眼，称为“扎谱子”，这种做法与彩画起谱子、扎谱子工序几乎完全相同；②将谱子摊在初步加工好的石构件上，用蒲包醮红土粉沿针眼拍打，将花纹拍打在石构件上，称为“拍谱子”，拍得的花纹如不清晰，再用笔按谱子勾画清楚，称为“过谱子”；③用錾子沿线“穿”一遍，使花纹变成浅刻痕留在石构件上；④根据“穿”出的图案进行大形雕刻，称为“打糙”；⑤在打糙的基础上，将需要细致刻画的部位进一步细画、细刻，称为“见细”；⑥修活，雕刻成形的部分进行仔细修理加工，最后成活。浮雕是多用于抱鼓石、滚墩石、陈设墩等石雕构件的主要图案，具有很好的艺术效果。

圆雕，石匠称其为“圆身”，即立体雕刻。圆雕的程序与前述两种有所不同。由于是做立体雕，所以首先要按设计“出坯子”“凿荒”。比如雕石狮子，先要按狮子的各部位比例尺寸下料并打出大的轮廓形状。待荒形打出来之后，在荒形上画出狮子的腿胯、头部、牙爪、胸脯、绣带、尾巴、铃铛等，并对这些具体部位进行初步打凿，称为“打糙”。在打糙的同时，对狮子的前腿、腹部的空当部分进行打凿，称为“掏挖空当”，这也是打糙的组成部分。打糙工作完成之后，狮子的各部具体形状均已出来，然后再进行细加工，包括对眉眼、口齿、舌头、毛发、绣带、铃铛、尾巴、绣球等逐一细画细雕，直至全部完工。用于四合院石活中的圆雕不多，主要是抱鼓石上的石狮。

石雕是传统民居“三雕”的重要组成部分，从纹饰到技法，都需要认真研究继承。

三、北京四合院的木雕艺术

木雕应用于传统建筑的历史，与石雕大体相当。《周礼 · 考工记》“梓人”篇所记，即包括木雕刻。战国时期，“丹楹刻桷”已成为宫廷建筑的常规做法。南北朝时期有关木雕的记载更为具体详尽。隋唐以后，雕刻已成为制度记载于《营造法式》中，并将“雕作”制度按形式分为四种，即混作、雕插写生华、起突卷叶华、剔地洼叶华，按当今的雕法，即为圆雕、线雕、隐雕、剔雕、透雕。明清时期又出现了贴雕、嵌雕等雕刻工艺，使木雕技术得到进一步发展。

木雕在传统民居四合院中的应用比较广泛，艺术价值也很高，但由于保护不力，当前的实物存留已很少。

1. 木雕的应用部位和雕刻题材

（1）用于宅门的雕刻——门簪、雀替、门联　门口上方，用以锁合中槛和连楹的门簪，是木雕的部位之一。门簪雕刻主要用在正面，题材有四季花卉——牡丹（春）、荷花（夏）、菊花（秋）、梅花（冬），象征一年四季富庶吉祥；有团寿字、“福”字或“吉祥”“平安”等吉辞，雕法多采用贴雕，雕好以后贴在门簪迎面上。

用于广亮大门、金柱大门的木雕刻，还有檐枋下面的雀替，其上雕刻内容多为蕃草，均采用剔地起突雕法。

门联也是宅门的雕刻内容之一，刻在街门的门心板上，通常采用锓阳字雕，属隐雕法，字体多为书法家手笔，有很高的艺术水平（图6—1—32）。

图6—1—32　宅门木雕刻举例

图6—1—32．1　宅门木雕刻举例（一）：门联

图6—1—32．2　宅门木雕刻举例（二）：门联

图6—1—32．3　木雕刻举例（三）：门簪

图6—1—32．4　木雕刻举例（四）：门簪仰视

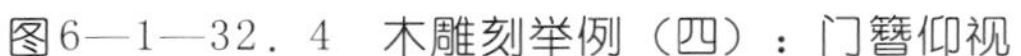

⑵ 用于垂花门的雕刻——花罩、花板、垂柱头　民居中的垂花门，罩面枋下少用雀替而多用花罩。用于垂花门的花罩，内容多为岁寒三友（松、竹、梅）、子孙万代（葫芦及枝蔓）、福寿绵长（寿桃枝叶及蝙蝠）一类世俗间常用的吉祥图案。也有极少数大宅门，采用回纹、卍字、寿字及汉文组成的万福万寿图案。

垂花门的垂柱头是进行雕饰的重点部位之一。垂柱头有圆柱头和方柱头两种形式，圆柱头的雕刻最常见的有莲瓣头，形似含苞待放的莲花，还有二十四气柱头（俗称风摆柳），头上的24条纹路象征二十四节气。方柱头一般是在垂柱头的四面做贴雕，雕刻内容以四季花卉为主。

在垂花门正面的檐枋（帘笼枋）和罩面枋之间由短折柱分割的空间内，还嵌有透雕花板，雕刻内容以蕃草和四季花草为主。垂花门上面的垂莲柱与前檐柱之间安装的骑马雀替或骑马牙子也做同样内容的雕刻。讲究的垂花门，包括月梁下的角背上面都有精美的雕饰，使垂花门格外华丽，富于观赏性（图6—1—33）。

图6—1—33　垂花门上的雕刻举例

图6—1—33．1
垂柱头及侧面雕刻

图6—1—33．2　花板和花罩雕刻

图6—1—33．3　花罩雕刻

图6—1—33．4　垂花门花罩雕刻——松、竹、梅

图6—1—33．5　垂花门花罩雕刻——子孙万代

图6—1—33.6　垂花门花罩雕刻局部

图6—1—34　隔扇、碧纱橱、风门裙板、绦环板雕刻举例

(3) 用于隔扇、碧纱橱、风门的雕刻　隔扇、风门以及室内装修碧纱橱上面的雕刻多用在裙板和绦环板上，通常是按照传统题材做落地雕或贴雕，内容以自然花草和吉祥图案为主，有的也雕些人物故事，如子孙万代、鹤鹿同春、岁寒三友、灵仙竹寿、福在眼前、富贵满堂、蕃草图案以及二十四孝图等（图6—1—34）。

(4) 用于室内外花罩的雕刻　传统民居中的大面积雕刻，主要见于室内外的花罩，如室内的栏杆罩、落地花罩，室外的花罩楣子以及前面提到的垂花门上的花罩等。

花罩为双面透雕，题材主要是自然花草及由此组成的富于文化内涵的内容，如岁寒三友，以松、竹、梅比喻为人正直、高洁；富贵满堂，以牡丹花、海棠花组成的图案借喻高贵富庶；松鹤延年，以松树枝叶及栖于其间的仙鹤隐喻延年益寿；福寿绵长，以蝙蝠、寿桃及其枝蔓缠绕的图案会意福寿长久不衰等等。在这种大面积的透雕中，有时还加进螺钿镶嵌等工艺手段，使画面更加丰富多彩，熠熠生辉（图6—1—35）。

图6—1—34.1　裙板与绦环板上的雕刻

图6—1—34．2　碧纱橱上的雕刻

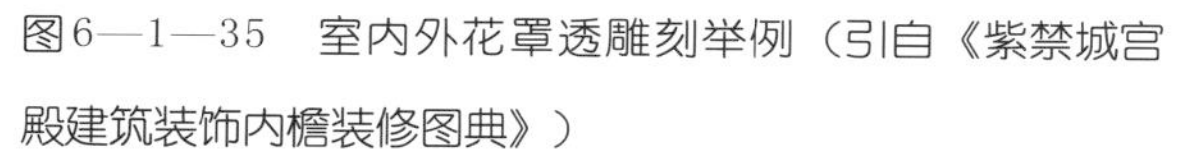

图6—1—35　室内外花罩透雕刻举例（引自《紫禁城宫殿建筑装饰内檐装修图典》）

图6—1—35．1　室内花罩透雕局部（一）：玉兰

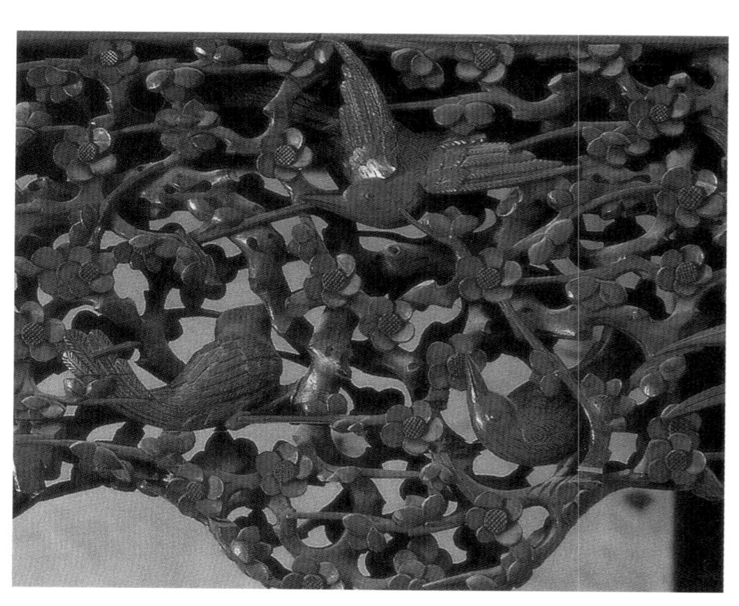

图6—1—35．2　室内花罩透雕局部（二）：喜鹊登梅

图6—1—35．3　室外廊子檐下花罩透雕（一）

图6—1—35．4　室外廊子檐下花罩透雕（二）

（5）用于室内外栏杆上的雕刻　室内栏杆主要用于栏杆罩，此外还有多宝格顶面上的装饰栏杆。室外栏杆主要指二层楼外檐的栏杆。

北京民居中的栏杆，采取杖杅栏杆形式者为多，其上的雕饰物件主要有镶在下枋和腰枋之间的花板、绦环板以及位于腰枋和杅杖扶手之间的净瓶。花板雕刻用于室内时以浮雕为主，用于室外时以透雕为主。净瓶雕刻的图案内容，应与栏杆罩的雕刻主题相一致。用于外檐的栏杆净瓶多用荷叶图案（图 6—1—36）。

（6）用于花格、棂条间的雕刻饰件　花格、棂条是隔扇、风门、支摘窗、楣子等木装修的格心部分，它们有步步锦、龟背锦、冰裂纹、盘长、豆腐块、卍字不到头等图案，这些图案中，都极少夹杂雕刻饰件；而灯笼框、万福万寿一类棂条花格，则夹杂着诸如卡子花、团花、团寿字、福寿字等吉祥图案。用于灯笼框棂条间的团花、卡子花，主要由自然花草图案组成，并有一定的寓意，如岁寒三友、梅竹、松竹、子孙万代、福寿（桃）等。团寿字、方寿字等图案则采自中国传统书法中的“福”字、“寿”字写法，经艺术加工，形成团花、卡子形状。这种类型的图案中，还有由蝙蝠构成的卡子花。这些团花、卡子花，除去美化窗格、表达吉祥寓意之外，还有连接加固相邻棂条的作用。

再一种用得比较普遍的雕饰，就是安装在倒挂楣子边梃间的花牙子。花牙子的雕刻内容以花草图案为主，也有的采用回纹、蕃草、夔龙、夔凤图案（图 6—1—37）。

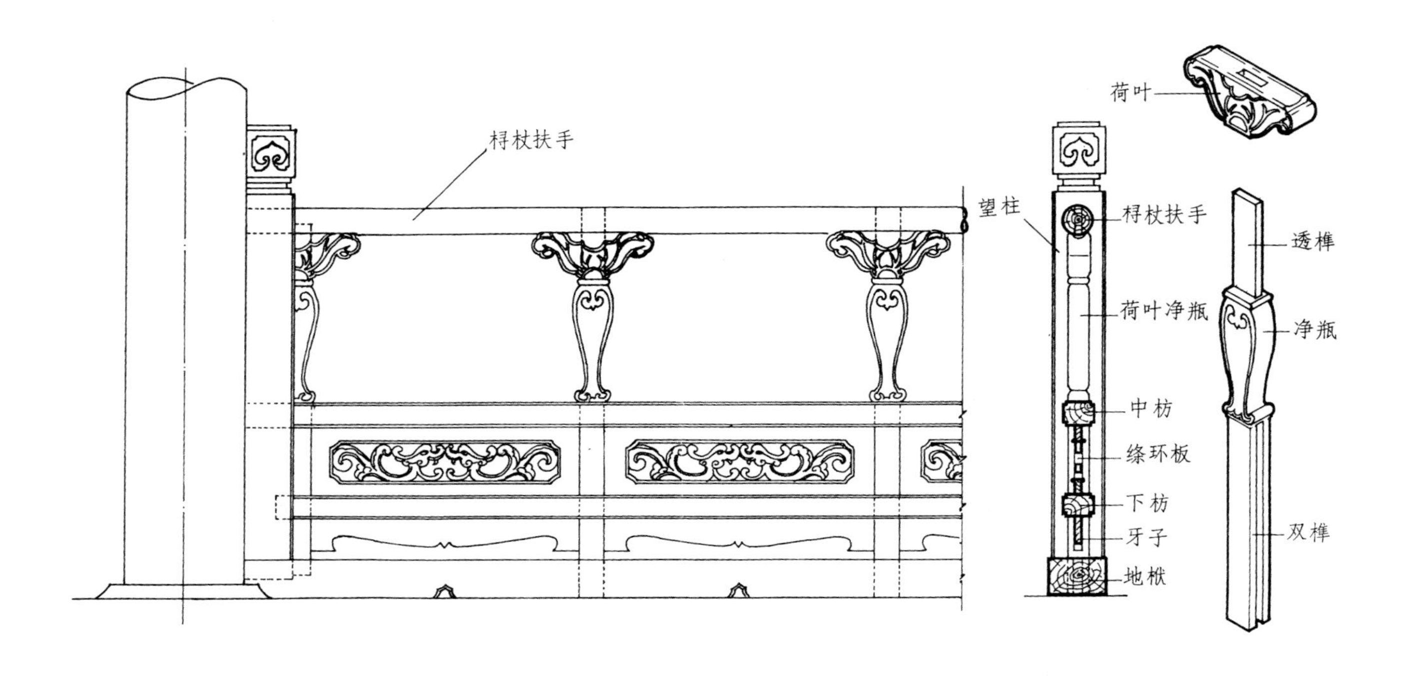

图6—1—36 用于栏杆的雕刻纹饰举例

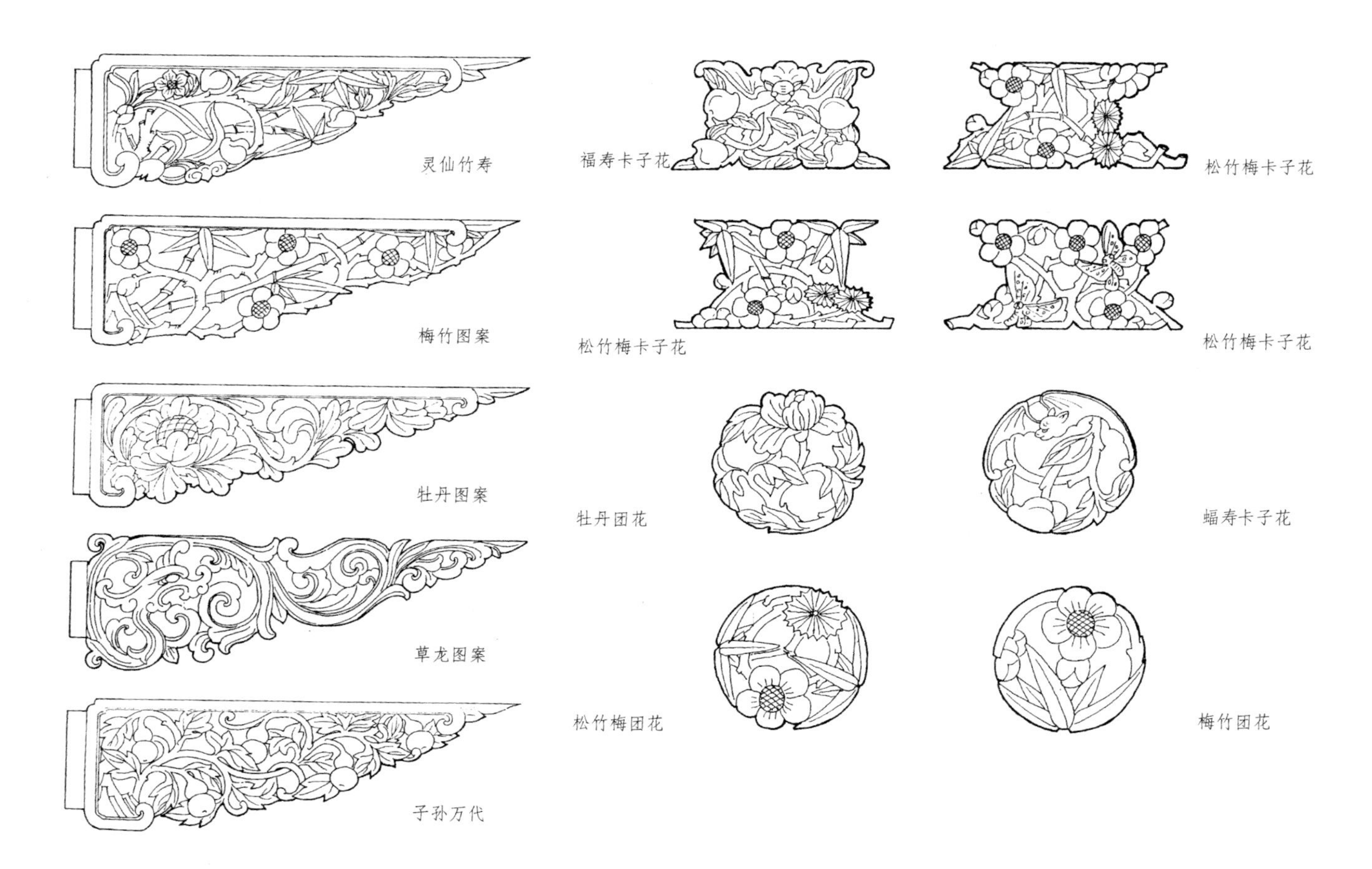

图6—1—37 团花、卡子花、花牙子

（7）其他雕刻 除去以上六部分雕刻外，其他如帘架边框上部的荷花拴斗雕刻荷花图案和下部的荷叶墩雕刻荷叶图案；落地罩下面的须弥墩雕成须弥座形状，讲究的做法还要在须弥座上雕出八达马、仰覆莲等花纹；在垂花门的木质博缝头上雕刻太极图案；在室内板壁、隔断、挂落上雕刻花草或吉祥图案；在倒挂楣子边梃端头雕刻“白菜头”图案；在匾联上雕刻回纹、汉纹、锦纹及花草图案等等。这些木雕刻有很大的随意性和灵活性，不再逐一叙述。

2. 木雕工艺技术简介

传统的木雕工艺技术有平雕、落地雕、透雕及贴雕、嵌雕等。这些木雕工艺，至今仍在古建筑、仿古建筑工程中采用。

（1）平雕 平雕是在平面上，通过线刻或阴刻的方法表现图案实体的雕刻手段。平雕常见的有三种刻法：一为线雕，这是一种用刻刀直接将图案刻在木构件表面的雕法，工艺类似印章中的阴纹雕刻，雕刻内容有兰草、梅花一类适宜用平雕表现的图案，其效果有如工笔画；平雕的第二种雕法是锓阳刻，这是一种通过将图形外轮廓阴刻下去，而反衬出图形本身的雕法，一般多用来雕刻门联、楹联、诗词等书法作品；平雕的第三种雕法是阴刻，是将图案以外的地子全部平刻下去，以托出图案本身。这种刻法，多用于回纹、卍字、丁字锦、扯不断等装饰图案。

（2）落地雕 落地雕在宋代称做剔地起突雕法，是将图案以外的空余部分（地子）剔凿下去从而反衬出图案实体的雕刻方法。

落地雕的程序如下，①画样，将要雕刻的图形按1:1足尺寸画在纸上。②过样子，将画好的样子贴在备好的木板上用刀凿将线条阴在木板表面，这道工序与砖雕中的“耕”和石雕中的“穿”具有同等作用。如果雕刻2块以上同样花纹的物件，还可从阴纹样的木板上反拓出多幅纸样（自从胶合板出现之后，已采用胶合板做样板代替纸样）。③凿活，用平凿、圆凿等雕刻工具沿图形凿至需要的深度（一般10 ～20mm，如一次不能达到这个深度时可分两次）。④起地子，用平凿、斜凿剔去空白部分，使之达到地子需要的深度，并用翘将地子修铲平整。⑤粗雕，将图案中枝叶大形及相互间的缠绕遮挡关系表达清楚，使大轮廓完整清晰。⑥细雕，在粗雕的基础上进一步细加工，并刻出花瓣、叶筋、花蕊等细部纹饰。落地雕不同于平雕，花头枝叶之间翻卷错落、缠绕遮挡关系必须清楚，要有立体感，雕刻要见刀功，棱角分明，线条流畅，要有艺术感染力。

（3）圆雕 圆雕是立体雕刻，首先要画样，并根据样子尺寸备料、落荒（做出大体形状），在将荒形修正得与所需形状很接近时，再在表面摊样（画样子），然后按样子进行糙刻、细刻，最后铲叶筋，刻细部纹饰（图6—1—38）。

图6—1—38 圆雕、贴雕、落地雕举例

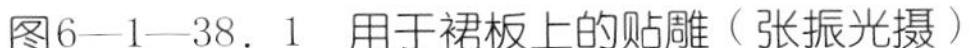
图6—1—38．1 用于裙板上的贴雕（张振光摄）

图6—1—38．2 贴雕举例

图6—1—38.3 落地雕举例（博古）

图6—1—38.4 用于室内栏杆罩上的花瓶（圆雕）

图6—1—38.5 贴雕一例

（4）透雕 透雕程序是：①画样；②过样；③锼活（将花纹以外的空地用锼锯挖掉）；④铲活，即粗雕，刻出枝叶花梗间的遮挡缠绕翻卷关系，透雕一般都是双面活，要两面雕刻；⑤细刻，在粗雕的基础上进行细刻，使之最后成活。做透雕时，对锼锯锼过的面还要进行修锉刻铲，以达到形象逼真、栩栩如生的效果（图6—1—39）。

（5）贴雕和嵌雕 贴雕是落地雕的改革雕法，兴于清代晚期，常见于裙板、绦环板的雕刻。方法是用薄板锼出花纹并进行单面雕刻之后，贴在裙板或绦环板上。它完全具备落地雕的效果，但在工料方面则要节省得多，雕出的效果也更好，尤其在地子平整、无刀痕刃迹方面非落地雕刻所能比。贴雕可通过使用不同质地和颜色的木料做地子和花纹达到特殊效果。

图6—1—39 透雕举例（引自《紫禁城宫殿建筑装饰内檐装修图典》）

图6—1—39．1 松树藤萝

图6—1—39．2 子孙万代

图6—1—39．3 喜鹊登梅

嵌雕是为解决落地雕中个别高起部分而采用的技术措施。如龙凤花板中高起的龙头、凤头，可在雕刻大面积花活时预留出龙头、凤头的安装位置，另外用木头单独雕刻并嵌装在花板上，从而可使雕作更加传神。

（6）木雕工具　与砖、石雕刻相比较，木雕工具更为讲究一些，它们主要有以下几种。

锼锯：由极窄的锯条或钢丝（带齿）制成，凭竹弓子绷紧，做透雕锼活之用。

平凿、斜凿：平凿是平刃的凿子，形似木匠用的扁铲；斜凿是斜刃的凿子。这两种都是用来起地子、修铲活的工具。

正口凿、反口凿：正口凿为凸面凿子，反口凿为凹面凿子，是专门用来剔凿修铲带有弧度部位的工具。

翘：形似锄头，刀刃部分弯曲成一定角度，主要用来修铲地子平面之用。

溜沟：刀刃一端做成沟槽形状，是专门用来雕刻花脉、叶筋、花纹、毛发的工具。

以上雕刻工具，除锼锯之外，每种都备有大小、宽窄不等的型号，以适应各种不同尺度和粗细的雕刻。

此外，还有木敲手，用枣木、檀木或其他硬木做成，头方柄圆是用来剔凿打凿柄的工具（以上均见图6–1–40）。木雕、砖雕和石雕构成了中国的传统建筑雕刻艺术，它们之间既有区别又有许多相通之处。通过三雕而创作出来的艺术作品，是中国传统建筑艺术中极为宝贵的组成部分，是人类珍贵的历史文化遗产。

图6—1—40　木雕工具

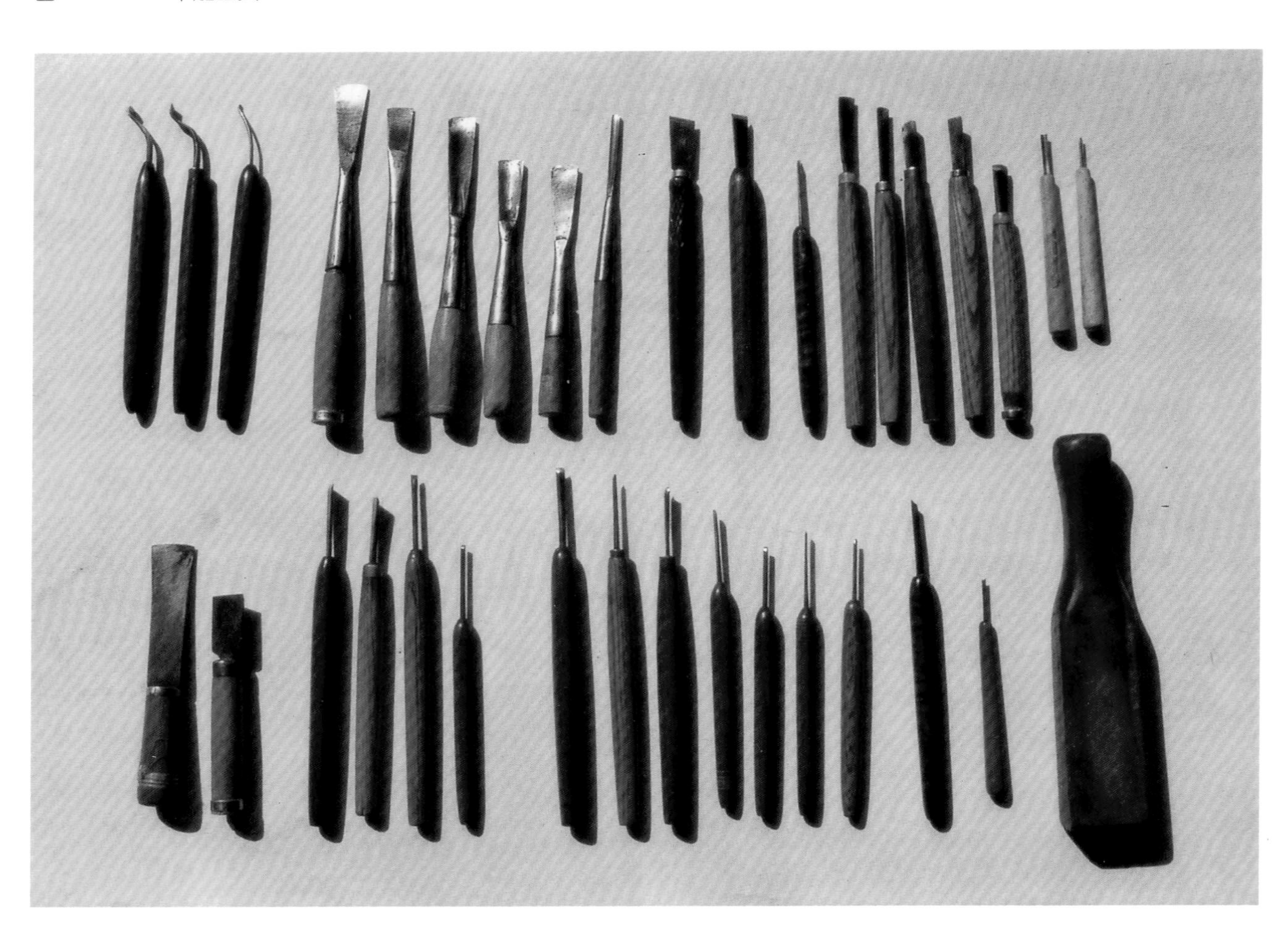

第二节 油饰彩画

一、古建油饰彩画的主要作用及等级差别

在木构件表面涂刷油饰色彩以利防腐并装饰建筑，是中国古建筑的传统做法。正如林徽因先生所指出的，在建筑上施用油饰彩画，“最初是为了实用，为了适应木结构上防腐防蠹的实际需要，普遍用矿物原料的丹或朱以及黑漆桐油等涂料敷饰在木结构上；后来逐渐和美术上的要求统一起来，变得复杂丰富，成为中国建筑艺术特有的一种方法”（引自林徽因《中国建筑彩画图案·序》）。

早期建筑上的色彩油饰，是没有明显区分的，它们都有保护木构件的作用，也都有色彩装饰作用。随着人类建筑活动的发展，油漆和彩画出现了明确分工，至明清时期，官式做法已有“油作”与“画作”之分，凡用于保护构件的油灰地仗、油皮及相关的涂料刷饰，被统称为“油饰”，而用于装饰建筑的各种绘画、图案线条、色彩被统称为“彩画”。

正如中国古代建筑历来具有森严的等级区别那样，作为建筑装饰的油饰彩画也具有严格的等级。据《春秋穀梁传注疏》载：“礼楹，天子丹，诸侯黝垩，大夫苍，士黈。”这说明，早在我国春秋时代，建筑所施色彩已经有了严格的等级制度。这种等级制度，历代沿袭不衰。《唐会要·舆服志》载：“六品七品以下堂舍……不得施悬鱼、对凤装饰……又庶人所造堂舍……仍不得辄施装饰。”可见唐代对各阶层人士之宅第装饰限制也是很严格的。

明代《舆服志》所载规定更为具体：“明初禁官民房屋，不许雕古帝后圣贤人物，及日、月、龙凤、狻猊、麒麟、犀、象之形。”洪武二十六年定制，“官员营造房屋不许……绘藻井，惟楼居重檐不禁，公侯前厅……用金漆及兽面锡环。家庙……梁栋斗拱檐桷彩绘饰，门窗枋柱金漆饰。一品二品厅堂梁栋斗拱檐桷青碧绘饰，门……黑油铁环。六品九品厅堂，梁栋饰以土黄，门……黑油铁环。品官房舍门窗户牖不得用丹漆……”“三品六品厅堂梁栋只用粉青饰之。”“庶民庐舍不过三间五架，不许用斗拱饰彩色。”

从以上关于油饰彩画的规定可以看到，在明代，对于象征皇权的龙凤等纹饰为皇家专用，其余官民绝对禁用。自皇家至士民的建筑装饰，至少划分了五个档次，并详细地规定了各自的装饰内容和色彩。可见明代对各阶层人士的房屋装饰规定也是非常严格的。

清代对此也有非常具体的规定。据《大清会典》载：“顺治九年定亲王府……正门殿寝……凡有正屋正楼门柱均红青油饰……梁栋贴金、绘五爪金龙及各色花草……凡房庑楼屋均丹楹朱户，其府库廪厨及祗候各执事房屋，随宜建置……门柱黑油。”“公侯以下官民房屋……梁栋许画五彩杂花，柱用素油，门用黑饰，官员住屋，中梁贴金……余不得擅用。”

从以上规定可以看到，清代关于王府公侯建筑与一般官僚及平民百姓的建筑在油饰彩画方面的差别是相当大的，但清代对公侯以下官民房屋的油饰彩画装饰较以往各个朝代明显放宽，如“梁栋许画五彩杂花”即是明证。

北京四合院是明清遗留下来的传统建筑，其油饰彩画内容、等级必然受到等级制度的严格制约。油饰彩画是建筑中最易风化剥落的部分，以现存建筑的油饰色彩去考察明清时期建筑的色彩等级，显然是不够的。了解历史上特别是明清时期有关色彩油饰等级的规定，对我们今天研究四合院的油饰色彩及恢复它原有的面貌是十分必要的。

二、四合院建筑的油饰

中国传统建筑的油饰分为油灰地仗和油皮两个层次，用于四合院民居的油饰也不例外。

1. 油灰地仗

油灰地仗（简称地仗）是由砖面灰（对砖料进行加工产生的砖灰，分粗、中、细几种）、血料（经过加工的猪血）以及麻、布等材料包裹在木构件表层形成的灰壳，主要起保护木构件的作用。由于在它的表面要涂刷油漆，所以它又是油漆的基层。清早期以前的地仗做法比较简单，一般只对木构件表面的明显缺陷用油灰做必要的填刮平整然后钻生油（即操生桐油，使之渗入到地仗之内，以增强地仗的强度韧性及防腐蚀性能）。清早期以后地仗做法日益加厚，出现了不施麻或布的“单披灰”，包括一道半灰、两道灰、三道灰及至四道灰做法，更讲究的则有“一布四灰”“一麻五灰”“一麻一布六灰”，甚至“二麻六灰”和“三麻二布七灰”等做法。讲究的四合院木构地仗、重点构件要做到一麻五灰，其余构件大多做单披灰地仗。王府建筑的地仗可厚于一麻五灰。

在木构件表面做地仗，在清代早期已形成制度。清代以来，木构地仗出现越做越厚的趋势，主要原因有二：其一，清代建筑多承自明代，因年久反复修缮，原有构件大多不太平直圆顺，棱角也不完整，只能通过加厚地仗，使麻糊布、过板闸线等工艺手段得以再现昔日光彩；其二，实践使人们懂得，很薄的地仗是不能长期抗御自然界各种侵蚀的，因此加大地仗厚度、加强地仗的拉力（糊布或使麻）也就成为必然。

2. 油皮及色彩

（1）四合院建筑油饰的色彩　涂刷在构件表面的油漆及涂料的色彩，对四合院整体环境的色彩构成起着决定性作用，所以历来备受重视。

传统的油饰色彩，一般都是由高级匠人将颜料入光油或将颜料入胶经深细加工而得。明清时期，适于古建油饰彩画的颜料是非常丰富的，仅清工部《工程做法则例》中所列的关于油饰的色彩，就不下20种，如朱红油饰、紫朱油饰、柿黄油饰、金黄油饰、米色油饰、广花油饰、定粉油饰、烟子油饰、大绿三绿及瓜皮油饰、香色油饰等等。涂料有广花结砖色、靛球定粉砖色、天大青及洋青刷胶、红土刷胶、楠木色等。由于当时已有如此丰富的油饰色彩，所以，用于建筑的油饰色彩也就十分丰富了。近年来出现的四合院油漆仅有红绿两色的现象，显然是不符合历史传统的。

（2）四合院建筑油饰色彩的一般做法及规律　关于四合院建筑油饰色彩的一般做法，可参见表（表6–2–1）。

（3）油饰色彩的变化及与建筑等级和装饰效果的关系　从表6–2–1可以看出，王公贵族居住的建筑，大多采用明亮鲜艳的紫朱油或朱红油进行装饰（多见于大门），以体现王侯“凡房庑楼屋均丹楹朱户”的非凡气派，以显示宅主人显要尊贵的社会地位；而一般官员、平民住宅只能用较灰暗的红土烟子油或黑红相间、单一黑色的油饰，这些正好符合建筑色彩庶民采用“黝”与“黑”的等级要求。

表 6-2-1 四合院建筑油饰色彩一般做法

建筑名称	做法编号	构件名称	油皮色彩	贴金	备注
房座	1	连檐瓦口	朱红油		见于王府建筑
		椽望	望板红土烟子油，椽红帮绿底油		
		梁枋大木	大多满彩画。少量局部彩画，余地紫朱油	分别按彩画制度贴金	
		下架柱框、槛框、榻板等	紫朱油	框线贴金	
		装修	各种扇活、门大边、边抹紫朱油，仔屉棂条三绿油	云盘线、涤环线及裙板花饰等雕饰贴金	
		坐凳楣子	坐凳面紫朱漆，楣子大边朱红油，棂条三绿油		
	2	连檐瓦口	朱红油		见于一般官员及平民住宅建筑（亦有用紫朱油代红土烟子油做法者）
		椽望	红土烟子油或红土刷胶罩油		
		梁枋大木	不彩画部位红土烟子油		
		下架柱框、槛框、榻板等	红土烟子油	高级做法者框线贴金，否则不贴金	
		装修	各种扇活、门大边、边抹红土烟子油，仔屉三绿油	裙板等雕饰，高级做法者贴金，否则不贴金	
	3	连檐瓦口	朱红油		用于一般官员及平民住宅建筑，但此类做法较少见
		椽望	红土烟子油或红土刷胶罩油		
		梁枋大木	不彩画部位红土烟子油		
		下架柱框、槛框、榻板等	烟子（黑色）油	高级做法者框线贴金，否则不贴金	
		装修	各种扇活、门、边抹烟子油，仔屉三绿油	高级做法者，裙板等雕饰贴金，否则不贴金	

续表6-2-1

建筑名称	做法编号	构件名称	油皮色彩	贴　　金	备　　注
游廊	1	连檐瓦口	朱红油		见于王府建筑
		椽望	望板紫朱油，椽红（紫朱）帮绿底油		
		梁枋大木	大多满做彩画。少量局部彩画者，其余地紫朱油	按彩画制度贴金	
		下架柱坐凳	大绿油		
		倒挂楣子、坐凳楣子	倒挂楣子朱红油大边，坐凳楣子大边朱红油，棂条三绿油	按彩画制度贴金	
	2	连檐瓦口	朱红油		见于一般官员及平民住宅建筑（特殊高官、富民的高级做法，亦有用紫朱油代红土烟子油做法者）
		椽望	红土烟子油或红土刷胶罩油		
		梁枋大木	彩画的所余部位或不彩画者，红土烟子油	做彩画者，按彩画制度贴金	
		下架柱坐凳	大绿油		
		倒挂楣子、坐凳楣子	倒挂楣子朱红油大边，余或苏装或饰三绿油；坐凳楣子朱红大边，余三绿油	做彩画者，按彩画制度贴金	
垂花门	1	连檐瓦口	朱红油		见于王府建筑
		椽望	望板紫朱油，椽红（紫朱油）帮绿底油		
		梁枋大木	大多满做彩画，少量局部彩画者，其余地紫朱油		
		博缝	朱红油或紫朱油	梅花钉贴金	
		下架柱框	梅花柱大绿油，门框紫朱油	框线贴金	
		装修	门、余塞板紫朱油，屏门大绿油撒金或纯油饰	框线贴金	见于一般官员及平民住宅建筑（特殊高官、富民的高级做法，亦有用紫朱油代红土烟子油者）
	2	连檐瓦口	朱红油		
		椽望	红土烟子油或红土刷胶罩油		
		梁枋大木	大多满做彩画。少量做局部彩画者，余地红土烟子油		
		博缝	朱红油或烟子（黑色）油或紫朱油（高级做法）	梅花钉贴金	
		下架柱框	梅花柱大绿油，框或红土烟子油或烟子油	框线贴金或不贴金	
		装修	见广亮门一般油饰栏及屏门一般油饰栏	框线贴金或不贴金	

续表 6－2－1

建筑名称	做法编号	构件名称	油皮色彩	贴 金	备 注
垂花门	3	连檐瓦口	朱红油		见于一般官员及平民住宅建筑
		椽望	红土烟子油或红土刷胶罩油		
		梁枋大木	檩、枋、梁满刷红土烟子油，花板、垂头等绿油		
		倒挂楣子	大边朱红油，棱条大绿油		
		博缝	朱红油或烟子油（黑色）		
		下架柱框	梅花柱大绿油，框或红土烟子油或烟子油		
		装修	见广亮门一般油饰栏及屏门一般油饰栏		
王府大门		连檐瓦口	朱红油		见于王府建筑
		椽望	望板紫朱油，椽红（紫朱）帮绿底（大绿）油		
		梁枋大木	大多满做彩画。少量局部彩画者，余地紫朱油	分别按彩画制度贴金	
		雀替	朱红色油地仗	按彩画制度贴金	
		下架柱框	朱红油或紫朱油	框线贴金	
		装修	大门、扇活大边、边抹同柱框，籽屉棂条三绿油	裙板等雕饰，门钉贴金	
广亮大门、金柱大门	1	连檐瓦口	朱红油		见于高官、富民住宅建筑，属高级做法
		椽望	望板紫朱油，椽红（紫朱）帮绿底（大绿）油		
		梁枋大木	大多满做彩画。少量做局部彩画者，余地紫朱油	分别按彩画制度贴金	
		雀替	朱红色油地仗	按彩画制度贴金	
		下架柱框	朱红油或紫朱油	框线贴金	
		装修	①大门、边抹、余塞板或朱红油或紫朱油 ②大门、边抹或朱红油或紫朱油，余塞板烟子油	框线贴金 门簪边框贴金	
	2	连檐瓦口	朱红油		见于一般官员及富民住宅建筑，属次高等级做法
		椽望	红土烟子油或红土刷胶罩油		
		梁枋大木	大多满做彩画。少量局部彩画或不彩画者，余地红土烟子油	按彩画制度贴金	
		雀替	朱红油地仗，若不彩画者，雕饰大绿油	按彩画制度贴金	
		下架柱框	红土烟子油	框线贴金或不贴金	
		装修	①大门、边抹、余塞板一律红土烟子油 ②大门、边抹红土烟子油，余塞板大绿油 ③大门、边抹红土烟子油（黑），余塞板红土烟子油	框线、门簪贴金或不贴金	

续表 6－2－1

建筑名称	做法编号	构件名称	油皮色彩	贴　金	备　注
如意门	1	门簪	或朱红色地或大青色地	边框和字贴金	见于一般民宅
		门框、门扇（无门联）	或一律红土烟子油（低等级做法者红土刷胶罩油）或一律烟子（黑色）油（低等级做法者烟子刷胶油）		
如意门	2	门簪	或朱红色地或大青色地	边框和字贴金	见于一般民宅
		门框、门扇（有门联）	门框、门扇烟子油（低等级做法者烟子油刷胶罩油），门联朱红油（低等级做法朱红刷胶罩油），文字或黑油或金字		
各类随墙门		无论有无门簪	同上述如意门做法1、2	同上述如意门做法1、2	见于一般官民住宅
各式屏门	1	门扇	门扇大绿色油，并做撒金，门扇上端，设朱红油斗方或圆形吉祥文字	斗方或圆形轮廓及文字贴金	见于王府及讲究的官民住宅
	2	门扇	门扇做单一大绿色油		见于一般官民住宅
各式什锦窗		边框 仔屉 棂条	烟子（黑色）油 朱红油 三绿油		见于各种等级住宅

说明：（1）传统朱红油，以一般名贵的“广银朱”色入光油（熟桐油）而得，色彩鲜艳稳重。
（2）紫朱油，即俗称之“二朱油”，由广银朱色加少许蓝色油而得，色彩鲜艳稳重带紫色调。
（3）红土，即广红土色，红土烟子油以红土为主，加少许烟子（黑色）色入光油而得，色彩近于土红色。
（4）烟子油即黑色油。

油饰色彩的多样化，对建筑色彩的变化和装饰起着非常重要的作用。

从王府建筑的紫朱油到一般民居的红土烟子油不仅体现了等级差别，而且说明古人是非常善于运用色彩的。紫朱油与红土烟子油，虽同属红色系列，但二色之间无论彩度、明暗度还是色相色温，都有许多细微差别。古人正是利用了这些差别，不仅避开了一般民宅用红与王府用红之间的忌讳，体现了等级差别，而且在色相运用上又保持了相互间的和谐与统一。

大多四合院广泛采用紫朱油或红土烟子油，还因为这两种颜色同属带紫色调的暖红色，它可以营造出一种亲切热烈的气氛，非常适合于四季分明的北方的居住环境。这种暖红色调，可与其周围的青绿彩画、大面积的青砖灰瓦产生冷暖对比，为建筑物带来盎然生机。

一般的民居四合院也运用高彩度的朱红颜色，但这种运用是有节制的，一般只用于建筑檐头的连檐瓦口、花门垫板及用来强调某些特殊部位、强调明暗对比的地方。

四合院油饰色彩的运用中，再一个常见而且具有浓郁地方特点的用法是，用黑色油（烟子油）与红色油（紫朱油或红土烟子油）相间装饰建筑构件，这种做法称为“黑红净”。如椽望用红色油，下架柱框装修用黑色油；大门的槛框用黑色油，余塞板用红色油；门扉的攒边用黑色油，门联地子用红色油。这些都属于黑红净做法。这种装饰可产生稳重、典雅、朴素而富于生气的效果（图 6–2–1）。

图6—2—1 四合院的三种宅门及其色彩效果

图6—2—1．1 随墙门

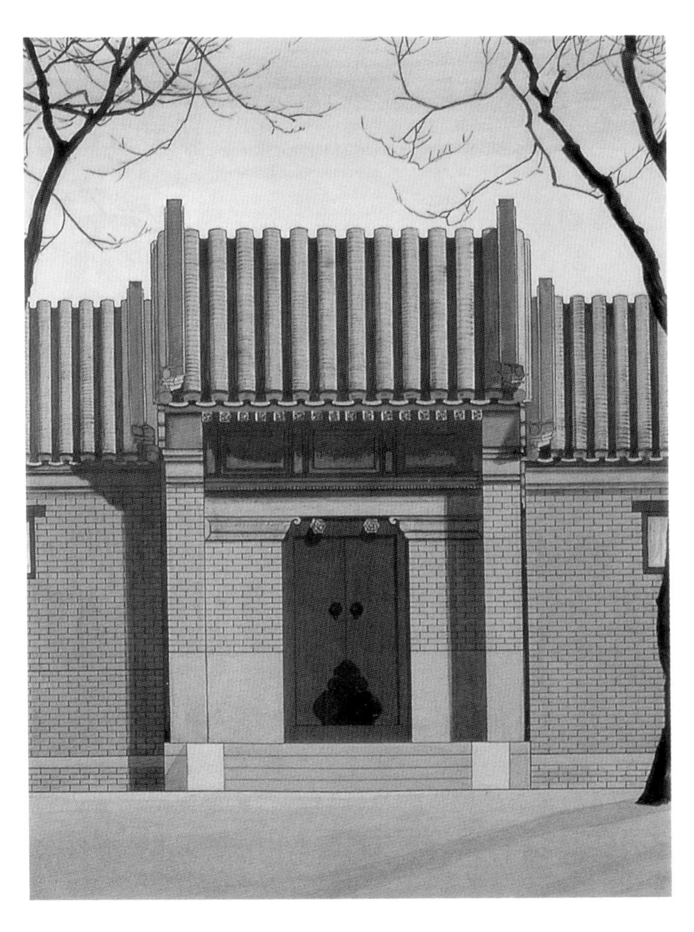

图6—2—1．2 如意门

图6—2—1．3 广亮大门

三、四合院建筑的彩画

1. 四合院建筑常见的彩画类别及其应用

彩画是四合院建筑的重要装饰手段，它运用鲜艳的色彩，通过在建筑构件上绘画达到装饰目的。四合院的彩画（上至王府下至一般民居彩画）涉及了清代两类建筑彩画，一类是“旋子彩画”，一类是“苏式彩画”。旋子彩画庄严肃穆，一般仅用来装饰王府。苏式彩画内容丰富，形式活泼，充满诗情画意，除用来装饰王府中一些次要建筑或园林建筑外，一般民居建筑也普遍采用。

彩画在四合院建筑中的应用大体有以下六种情况。这六种情况也可以代表六种不同等级。

（1）大木满做彩画　即檩、垫、枋等大木构件，或满做旋子彩画，或满做苏式彩画（椽柁头、三岔头、穿插枋头、雀替、花牙子、花板、天花、倒挂楣子等也做与大木相配的彩画）。

（2）大木做“掐箍头搭包袱”的局部苏式彩画　即檩、垫、枋大木构件端头做带状活箍头、副箍头，构件中段做包袱图案。包袱内饰各种绘画内容（椽柁头、雀替、牙子等构件做与大木相配的彩画）。

（3）大木做“掐箍头”的局部苏式彩画　在檩、垫、枋大木构件的端头做各种活箍头及副箍头（椽柁头、雀替、牙子等构件做与大木相配的彩画）。

（4）只在椽柁头部位做彩画，其余全部做油饰

（5）只在椽柁头迎面刷颜色　一般在飞椽刷大绿色，檐椽头和柁头刷大青色，其余部位做油饰。

（6）所有构件全部做油饰

以上六种做法，在不同等级和四合院中均有体现。

在古代，人们普遍重视对宅门、二门（垂花门）的彩画装饰，这些部位的彩画要比宅院内其他建筑的彩画高一个等级。如内宅正房、厢房做“掐箍头搭包袱”彩画，那么该院的大门、垂花门则要满做苏画（图 6–6–2）。

2. 四合院彩画的题材、内容、构图及做法特点

（1）旋子彩画　这种广泛用于王府建筑的彩画，其基本构图特点为，檩枋大木两端绘箍头，开间大的檩枋内侧还要加画盒子或多画一条箍头，檩枋中段占构件 1/3 长的部位画方心，方心与箍头之间的部分画找头。体现旋子彩画主要特征的旋花图案，主要在找头部位得到充分表现。

找头图案的旋花画法采用“整破结合”的方式。旋子彩画的主题纹饰，主要在檩枋彩画的方心内得到表现。王府旋子彩画的方心，一般采“龙锦方心”和“花锦方心”。

旋子彩画从纹饰特征、设色、工艺制作方面分，大致有八种做法：混金旋子彩画、金琢墨石辗玉、烟琢墨石辗玉、金线大点金、墨线大点金、小点金、雅伍墨和雄黄玉。王府建筑对旋子彩画的运用最多的是金线大点金和墨线大点金，其中个别重要的建筑（如大门等），亦有用金琢墨石辗玉做法的，值房类等附属建筑一般用小点金或雅五墨彩画。

旋子彩画的设色具有固定的规制，彩画中用金面积的大小，直接反映着该彩画的做法等级，上述八种旋子彩画也正体现着其用金方面的差别。旋子彩画用色，是以青绿二色为主，其设色的主要特征，是按图案划分部位，按“青绿相间”的原则分布色彩。这种方法，可使构成图案的色彩协调匀称。

图6—2—2 垂花门苏式彩画与纹饰举例

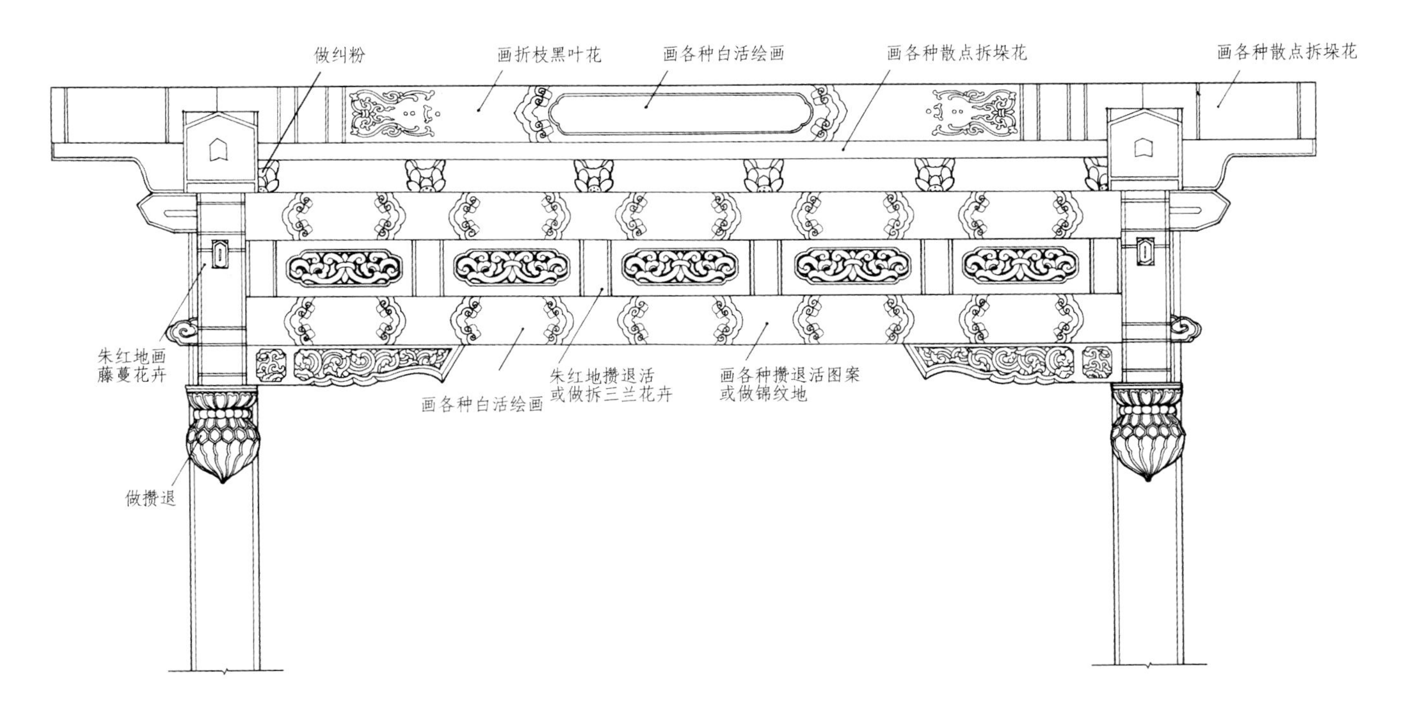

图6—2—2.1 垂花门正立面苏画纹饰示意图

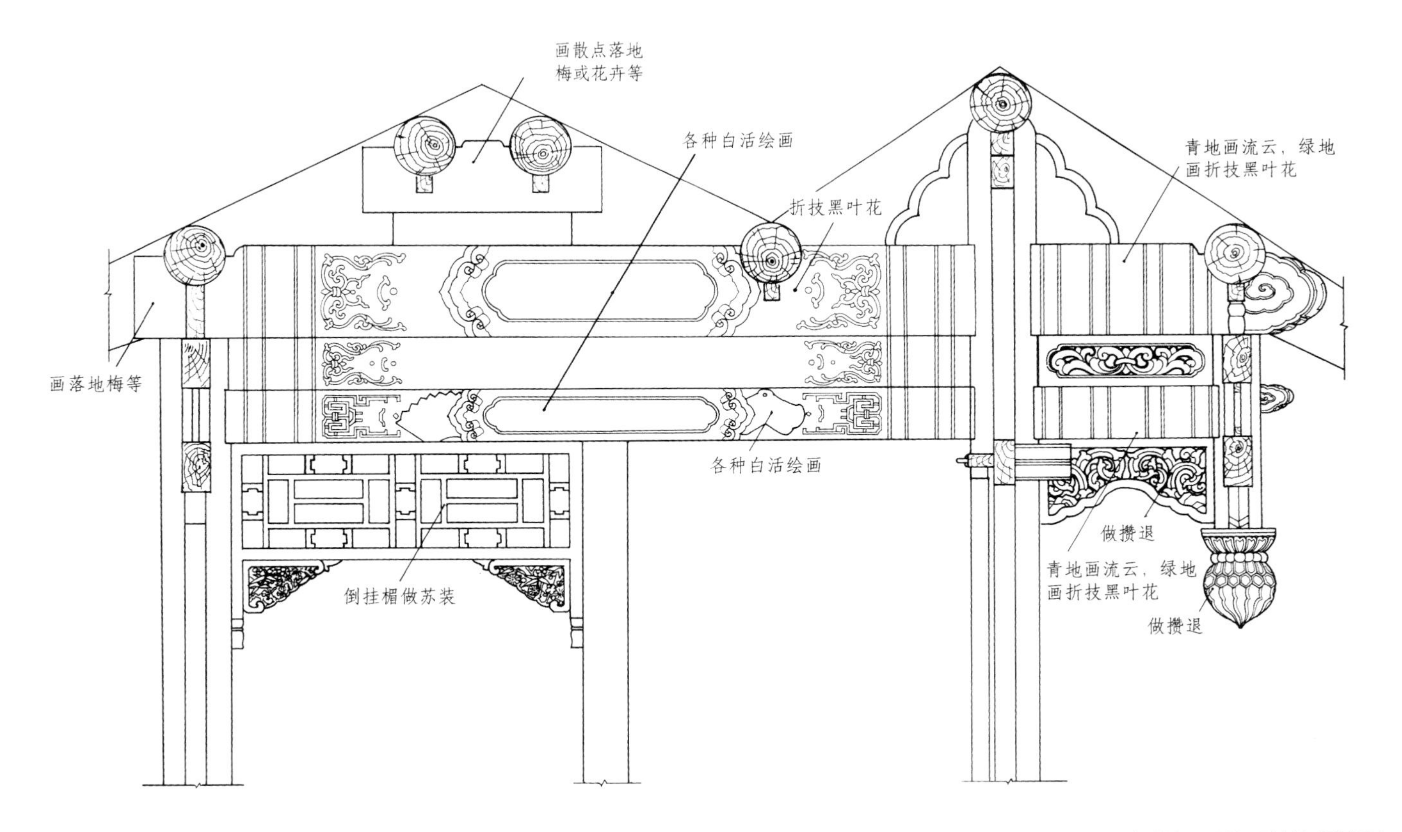

图6—2—2.2 垂花门侧立面苏画纹饰示意图

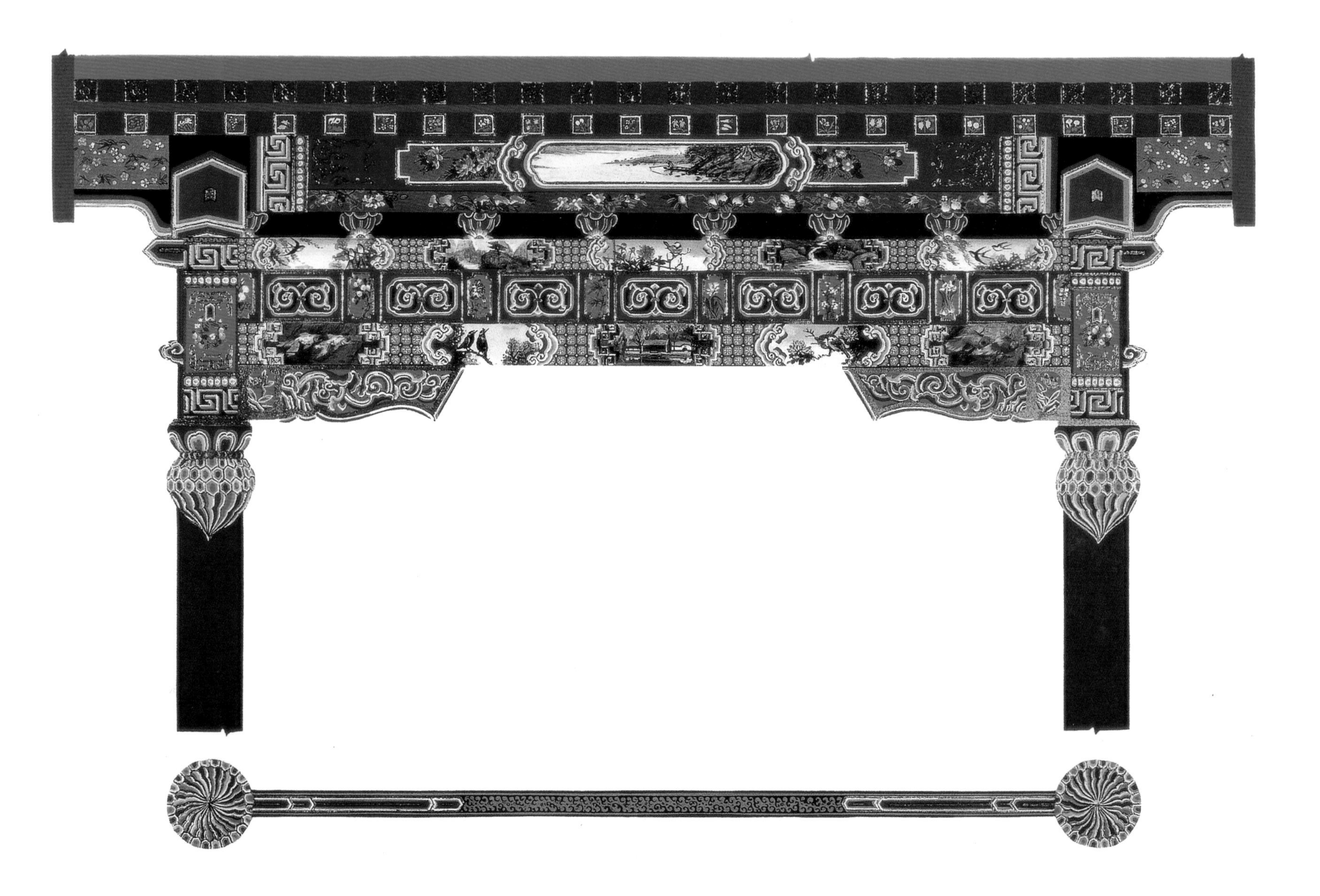

图6—2—2.3 垂花门苏式彩画

（2）苏式彩画　苏式彩画分三种主要表现形式，即包袱式、方心式和海墁式。四合院的苏画装饰，几乎全部用到了这三种形式。

清代晚期的苏式彩画基本分为三个等级做法：高等级者称“金琢墨苏画”，中等级者称“金线苏画”，低等级者称“墨线（或黄线）苏画”。但从北京城区现存清晚期民居彩画遗迹看，建筑只要有装饰彩画的，绝大多数都要贴金，极少见有墨线苏画。

苏式彩画的施色，与旋子彩画基本一样，也是以青绿二色为主，但某些基底色，较大量地运用了各种间色，比如石三青、紫色、香色等等。所以，这类彩画可给人以富于变化和亲切的感受。

苏画细部题材的表现是多方面的，有各种历史人物故事画，有百态千姿的花鸟画，有表现殿堂楼阁的线法风景画，有笔墨酣畅的水墨山水画等等。这些趣味活泼的绘画内容，在包袱、池子、聚锦内都得到了充分表现。由于这些画题与人们的生活及周围环境紧密相关，为人们所喜闻乐见，所以特别适于宅第四合院建筑的装饰（图 6—2—3）。

（3）椽柁头彩画　这是用于王府和一般民居的椽头彩画。飞椽头常见的有“沥粉贴金万字”“阴阳万字”“十字别”和“金井玉栏杆”等；檐椽头常见有片金或攒退做法的“方圆寿字”，作染或拆垛做法的“福庆”“福寿”“柿子花”“百花图”等（图 6—2—4）。常见的柁头彩画有“作染四季花”“线法及洋抹山水”“作染或洋抹博古”“攒退汉瓦”“攒退活图案”等。

图6—2—3　掐箍头搭包袱苏画纹饰举例

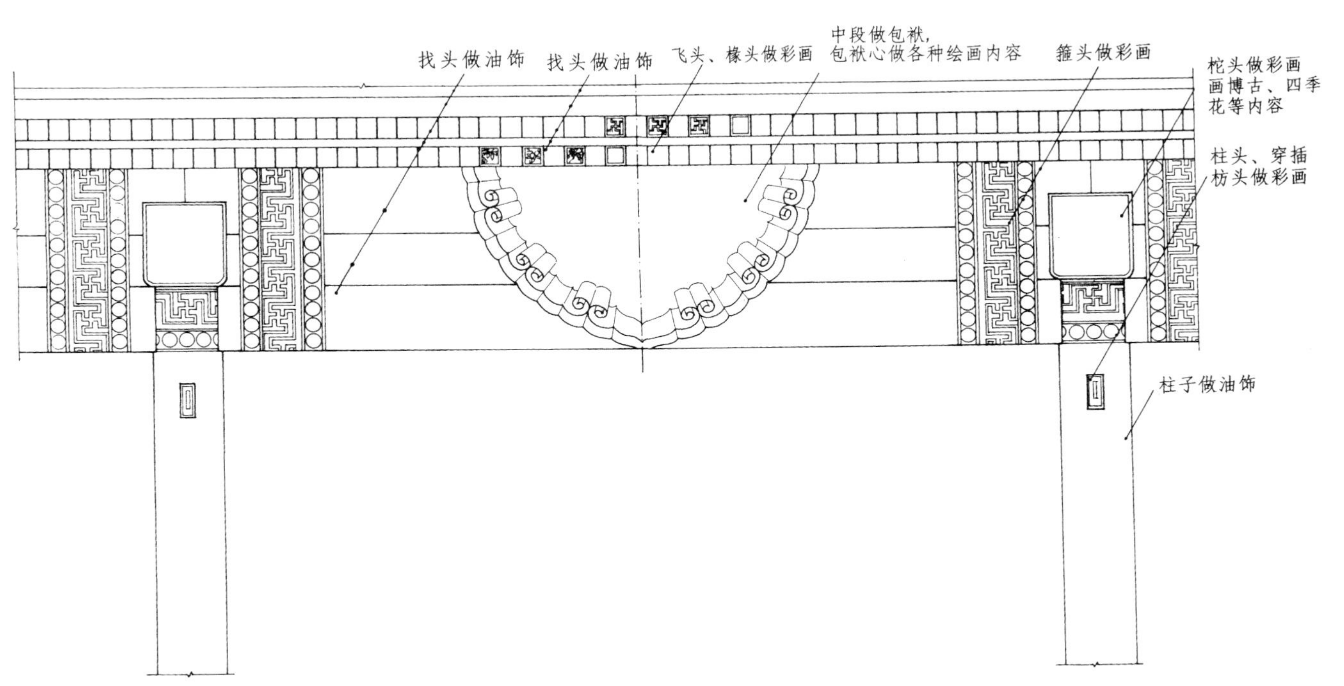

图6—2—3．1　苏画掐箍头搭包袱纹饰示意图

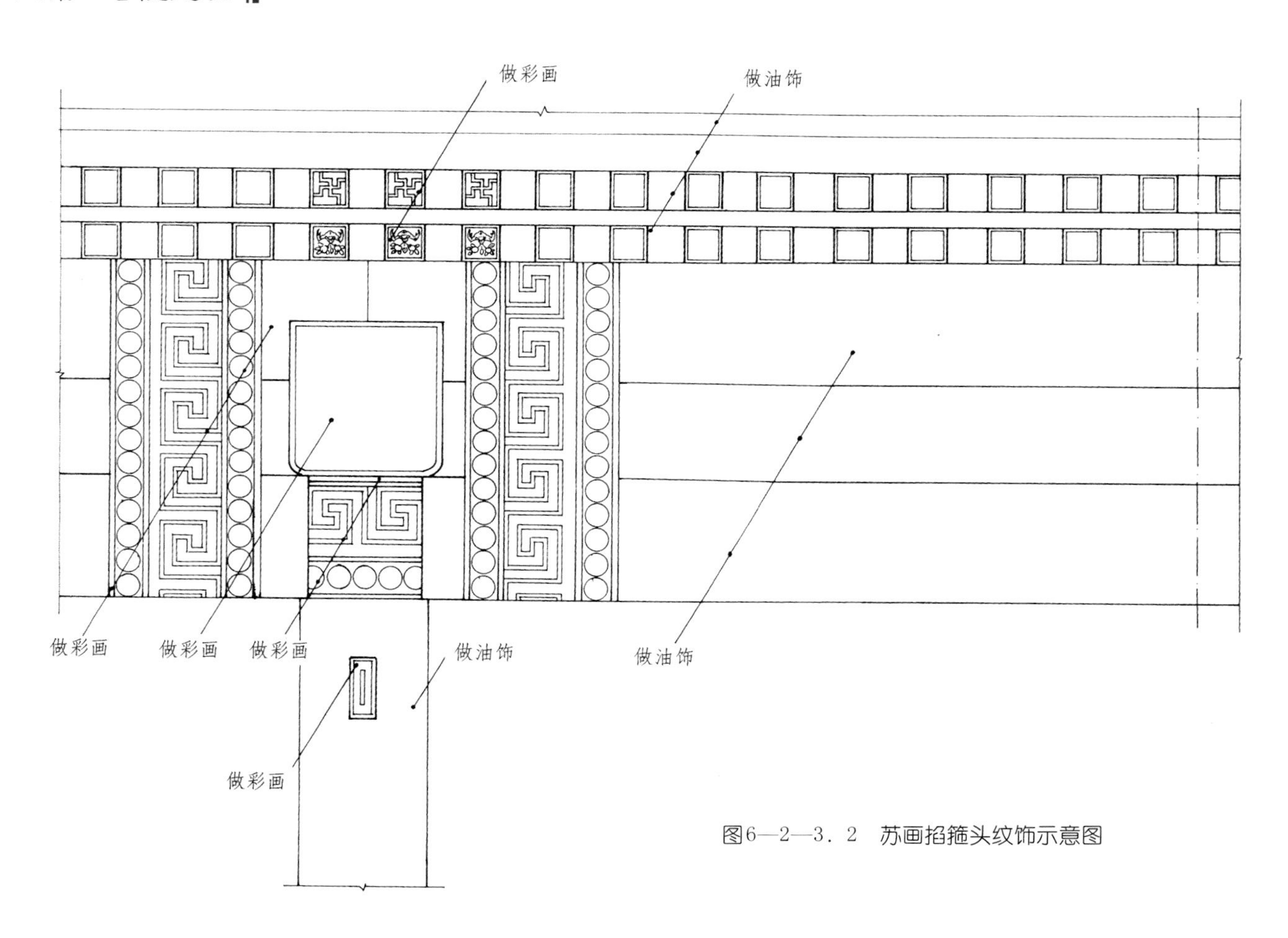

图6—2—3.2　苏画掐箍头纹饰示意图

图6—2—4　四合院常见椽柁头彩画、椽柁头刷画、椽头纹饰图案举例

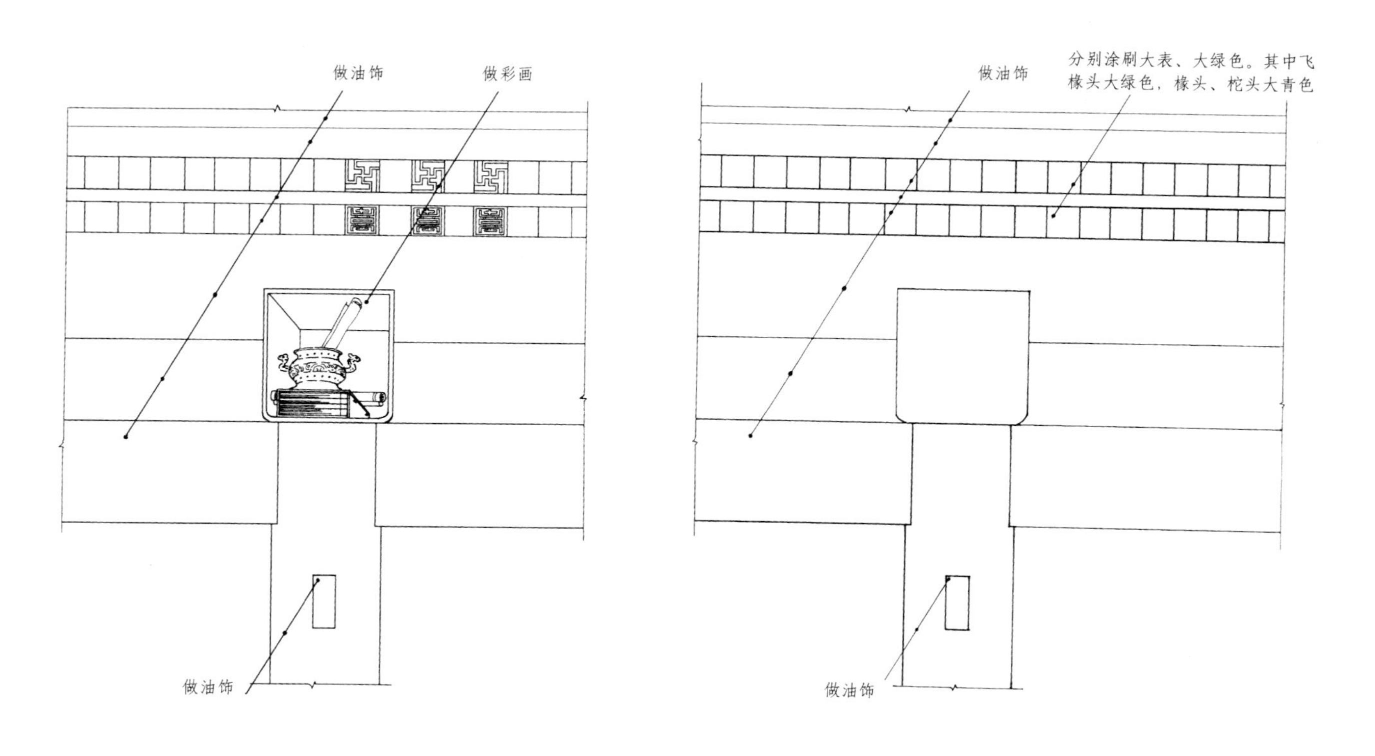

图6—2—4.1　椽柁头彩画示意图

图6—2—4.2　椽柁头涂刷色彩示意图

图6—2—4．3 椽头纹饰举例

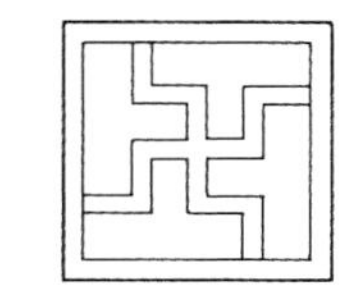

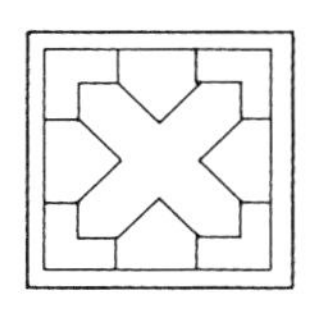
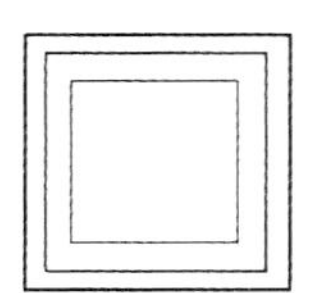

飞檐椽头纹饰举例

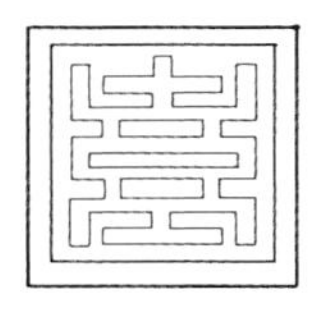

檐椽头纹饰举例

（4）*天花彩画* 天花彩画常见的有“片金龙天花”（仅限于王府）、“作染团鹤天花”“攒退活图案天花”“作染百花图天花”等（图6–2–5）。

（5）*倒挂楣子彩画* 多见于清晚期的“苏装楣子做法”。

3. 四合院彩画纹饰的寓意

四合院的许多彩画纹饰都有一定的象征意义和吉祥寓意。如龙纹是专用来象征皇权的，所以建筑制度限定，只有帝王之家才可运用，庶民是绝对禁用的。又如飞椽头用的“万”字，椽头用的“寿”字，加在一起称

图6—2—5 四合院常见天花彩画图案举例

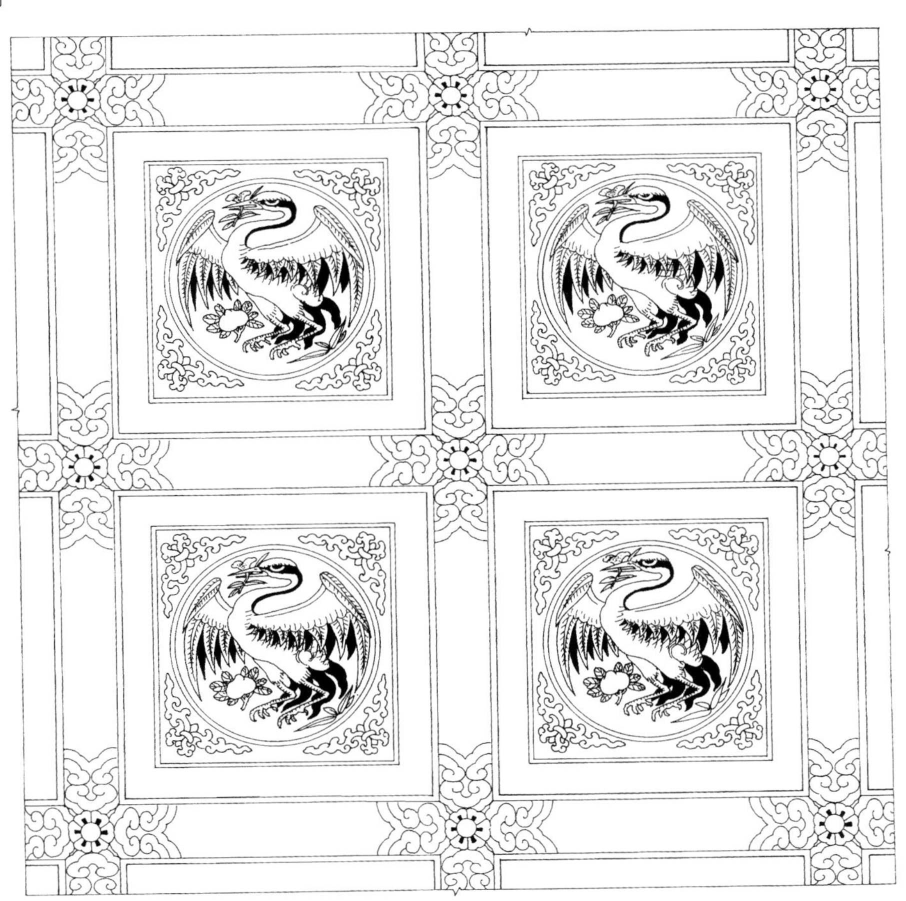

图6—2—5．1 团鹤天花纹饰

图6—2—5．2　百花图天花纹饰

为“万寿”，寓意长寿。如飞椽头用“万”字，椽头用“蝠寿”，则寓意为“万福万寿”。再如画牡丹和白头翁鸟，寓意富贵到白头；而画博古，则寓意主人有文化、有才学、博古通今，不同于凡俗之辈；画灵芝、兰花和寿石，寓意“君子之交”……彩画纹饰含有吉祥寓意的例子不胜枚举。这些图案绘画主题鲜明、构图巧妙、寓情于景、情景交融，不同程度地代表着各个宅主人对幸福、长寿、喜庆、吉祥、健康向上的美好生活的向往和追求。

第三节　室内家具与陈设

明清两代不仅是传统住宅发展的鼎盛时期，也是传统家具发展的黄金时代。这个时期，家具和建筑产生了更加密切的联系，与花罩、碧纱橱、多宝格等室内装修风格一致、色泽协调。高雅考究的室内家具，除满足人们生活起居需要之外，还同文玩字画、古董摆设在一起，共同构成具有极大艺术魅力和收藏、观赏价值的文化艺术瑰宝。

一、传统家具与陈设的基本内容和艺术价值

室内家具主要供人们生活起居之用，如依其功用划分类别，大致可分成椅凳、桌案、床榻、柜架和其他家具五类。

1. 椅凳类

椅凳类家具主要包括杌凳、坐墩、交杌交椅、长凳、椅、宝座等。杌凳是无靠背的坐具的统称，有无束腰和有束腰、直腿、弯腿、曲枨、直枨等多种造型。坐墩又名绣墩，是一种鼓形坐具，有五足、八足、直棂和四开光、五开光等多种造型。交杌又俗称马札子，其构造和形状来自古代的胡床，是一种可折叠的简易坐具。长凳是供两人以上安坐的凳子，有案形（形似案子，窄长，四腿八乍，民间俗称板凳）和桌形（形似桌子，直腿或弯腿）两种。椅是人们所熟悉的坐具，传统的椅有靠背椅、扶手椅、圈椅、交椅四种。靠背椅是只有靠背没有扶手的椅。扶手椅既有靠背又有扶手。圈椅又称圆椅、马掌椅，形状圆婉美观。交椅是交杌（马扎）与圆形靠背椅的结合。宝座是专供帝王用的坐具，民间不能使用（图 6–3–1）。

2. 桌案类

桌案类家具主要包括炕桌（炕几、炕案）、香几、酒桌、半桌、方桌、条几（条桌、条案）、书桌（书案、画案）等。炕桌（炕几、炕案）是在炕上使用的矮脚家具。炕桌长宽之比为 3∶2，用时放在炕中间，炕几、炕案较窄，用于炕的侧端。香几为放置香炉的家具，以圆形三足为多，腿足弯曲夸张，造型优美秀丽。酒桌是一种较小的长方形桌案，因古代常用它作酒宴之用而得名。半桌是相当于半张八仙桌大小的桌子，当一张八仙桌不够用时可用其拼接，故又名“接桌”。方桌， 是应用最广的一种家具，有大、中、小之分，分别称“八仙”“六仙”“四仙”。条几（条桌、条案）都是窄而长的家具，大小不等。其中条案最长，可达丈余。条案的形式有平头案、翘头案，造型各异。书桌书案、画桌画案，是比较宽大的长方形家具，其形状往往与条桌、条案相同，但宽度较大以便于阅读书画（图 6–3–2）。

3. 床榻类

床榻类家具主要有榻、罗汉床和架子床三种。只有床身，没有后背、围子和其他任何装置的称为榻。床上有后背和左右围子的称为罗汉床。这种床的后背、围子形状与建筑中的罗汉栏板十分相近。架子床是带床顶的床。床顶由四根或六根立柱支撑，架子床四周可以安装床围子，是一种很讲究的卧具（图 6–3–3）。

4. 柜架类

柜架类家具包括架格、亮格柜、圆角柜、方角柜几种。架格又称书格或书架，其上可放书籍或其他器物。亮格柜是分成上下两段的一种家具，上部是亮格，下部是柜子，它是二者相结合的产物。圆角柜是一种带柜帽子的柜子，柜帽转角处做成圆形。这种家具多用在炕上，故又称“炕柜”。方角柜无柜帽，上下等大，体量不一，大者高达两米或三四米，小的一米五左右，可以用在炕上（图 6–3–4）。

图6—3—1 椅凳类家具举例

图6—3—2 桌案类家具举例

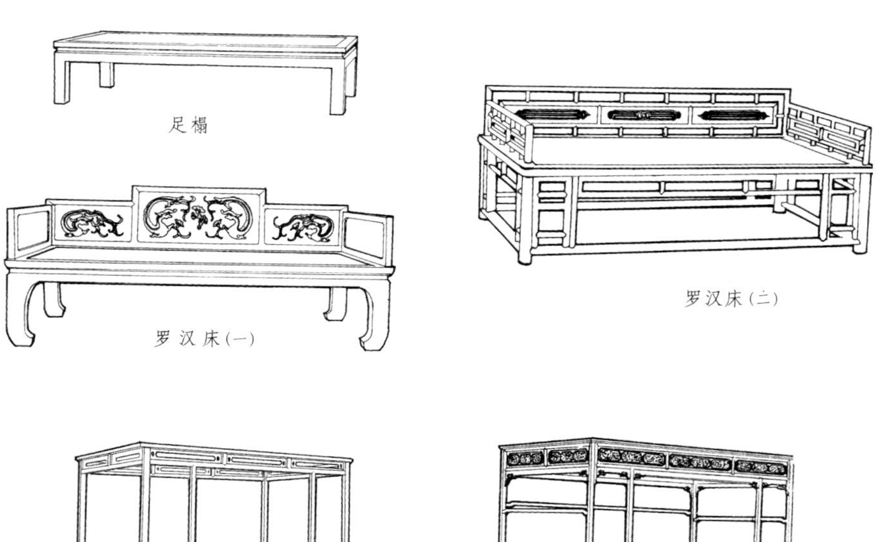

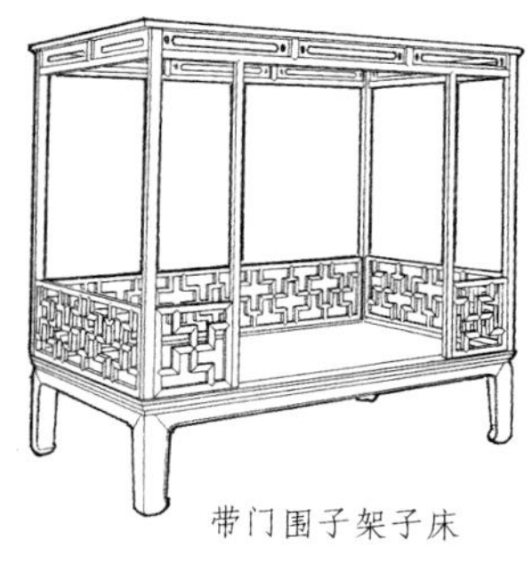

图6—3—3　床榻类家具举例

图6—3—4　柜架类家具举例

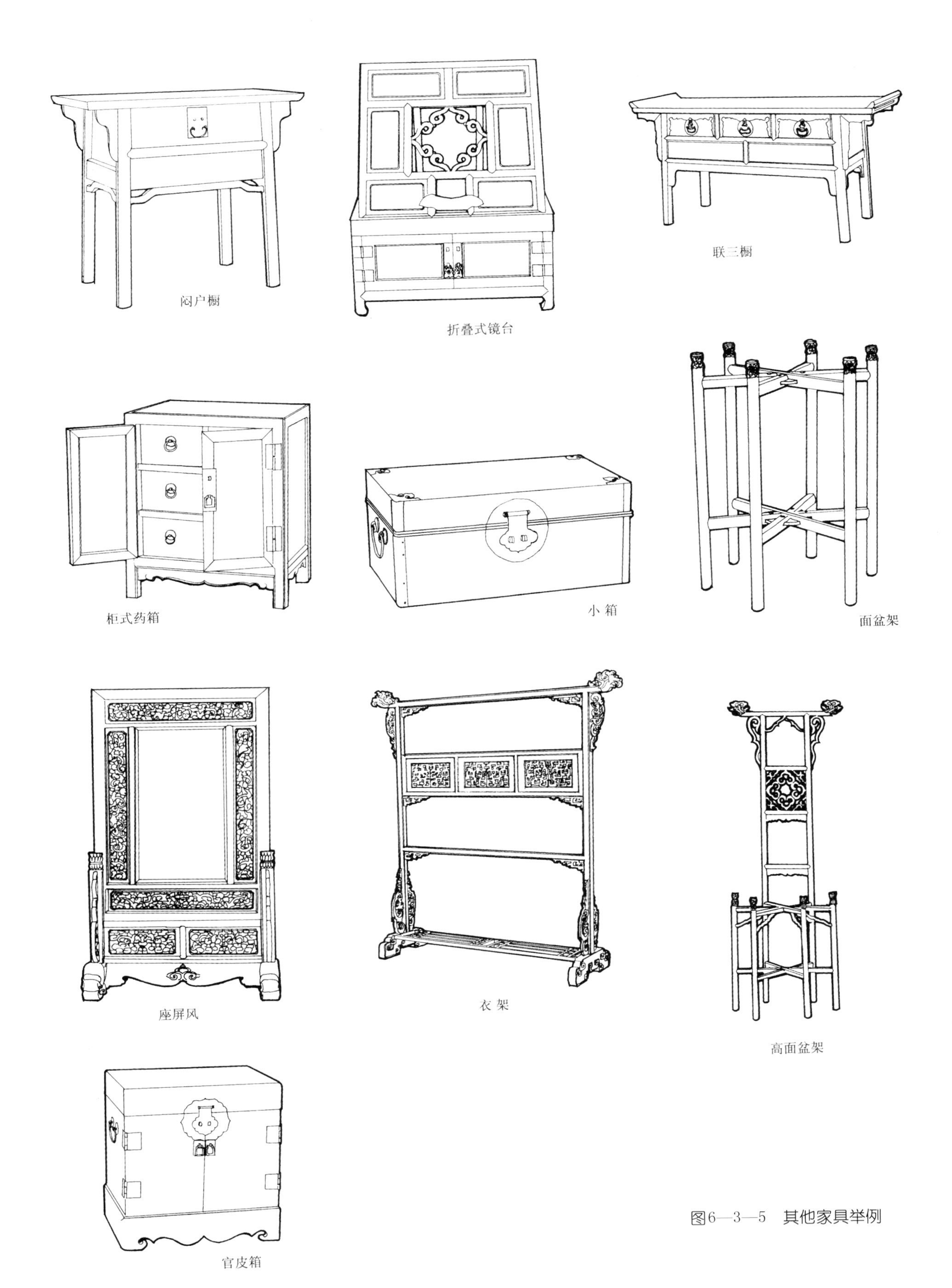

图6—3—5 其他家具举例

5. 其他类

凡不能包含在上述四类的，都可称为其他家具。其他家具品种很多，有屏风、闷户橱、箱、提盒、镜台、官皮箱、衣架、面盆架等等。屏风是屏具的总称，由多扇组成，有可以折叠的围屏和带底座的座屏。座屏有独扇、三扇、五扇等不同形式，这种家具多见于宫殿府邸之中。闷户橱是一种带有抽屉和闷仓的家具，形似翘头案，案面上可以摆放器物陈设，抽屉和闷仓内可以储藏物品。箱是一种有底盖可以储存物品的家具，依功能不同可分为衣箱、药箱以及存放金银细软的小箱等。提盒是带提梁的分层的长方形箱盒，有大、中、小几种，大的须两人抬，中的可一人挑（用扁担挑，一头一个），小的可用手提。镜台即梳妆台，多在居室中使用。官皮箱是镜箱，也可以说是一种可移动的梳妆箱。衣架是专门用来搭挂衣服的架子，多用于室内床榻两侧。面盆架是用来放置脸盆供人洗脸用的架子，有高、矮两种，矮者高65~70cm,有三足、四足、六足等不同形式，可以折叠；高者是在矮架基础上，将其后面二足升高至1.7m左右，上安挂牙、搭脑，可搭置巾帕（图6–3–5）。

陈设是摆放或悬挂在室内供人品玩欣赏的艺术品的总称，它主要包括青铜器、瓷器、玉器、竹木雕刻、漆器、刺绣、字画等。我国的青铜器在商代即已进入成熟阶段，早期的青铜器以实用为主，包括炊具、酒器、容器、农具、武器、乐器、车马器、玺印、符节等。后来随着铁器、陶瓷和木器的发展，青铜器逐渐不再作为用具，而成为载有古代历史文化信息的文玩而流传于世。

陶瓷器具一出现，就是一种实用品与艺术品的结合体，在室内陈设中占有重要地位。不同朝代的陶瓷，无论从造型、釉色、胎质、装饰、款式上，都带着不同时代的艺术特点和时代 特征。如唐代的青花和唐三彩，明代的青花红彩，清代的康熙五彩、粉彩、珐琅彩等，都具有独 特的艺术性和鲜明的个性，是供人使用和玩味的艺术佳品。

其他，如玉器、雕刻、刺绣、漆器、钟表等，或是纯粹的艺术品，或是具有艺术欣赏价值的器物，是室内重要的装饰陈列品。

传统家具和陈设，是建筑艺术不可缺少的组成部分，是中国几千年文化、艺术成就的积淀，是世代人民聪明才智和艺术才华的结晶，是中国传统文化的精粹。它们不仅是中国的宝贵财富，也是全人类共同的宝贵财富，具有极高的艺术价值和收藏价值。

二、各类房间家具陈设的内容与配置

中国传统民居的家具与陈设是非常讲究的。不同功能的房间，其家具陈设的内容、形式、风格、特点各不相同。现以堂屋、居室、书房、厅堂为例分别介绍。

1. 堂屋

正房的明间为堂屋，堂屋具有起居、会客和礼仪的综合功能，因此既要规矩、肃静，又要有一定文化和生活气氛。堂屋对内可以闲坐谈心，行各类家礼，如祭神、祭祖、团拜、拜堂、行家规等；对外可用于族内的接待和礼节性拜访。堂屋家具陈设的中心是靠墙的翘头案，案前放八仙桌，桌两侧配圈椅、扶手椅或太师椅。翘头案上面的陈设根据堂屋的使用内容不同而异。纯作起居、会客和行礼仪的，一般两侧配青花、青花釉里红或哥釉类釉色素雅深沉、器型较大的瓷器，如天球瓶、方瓶等，以取平衡稳定感并突出古色古香的气氛。案子中央宜置重力感强的摆设，如青铜鼎、炉类，器型不宜太大，但与使用功能结合较好，亦可用较粗犷的石雕、玉雕等。八仙桌上只宜放果盘或茶具。墙正中可挂中堂字画，两侧配条幅。色彩不可过艳，内容宜选治家修身名言为好。如堂屋兼作佛堂，则翘头案正中应设佛龛，案上配置香炉、蜡扦、花筒等五供。此类堂屋桌前常挂刺绣桌围，下摆蒲团。堂屋与次间常有隔断或落地罩分隔，并适当摆设坐椅用以待客或供家中晚辈坐用。如堂屋较大，还应配置左右对称的高几和宫灯。但堂屋布置不宜奢华，不摆放精细贵重的陈设（图6–3–6）。

图6—3—6 堂屋家具陈设平面、立面图

图6—3—6．1 堂屋家具陈设立面

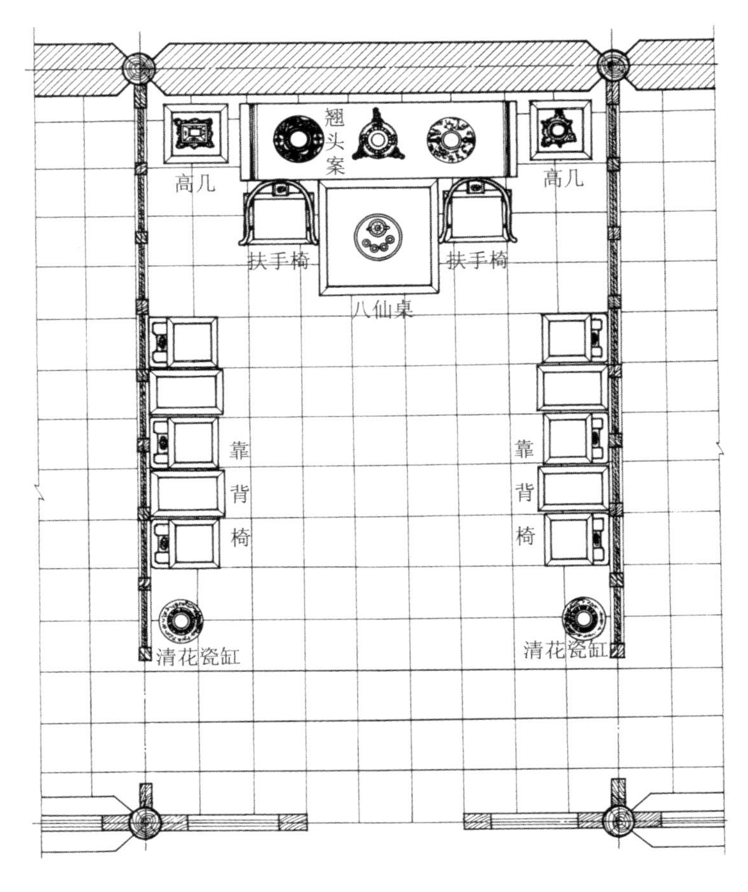

图6—3—6．2 堂屋家具陈设平面

2. 居室

传统建筑居室一般设在正房的次间或套间，也可设在厢房或耳房。居室是供休息及少量起居活动的地方，门窗隔扇和顶棚不做过多雕饰，以四白落地最显清雅安谧。由于居室活动有小范围的功能划分，因此可以形成几个装饰陈设的重心。榻、架子床或炕是居室的功能重心，北方民居的寝具多为炕，它既符合北方气候条件，又经济适用。榻或炕上一般设炕桌，炕桌配茶具。除此便只是被褥和炕箱之类。居室内装饰陈设的重点为北墙的连二橱、连三橱或闷户橱。此类橱长度较大，中间常放置帽镜。帽镜两侧依次为茶叶罐、帽筒和胆瓶，其间夹放梳头匣或拜匣，案前放茶具或果盘。家境不同，摆设的档次和品位也不同。墙上可配挂屏或字画，内容应根据主人的性别、气质和爱好而定。单身男性一般不设镜台而设多宝橱，陈放文玩和书籍，绣房则应配置镜台用于梳妆。古代居室无卫生间，室内应设面盆架。如无镜台也可在闷户橱的帽镜前放镜支用于梳妆和放化妆用品。居室的陶瓷宜采用色泽较鲜艳、器形较活泼流畅的品种，如斗彩、粉彩、五彩，亦可用少量珐琅彩和矾红彩等。纹饰内容要以吉祥喜兴为好，如“富贵白头”“燕喜同和”“长春同庆”等，内容多用花鸟和婴戏等。居室中可加少量画龙点睛的小巧摆设，如翡翠或玉雕盆景、粉桩、花瓶、漆盒、抿缸等，亦可在适当位置装饰刺绣、剪纸、布饰等。总之，居室要装饰陈设出安静、温馨、喜兴、和谐的气氛，以增加生活气息（图 6–3–7）。

图6—3—7　居室家具陈设布置立面

3. 书房

书房又称书斋，是读书的专用房间，兼有琴、棋、书、画等活动功能，亦可会密友。因需要安静，常设在较僻静的位置，如套间或跨院中另建单独的房间。书房需要有文化气息，要安适、沉静幽雅，不宜过于呆板和肃整。因此，家具陈设一般不用对称而用平衡，布置较为灵活。但也应分清主次，不可杂乱无章，也不能平均落墨。书房的视觉重心和功能重心均为书桌，因此主要功能性和装饰性陈设均在书桌及其临近的视觉范围内。读书需要良好的光线，故书桌应放在窗下，并配以圈椅或扶手椅，以供较长时间读书写字时中间歇息。

书桌上一般不放纯装饰性陈设，避免形成视觉重心而分散精力，应以实用且具有装饰色彩的物品为宜，一般可放置书匣、笔筒、笔洗、砚台、笔架、印盒等。书房宜设书架、书格或书柜，亦可设多宝书柜兼放书籍和文玩。在距离书桌较近处置琴几用于安放古琴或古筝；书桌较远处设棋桌用以闲时对弈；在适当位置设方桌式鱼缸，用于观赏并增加室内生气。这种有层次的陈设使书房内形成以书桌为主体的数个功能子系统，从而组合成全部书房的装饰陈设群。书房的书画陈设依主人的情趣而异。一般绘画以立轴绢本设色或浅设色山水画为宜。这种山水画不仅显示古朴深远，烘托层次较深的文化气息，而且可以延伸视觉焦点，从人们的心理上扩大室内空间。书法陈设忌平淡呆板，或神采飞扬、雄肆奔放，或俊逸跌宕、耐人玩味。一幅无神的书法陈设会成为书房的败笔。书房可在多宝格或书架上适当放置精致的青铜、玉雕和陶瓷等文玩摆设，但形体不宜过大以防喧宾夺主。瓷器可用青瓷、甜白、哥釉等素雅品类，也可用青花、釉里红、斗彩和五彩等较活泼鲜艳的品类，最好不用大红大绿的硬彩，如用珐琅彩，其器形要小些，色彩要淡些。家具以黄花梨最为相宜，它可表现文人内心世界的沉静、安恬、含蓄及谦和（图6—3—8）。

图6—3—8　书房家具陈设平面、立面图

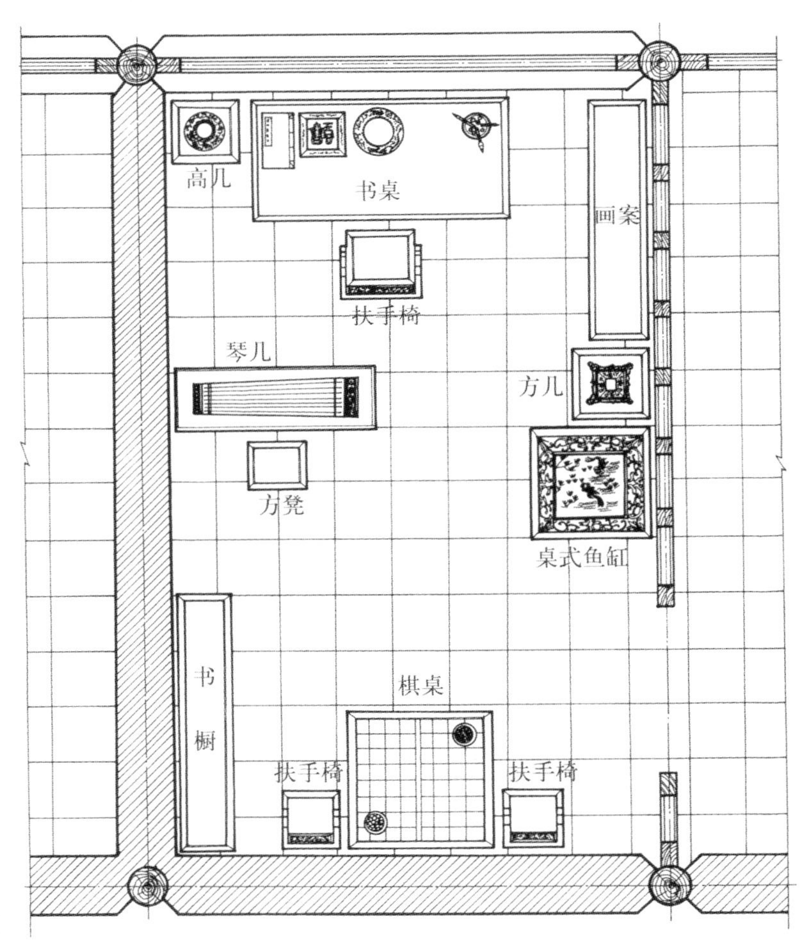

图6—3—8．1　书房家具陈设平面

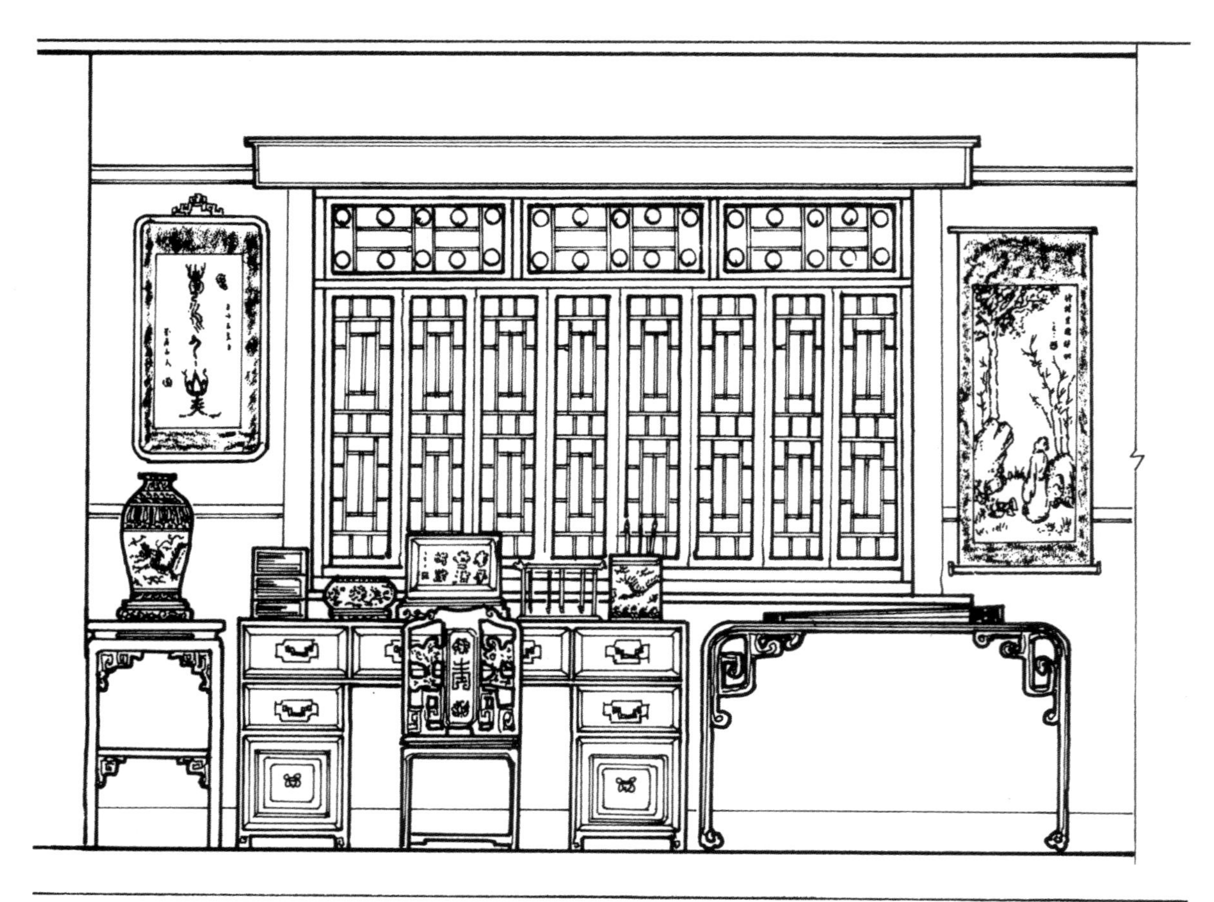

图6—3—8．2　书房家具陈设立面

4. 厅堂

厅堂是住宅内较大的公共活动场所。私家厅堂一般设在第一进和第二进院之间的过厅，其位置便于内、外活动。厅堂类型较多，如祭祀用的祠堂、议事用的议事厅等。功能不同则气氛不同，祠堂应肃穆神圣，议事厅应整洁庄严。大厅的陈设应以应用性物品为主，议事厅家具主要是桌椅，祠堂则摆设供桌牌位和香蜡用品。一般厅堂为使空间开阔常不做顶棚而彻上露明，以结构展示建筑的造型美，并可在结构构件上做彩画或油饰。各类大厅均因面阔进深较大需照明而挂宫灯。大厅家具一般应平素，不宜多加雕饰，应整齐摆放并留出明显的走道。如大厅由数间组成，可用落地罩分隔以保持隔而不断。这种分隔还可以形成人走景移的长卷画效果（图 6–3–9）。大厅的主要轴线上以翘头案、八仙桌和扶手椅为主，即如前所述的堂屋的摆设，但其规模和空间比堂屋要大，应设较多的椅凳和榻类坐具，供多人活动使用。厅堂家具以深色为多，但也可用红木和黄花梨的浅色家具。大厅常在侧门处置屏风便于出入和服务。大厅的书画常用横幅水墨山水或书法，风格应粗犷以适应视距较远的观赏。其风格宜凝重浑厚、刚正醇和，具有大家气派。在厅的适当角落可置高几，陈设器形较大的装饰性陶瓷器，如天球瓶、赏瓶、方瓶、尊等，而在比较显著的位置则应陈列大型瓷器，如瓶、罐、缸之类。有的厅内需放柜橱类家具以存放小型用品。厅内盆花一般选较高大的花株，以求与空间、家具的平衡和协调。大厅有堂名时还可以设匾额，并加配楹联。楹联常表达主人的志向、品格、德行或家风、家教。楹联常设在檐柱上，也可以条幅的形式悬在中堂两侧。

在传统民居房间中，常出现隔而不断的房间关系，即两间不同功能的房间用落地罩、局部隔扇做分隔，形成一种既分隔又沟通的连续性功能布局。如书房和居室的连通，不但增加了空间尺度，也增加了视距深度，从房间中心望去，室内家具秀美高雅、错落有致，另有一番情趣（图 6–3–10）。

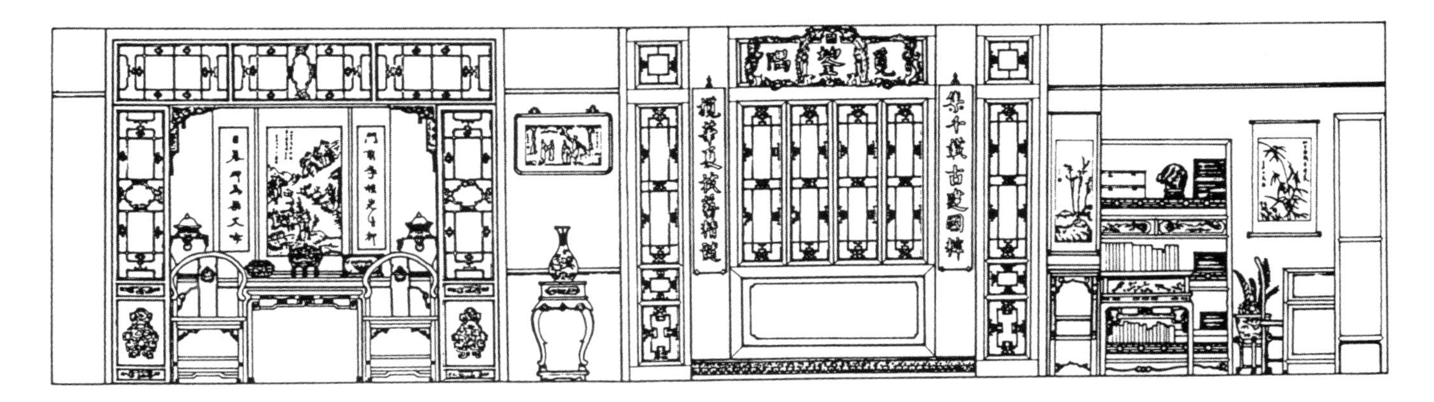

图6—3—9　室内陈设的长卷画效果

图6—3—10　房间连通的透视感

三、房间风水、习惯心理与家具陈设的关系

1. 家具陈设与房间风水

房间风水是风水理论在室内的应用，它涉及室内气场定位，影响到家具布置。按照中医理论，气是生理功能或动力。气有化生、推动与固摄血液、温养全身组织、抗拒外邪、推动脏腑组织活动之作用。人生活在自然界，人的气要受到自然界的影响，自然界的风、湿、暑、寒、燥、火等都会影响人的生理功能。根据《素问 · 五藏真言论》的“五藏应四时”的理论，人与天地相应，生命的节奏协和自然。室内也是自然的一部分，它的不同位置会形成不同的小气候，还会受重力、磁场、微波、电磁波的多种影响和干扰，家具位置显然也应针对这种影响而选择适当的合理位置。如床应放在生气的位置，忌面向西窗。西窗是人们最嫌弃的，夏日酷热、冬日严寒多来自西窗，床作为居室最主要的家具理应避开西窗。又如床不对门，是说应避免当风而卧；床不正放，是说要避免南北磁场对人体磁场的干扰；床不对镜，是怕反射光影响视觉，干扰休息；床靠实墙，会给人一种安全感；室内勿两镜相照，是为避免多次反射造成幻影与炫光惊扰休息；书桌应为矩形，是为了充分利用空间，便于阅读、书写、摆放文具；书房左面应靠窗，是因左前方光线最便于书写和阅读；浴室不正对卧室，是为避免目光干扰……总之，室内风水的许多内容都是有一定科学道理的，可以在室内装饰陈设中参考。

2. 家具陈设与习惯心理

室内装饰陈设除了考虑人的生理需求外还要考虑心理适应。人们在长期生活中形成了各种心理适应与不适应因素，如室内色调，建筑物柱子颜色皇室为红、诸侯为黑、大夫为青、百姓为黄，久而久之就形成了某些地区、某些人群的惯用色彩，家具陈设亦须与之协调。又不同颜色对应五行，即金白、木青、水黑、火赤、土黄，五行又与人的属相有联系，于是就会出现色彩与人的相生相克，虽无多少科学道理，但却影响人的心理。又如室内饰品的题材内容也常常与人的心理因素有关，有些吉祥、桃符、字画之类饰品，不但有一般装饰作用，还对部分人有心理诱导作用。传统室内装饰陈设物，如家具、香囊、绣袋、珠宝等，可通过气味释放或分子扩散方式影响人体的生理功能，使不同的人出现对某些装饰陈设的特异选择。根据《本草纲目》可知，紫檀可消风热肿毒，楠木治转筋足肿，柞木催生利窍，杨木治暑月生疖，琥珀清心明目，翠玉治面身瘢痕，珊瑚镇心明目，玛瑙宝石去翳明目……尽管装饰陈设的这类作用未必明显，但它可以在长期使用中产生心理的适应或不适应。这些心理因素，也是现代室内装修和家具陈设中应当注意的。

第七章
私家园林和庭院绿化

中国的古典园林包括皇家园林、私家园林、寺庙园林和衙署园林，它们是伴随中国古典建筑的发展而发展起来的。最早开始出现的是皇家园林。据记载，早在商周时期就有了营建宫苑的活动，从最早的供帝王狩猎的纯自然的山林，逐步转向仿照自然营造山峦、水泊以点缀建筑。从秦汉至隋唐至宋元明清各朝，都营造过许多著名的帝王宫苑。

私家园林是在皇家造园活动的影响下逐渐发展起来的。一些有权势的官僚贵族步皇家之后尘，率先营造私家园林，使私园得以发展。魏晋南北朝的三百余年间，由于社会动荡，战乱纷起，士大夫阶层逃避现实，洁身远祸，纷纷过起隐居生活。他们寄情于山水之间，吟诗作画，陶冶性情，逐步确立了自然审美观，从而大大推动了造园艺术的发展。这个时期的造园之风极盛，私家园林如雨后春笋。此后的一千多年间，营建私家园林已经成为官僚、贵族和文人雅士的高雅时尚，并且这种时尚随着文学艺术，尤其是诗词、绘画艺术的发展而不断升华。历代著名的山水画家和诗人（如郭熙、李成、米芾、苏轼等）的绘画艺术理论，对造园艺术产生着重大影响。至明代，终于产生了专门用于指导造园的理论著作——《园冶》。

明清两代，是我国历史上造园的最后一次高潮。这时期不仅皇家造园之风极盛，私家园林也高度发展，并且达到了空前绝后的水平。历史上著名的勺园、梁园、李园（清华园）、官园、礼王府花园、成王府花园、恭王府花园，以及苏州、扬州一带的拙政园、留园、艺圃、个园等等，都是这个时期的杰作。

第一节　北京的私家园林

一、明清时期的北京私家园林

明代的北京私家园林见于记载的，有位于海淀的李园（为李候的别业，即今之清华园前身）、勺园（明米仲诏所建，与李园相邻）、湛园（亦为米仲诏所建，位于米宅之左），有位于德胜门附近的定园（明初大将徐达后代之园），有位于阜成门内的宣家园，位于西直门右侧的齐园等等。明代《帝京景物略》记载，李园“方十里，正中‘挹海堂’。堂北亭，置‘清雅’二字，明肃太后手书也。亭一望牡丹，石间之，芍药间之，濒于水则已。飞桥而汀，桥下金鲫，长者五尺，锦片片花影中，惊则火流，饵则霞起。汀而北，一望又芙蕖。望尽而山，剑铓螺矗，巧诡于山，假山也。维假山，则又自然，真山也。山水之际，高楼斯起。楼之上斯台。平看香山，俯看玉泉，两高斯亲，峙若承睫。灵碧、太湖、锦川百计，乔木千计，竹万计，花亿万计，阴莫或不接”。从以上记载可以窥见当时李园园容之盛，景致之美。而与之相邻的勺园，则又是另一番景象：“入路，柳数行，乱石数垛。路而南，陂焉。陂上，桥高于屋。桥上望园一方皆水也。水皆莲，莲皆以白。堂、楼、亭、榭，数可八九，进可得四。覆者皆柳也，肃者皆松，列者皆槐，笋者皆石及竹。水之，使不得径也；栈而阁道之，使不得舟也。堂室无通户，左右无兼径，阶必以渠，取道必渠之外，廊，其取道也，板而槛，七之；树根槎枒，二之；砌上下折，一之。客从桥上指，了了也……”明代叶向高评论说：“李园壮丽，米园曲折。”两园风格迥然不同，但由于海淀多水，两园皆以水取胜。

梁家园也是明代著名私园之一。梁家园的创建者梁梦龙，是嘉靖年间进士。该园选址于湖水之畔，在林木掩映之间建有疑野亭、半山房、警露轩、看云楼、朝爽楼等建筑。梁家园“亭榭花木极一时之盛”，曾是达官贵人、文人墨客饮宴娱乐之所。王横云《招饮梁家园诗》云：“半倾湖光摇画艇，一帘香气扑新荷。”可见当时园景之美。梁家园也是以水取胜的园子。它将凉水河引入园中，使死水变活水，创造了名噪一时的私家园林。

清代的私园不少，但有详细文字记载的不多，其中有成国公园、宜园、曲水园、礼王府花园、成王府花园、官园、贾家花园、前孙公园、后孙公园、和坤宅园、怡园、自怡园、朱彝尊故宅、吴三桂故宅、李渔故宅等数十处。这些私家园林，历经百余年风雨飘摇，社会动荡，至今大部分已被毁，仅有清恭王府之萃锦园、醇王府之鉴园等极少数私园还基本保存着原来面目。至于清代不甚有名的可园等几处小型私家园林，如今已成为我们寻迹觅踪的唯一借鉴了（图 7—1—1）。

图7—1—1　摄于1922年的成王府花园、礼王府花园、僧王府花园旧影　（引自《帝京旧影》）

图7—1—1．1　成王府花园

图7—1—1．2 礼王府花园

图7—1—1．3 僧王府花园

图7—1—1．4 礼王府花园水榭

图7—1—1．5 成王府花园来声阁

图7—1—1．6　成王府花园净真亭

图7—1—1．7　成王府花园跨虹亭

图7—1—1．8　成王府花园一景

图7—1—1．9　成王府花园一景

图7—1—1.10　僧王府花园一景

图7—1—1.11　成王府花园一景

图7—1—1．12　僧王府花园一景

二、北京私家园林的一般特点和造园手法

本书所谈的北京私家园林，主要指附属于四合院建筑的较小型园林。上文提到的李园、勺园、梁家园以及恭王府萃锦园、醇王府褴园等都是规模较大的私园，它们属皇亲国戚所专有，远非士民阶层所能望津。而较小型的私园，则多为一般官僚及文人富商所拥有，因此更具代表性。

北京的私家园林大多建于城区内，与主要的居住建筑相毗邻，用地范围十分有限，更没有真山真水可供借用。在有限的空间内，要将园子建得小中见大，富于层次和空间感，一般都采取内向型布局。这种布局是以集中的水面为中心，并环绕中心来布置厅堂、回廊、亭榭等。这种布局的优点是，在极其有限的用地范围内可以布置较多的建筑，且不会造成局促拥塞的感觉。由于建筑都沿园子周边布置，而且都背外面内，不仅使园子的中心得到充分的强调，而且建筑之间还可互成对景，互为因借，相映成趣。

北京现存的可园，就是这样一座比较典型的私园。该园位于北京东城区南罗鼓巷帽儿胡同，原为荣源府邸所属的花园。其西侧为一大型四合院，有五进院落，纵深七十余米，花园部分与宅院部分占地面积大体相当，有两千余平方米。可园占地狭长，不利造园，于是，以园中的主要建筑大花厅为界，将整个花园分成前后两部分。前园为主园，约占全园3/5，后园为辅园，约占全园2/5。主园以水池为主体，位于园子中心。水池南面为一座规模较大的假山，由太湖石叠成，玲珑而不失雄伟，池水由假山山脚蜿蜒而出，似由此发源。假山之上花草扶疏，树木繁茂，一六角小亭立于假山之巅，秀雅玲珑，为园子南端的主要景观建筑，登亭可俯瞰全园。水池迤北，为较开阔之地，遍植花草树木，园中主要建筑大花亭位于主园之北，面阔五间，硬山形式，体量宏大，是全园视觉中心。花厅一池水一假山沿主轴线顺序排开，构成全园南北方向三个主要景观。水池西侧有敞轩三楹，东侧有四角亭一座。亭北面十余米处有悬山式敞轩三间，为大花厅之陪衬。园中建筑，均沿周边布置，由游廊相联接，形成比较完整的空间序列。可园的后半部分，以假山为主体，位于园子中心部分，用太湖石叠就，峰、岭、洞、壑皆备，构成自然态势，沿小径经假山北行便见平地，顿觉疏朗，各色树木花草掩映之中，有厅堂坐北面南而建，园东侧为一座三开间歇山式敞厅，建于岗阜之上，为全园的制高点和视觉中心，可凭栏观赏全园。敞厅迤南，一座八角半亭串联于游廊之间，成为园子东南角之景点建筑。与半亭相对的，是西南角的三间厢房。园中建筑由抄手游廊连接贯通形成序列。

可园虽小，但巧用了对景、隔景、藏露、因借等造园手法，使园内的山、水、树木花草、建筑以及峰石、小桥等景物成为有机的整体，满足了园林可行、可望、可游、可居的功能要求，是北京现有小型私家园林中较好的一例（图 7—1—2）。

图7—1—2　可园园景及鸟瞰

图7—1—2．1　大花厅前山石花草

图7—1—2．2 从水池北望

图7—1—2．3 后园歇山式敞厅

图7—1—2．4　大花厅东侧敞厅

图7—1—2．5　假山上六角亭

图7—1—2．6　可园后园叠石

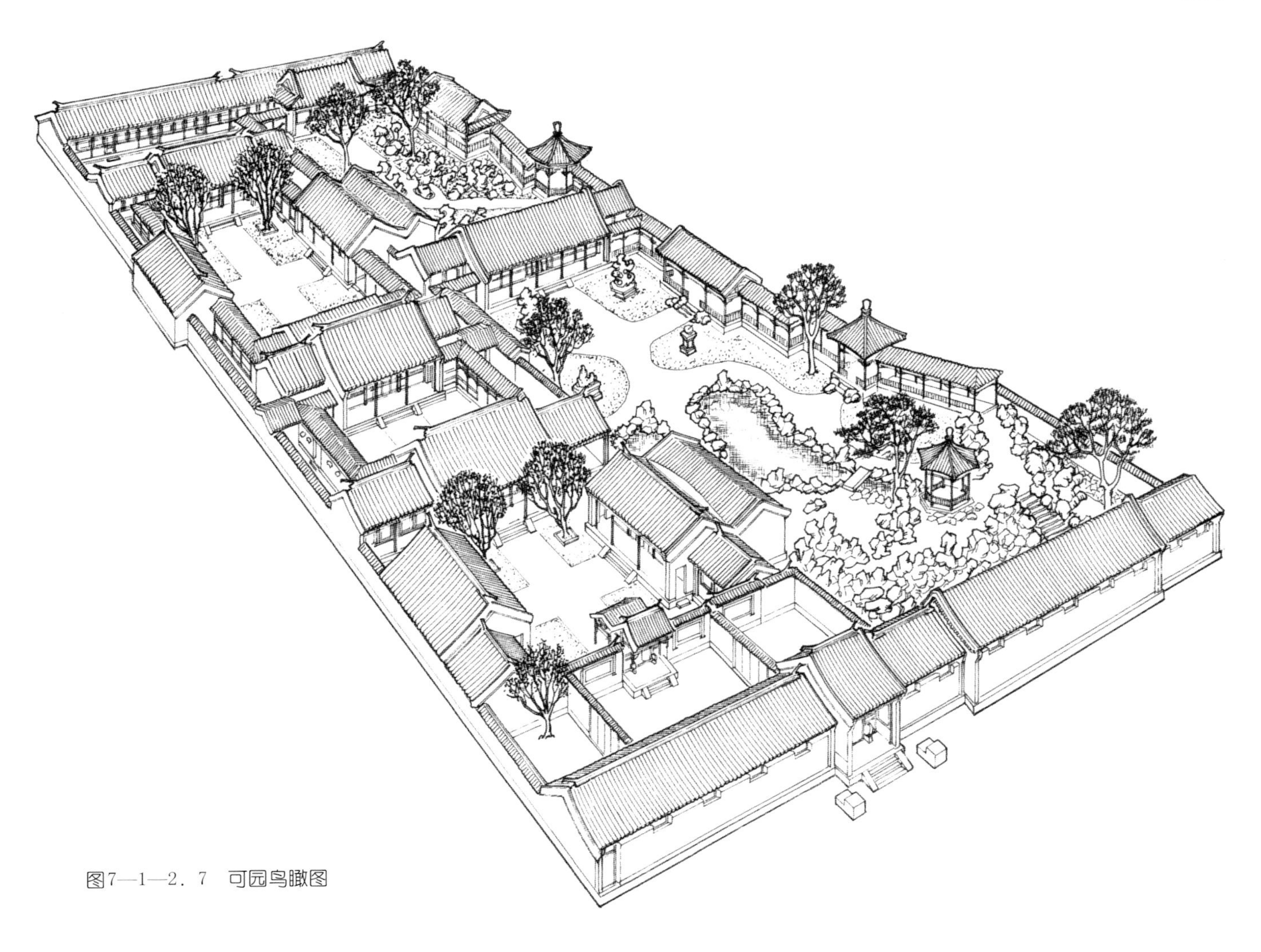

图7—1—2．7 可园鸟瞰图

建于城市内的小型私家园林由于缺乏真山真水衬托，空间呆板狭小，所以，无论道路、建筑、池水、泊岸都应力求曲折多变，使之更加贴近自然，使山回路转，曲径通幽，步移景迁，小中见大，虽由人造，宛自天开，令人百游不厌，玩味无穷。

由于北京缺水，北京的小型私家园林引水入园的不多，而以叠石为主的园子相对较多。这应是北京小型私园的一个特点。比如颇有些名气的北京牛排子胡同的半亩园，就是一座以叠石和建筑为主体的花园。秦老胡同某宅花园也是由叠石、游廊、花草树木组成，布局简洁、灵巧，以小巧取胜。

小型私家园林宜有明确的主题。造园好比作文章，应当有命题，文章要按命题展开，围绕一个中心，从不同角度进行阐释，将内容写深写透。园子的主题要由主人或设计者而定，同时还应考虑周围环境和条件。前面所谈的秦老胡同某宅花园题额为“西园翰墨”，说明它是一个以书房为中心的具有浓郁书卷气的文人花园。

造园手段除掇山、理水、开池、建筑之外，还有植树、种草、栽花、点竹、饲鸟、养鱼等等，通过各种手段美化环境，使三季有花，四季常青，春赏鸟啾，夏听蝉鸣，精心营造美好的环境气氛（图 7—1—3）。

图7—1—3　私家园林园景举例

图7—1—3．1　西四北三条某宅花园

图7—1—3．2　后圆恩寺某宅花园

图7—1—3．3 后海某宅花园月洞门

图7—1—3．4 帽儿胡同某宅花园假山

图7—1—3．5
西四北三条某宅
花园爬山廊

图7—1—3．6
西四北三条某宅花园山石及绣楼

图7—1—3．7
秦老胡同某宅花园假山

图7—1—3．8
后海某宅花园叠石

第二节　四合院的庭院绿化

在传统住宅当中，庭院泛指房屋围合成的室外空间。实际上，在古代庭和院是有区别的。古代称最小的户外空间为庭。“堂下至门，谓之庭”，“庭，堂前阶也”。可见，古人心目中的庭是指住宅正厅前面的小块空地，北京地区称这种室外小空间为“小天井”。“院”则是指由四面房子围合起来形成的较大空间。由于院子面积大，所以，它不仅能满足住宅通风、采光的要求，还可以栽植花木以点缀环境，创造优越的户外生活条件。

在传统民居四合院中，庭院是宅人活动的主要场所。即使带花园的住宅，人们的活动场所也是以庭院为主，以花园为辅。因此，人们历来十分重视庭院的种植和美化，通过绿化庭院去创造优美的生活环境。

庭院种植不仅可以满足人们视觉、听觉的需求，而且还能给人以极其丰富的感受。清代画家郑板桥曾对一个小院落的种植作过这样的描写：“十笏茅斋，一方天井，修竹数竿，石笋数尺，其他无多，其费亦无多也。而风中雨中有声，日中月中有影，诗中酒中有情，闲中闷中有伴。非惟我爱竹石，即竹石亦爱我也。彼千金万金造园亭，或游宦四方，终其身不能归享。而吾辈欲游名山大川，又一时不得即往，何如一室小景，有情有味，历久弥新乎！对此画，构此境，何难敛之则退藏于密，亦复放之可弥六合也。”（《郑板桥集·竹石》）一个小院，几竿翠竹，数尺石笋，就能给画家带来如此丰富的心理感受，可见营造环境对人的心理影响之大。

四合院的庭院绿化，首先是种树。树木一般都种在内宅的十字甬路与房屋之间的空地上，可以每块地各种一棵。也可在北房阶前两侧各种一棵。树木的品种有海棠、丁香、石榴、榆叶梅、玉兰、枣树、槐树等。

海棠是人们喜爱的树种之一。海棠春可赏花，秋可结果，春华秋实，不同季节给人以不同的美感。丁香也是人们喜爱的树种，鲁迅先生的日记中就有“晚庭前植丁香二株”的记载。过去，曾见过很多四合院中种植丁香的，尤其是生长多年的老丁香树枝杈茂盛，花开时满庭芳香，醉人心脾。玉兰是讲究人家喜欢选用的树种。玉兰不仅花型美观，气味清香宜人，而且寓意也极好。它同海棠一起，常被人称为“玉棠富贵”，因此格外受人青睐。北京人还十分喜爱栽种石榴。石榴花开时如火如荼，结果时硕果累累，极富观赏性，不仅有春华秋实之美意，而且还因石榴多子，常隐喻“多子多福”。如果再加上葫芦藤蔓，就更是“子孙万代”了。除此之外，夹竹桃、榆叶梅、山桃等，也都是颇具观赏价值的树种，为广大居民所喜爱。

北京人还喜欢种枣树。枣花开时，幽幽清香招来蜜蜂采蜜授粉，给四合院带来勃勃生机。秋季来临，红枣挂满枝头，不仅显出秋色佳美，更带给人们收获的喜悦。槐树也是四合院常见的树木。槐树树形好，树冠大，遮阴凉，尤其夏日酷热之时，香喷喷的槐花绽开，浓绿中泛着翠绿色，空气中飘着幽香，此伏彼起的蝉鸣声给人们的是又一番感受。有些树种是不宜种在四合院里的，如松柏树，一般都种在墓地。“白杨萧萧”，杨树也不宜种在院内。桑树的“桑”字与“丧”字谐音，更不可取。倒是有人喜欢在院内种几棵椿树。香椿可供宅人美美地吃上几顿椿芽。臭椿呢？据说臭椿是树中之王，哪里有臭椿，哪里的树木都长得茂盛。

四合院里除去种树就是种花种草了。花草有地上种的，还有盆栽水养的，这在四合院中都是十分讲究的。“天棚、鱼缸、石榴树，老爷、肥狗、胖丫头”，这是概括老北京小官吏家庭生活的两句很生动的话。这石榴树和鱼缸就是指的盆栽、水养。北方冬季寒冷，石榴树种在大木桶内，春夏秋三季搬出来，冬天抬到室内过冬。采用盆栽的花木还有夹竹桃、杜鹃、栀子、银桂、金桂等品种。顺便说及，讲究人家培植的盆景，摆在廊下室内，是极富观赏价值的。

四合院里的水养植物主要有荷花、睡莲、菱角、水葫芦、茨菇等。这些养在荷花缸里的水生植物，滋润了缺少水景的四合院，缸内往来翕忽的游鱼和盘桓在花间的蜻蜓，更给人们带来江南水乡的意境。

草花也是四合院绿化的主要内容。北京人喜欢的草茉莉、凤仙花（俗名指甲草）、西番莲、喇叭花、丝瓜花、扁豆花，常生长在老百姓四合院的房前屋后，点缀着人们的生活环境。而名贵的牡丹、芍药、大丽花、秋菊花，则是供讲究人家观赏的上品了。

绿化环境，美化生活，是我国传统居住文化的重要内容。前人在这方面留给我们的许多宝贵经验和做法，值得认真研究、继承（图 7—2—1）。

图7—2—1　北京四合院庭院绿化种植举例

图7—2—1．1　某名人故居院内种植的海棠树

图7—2—1．2　后圆恩寺某宅庭院绿化

图7—2—1．3 石榴树硕果满枝头

图7—2—1．4 庭院里的盆栽荷花及其他花草

图7—2—1．5　后圆恩寺某宅庭院绿化

图7—2—1．6　后圆恩寺某宅院内种植玉兰、月季

第八章

北京四合院的设计、施工与修缮

第一节　四合院的设计

一、传统的设计方法和程序

据有关史料记载，明清时期传统建筑的设计已经有当代建筑设计的萌芽，如宫廷建筑设计有样房，通过画图样、烫样（相当于今天的做模型）来对建筑群体进行规划，然后再根据批准的图样进行施工。对于复杂的建筑，在正式施工前还要“扎小样”，即做1/10比例的模型，待将构造真正弄清后再行施工。

一般的民居建筑，由于构造相对简单，所以往往省去了绘图样、烫样和扎小样的程序，由工匠中的主角儿——木匠头儿与业主共同商量，排定尺寸，确定房子的面宽、进深、柱高、举架等，然后根据已成定规的权衡比例关系确定木构架各部分的详细尺寸和具体做法。瓦、石、土各工种则随木作的规矩和约定俗成的尺度做法进行砖、石工程和地面排水工程。

在传统的设计过程中，风水师发挥着相当重要的作用。首先，房主买地，要请风水先生看风水，以判别吉凶。地皮定下后，在确定院落房间的位置，特别是确定正房和宅门位置时，还要以罗盘校正方位，并用五行八卦、阴阳学说及房主的生辰八字来定出房间朝向。如果经过这些工作，仍然存在缺陷和不足，则采取“避让”“改造”“符镇”等法进行调整，使宅院处于吉利之位（详细内容可参见第三章）。

二、现代的设计方法和程序

我们近年来遇到的四合院设计，多数属于单位院落的翻建或改建。建设单位（俗称甲方）一般是个人或开发商，也有的属于机关单位（例如外省市驻京办事处）。他们买下的旧院，房子都已十分破旧，难见原来面目，基本都是拆掉旧房，重建新房。虽称翻建，但绝不是原有房子的就地重建，而是要按照功能要求重新进行设计。其内容、设施都要符合现代居住（或办公）的要求，只不过外形是传统的四合院而已。

一般的设计程序是，由甲方提出具体要求，设计者根据甲方的要求，做出第一轮方案。第一轮方案的重点是排平面，在平面上将甲方意见具体化，然后交给甲方进行讨论。针对第一轮方案，甲方还会提出进一步的修改意见，设计人要在同甲方的探讨中阐明自己的设计意图，以便使双方产生共识。在做方案的过程中，设计人除要贯彻甲方的意图外，还应在设计中贯彻国家有关的法律法规和设计规范，使设计方案成为既能体现使用者的意图，又符合国家政策法规的合格方案。

方案确定后，画正式报建图（报建图要做总平面图、建筑平面图、立面图、剖面图、基础平面图），由甲方呈报规划部门办理报建手续。待报建手续完备后，再进行施工图设计。在现代的四合院规划设计中，对风水问题也要给予充分关注。除去用一般的风水理论和判别吉凶的标准对地形、环境进行勘察外，有时还需专门聘请风水方面的专家进行实地考察，以做出正确判断，满足房主对风水方面的要求。

施工图设计时建筑专业要先走一步，然后依次为结构、设备，如有室内装修装饰，要与设备配合。各专业设计都要严格执行国家规范，以确保做出合格的设计作品。

近些年，我们还不止一次地遇到过成街成片的四合院翻建、改建工程。成片改建一般有两种情况。一是成片院落全部拆除，重新规划建设。这种情况比较单纯，除要考虑保护原有地面上的古树名木之外，其他均可按照甲方的要求，结合国家的有关规范进行规划。另外一种情

况是在要拆除的建筑中，夹杂有文物保护单位或有文物价值的建筑物、构筑物。遇到这种情况时，不仅要认真贯彻执行国家有关文物保护的方针政策，还要使新四合院街区的规划设计与原有建筑协调一致，融为一体。所做方案要得到文物部门和规划建筑部门的共同认可，同时还要满足建设单位的合理要求。

三、四合院设计的指导思想和原则

通过多年的设计实践，我们认为，以下几点是四合院设计的基本指导思想和原则，应当在设计中认真贯彻。

1. 巧用法则，因地制宜——四合院平面布置的原则

由三面或四面房屋围成院落，院落可向纵深或两侧发展，每个院子都应有正房、厢房……这似乎是通用于一切四合院的平面布局原则。进行四合院设计，应贯彻这些原则。但是在实际中遇到的地形并不都是方方正正、规规矩矩。尤其单体院落的占地，有各种形状、各种尺寸，要在这不甚规则、不同形状的房基地内作文章，绝对不能死套规矩，应当巧用法则，因地制宜。例如，我们曾为东城区秦老胡同做过一个院子的设计。该院基地呈“⌈”形，前宽15.4m，后宽22.8m，进深48.8m，占地面积855.36m^2。要在这样一块异形基地上做出方方正正的理想院落是不可能的。甲方要求做出800m^{2}以上的房子，其中包括要在后院做地下室（后院还有一棵长势茂盛的大槐树）。

根据地形情况，我们把它分成前后两部分，前部院落较窄，宽15.4m，进深32.5m，可以做出不带抄手廊的两进院落；后院向东侧宽出，其南北中轴线与前院偏差3.5m。根据这一情况，我们在平面布置上进行灵活处理，后院随偏就偏，在东、西、北三面做成“⊓”形建筑，一则扩大面积，二则为后院地下室的安排提供了条件。当地下室与槐树发生矛盾时，我们适当减小地下室面积，留出槐树保护范围，使这一矛盾得以解决。这项设计，既满足了建设方多出面积的要求，又达到了规划部门关于绿地不少于35%的指标，还有效地保护了原有的树木（图8–1–1）。

再一个例子是东四八条某号。房主是一外国公司驻京机构，买的是一个三合院，占地面积271.5m^2。原有北房三间，耳房各一小间，东西房各二间，共计九间。南面偏东是一座随墙门，院内靠近南墙有一棵大槐树，枝繁叶茂，荫及全院，虽不是古树名木，但有保留价值。在接受这项设计时，房主人不仅提出了多建房以满足办公、住宿的要求，还要求将此树保留下来。在这样一个面积仅271.5m^{2}的基地上，要盖7间办公室、一套主人卧室、一套客人卧室、一个接待室，还要有锅炉房、厨房、公共卫生间等，不仅房间数大大超过原有数量，而且还要保护古树，宅门还要做成气派的屋宇式门，要求是相当高的。根据这个情况，我们采取了三项措施。首先，改三合院为四合院，原来的临街一面增建大门、锅炉房、公共卫生间和厨房。第二，大门做成浅进深、半坡式屋宇大门。从街面看，是屋宇式金柱大门，足够气派，但其进深仅有2.8m，大门的外廊占去1.2m，门内仅剩1.6m的空间。我们在大门内东厢房的南山墙上贴砌坐山影壁，将该公司徽记作为中心花刻在影壁上。在这极有限的空间内，完善了宅院大门的一系列内容。第三，临街南房居中一间留给槐树，这间房做成半坡屋顶：南坡做屋顶，北坡留洞口，槐树由屋顶洞口伸出。在做这一设计之前，我们仔细测量了槐树的准确位置和倾斜的尺寸，并以此为据进行半坡屋面设计，使树正好从北坡屋面伸出，南坡保持了屋面的完整性。从街面看，是一排完整的倒座房；而从院内看，槐树从屋面方洞中伸出。为了借助槐树增添情趣，我们又在槐树北面设一月洞门，门可启可闭，上刻横额，取名“槐园”。

图8—1—1　东城区秦老胡同某号四合院翻建设计举例

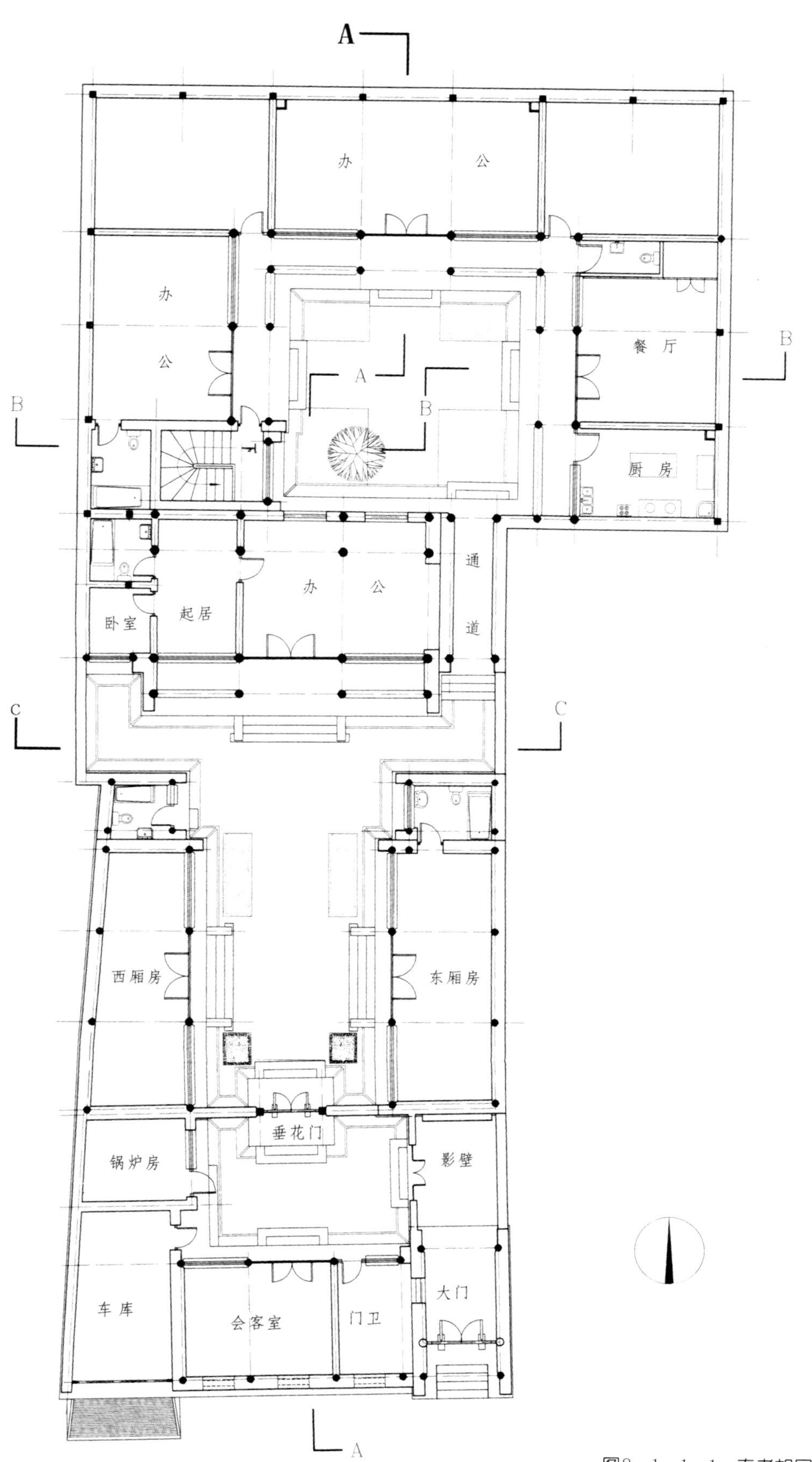

图8—1—1.1　秦老胡同某号四合院翻建设计平面图

图8—1—1.2 秦老胡同某号四合院翻建设计剖面图、立面图

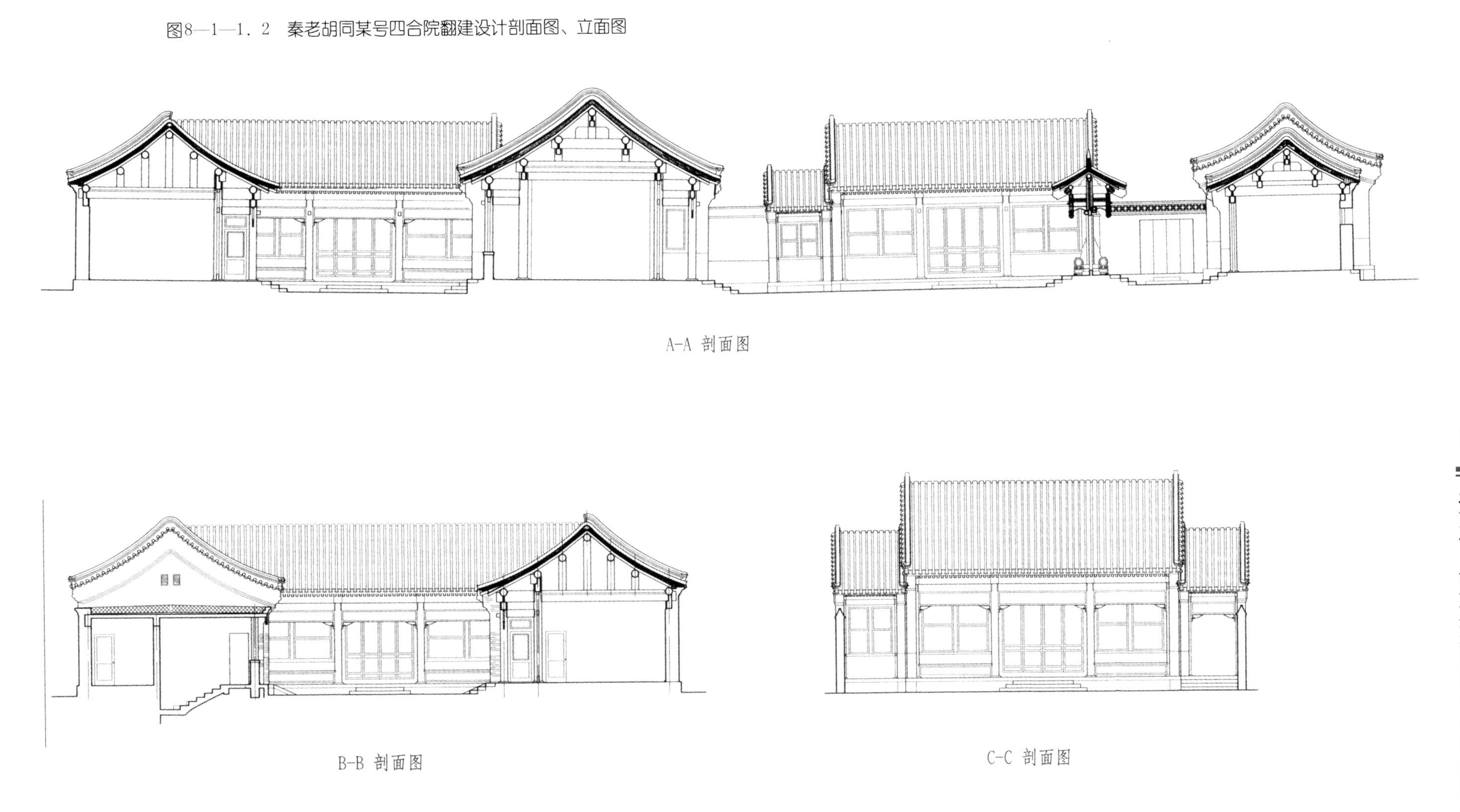

我们在占地仅有271.5m²的地块内，做了大小共16间房，不仅满足了房主人的使用要求，而且还解决了建一个体面的宅门、保持完整的沿街立面、保护槐树为院内增加趣味景致等一系列问题。这个设计虽小，但它体现了巧用法则、因地制宜、巧妙构思、精心设计的原则，取得了非常理想的效果（图8–1–2）。

图8—1—2　东四八条某号四合院翻建设计举例

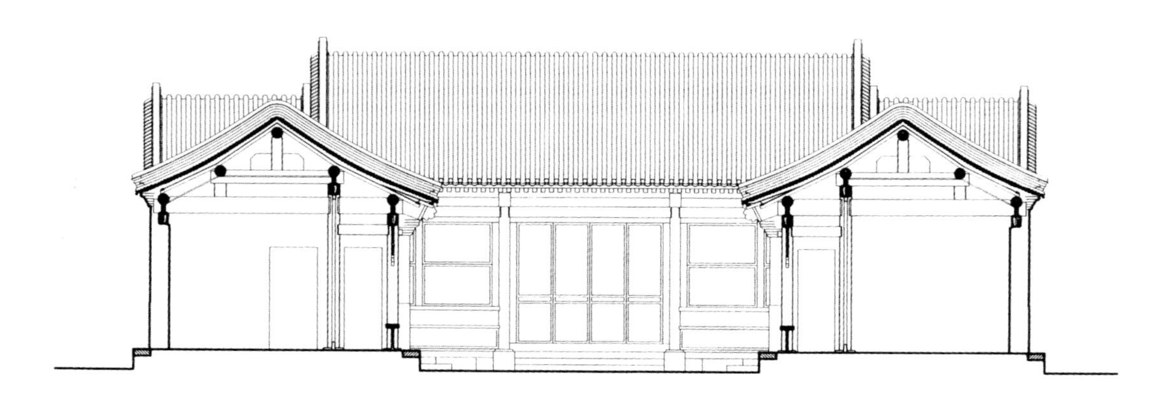

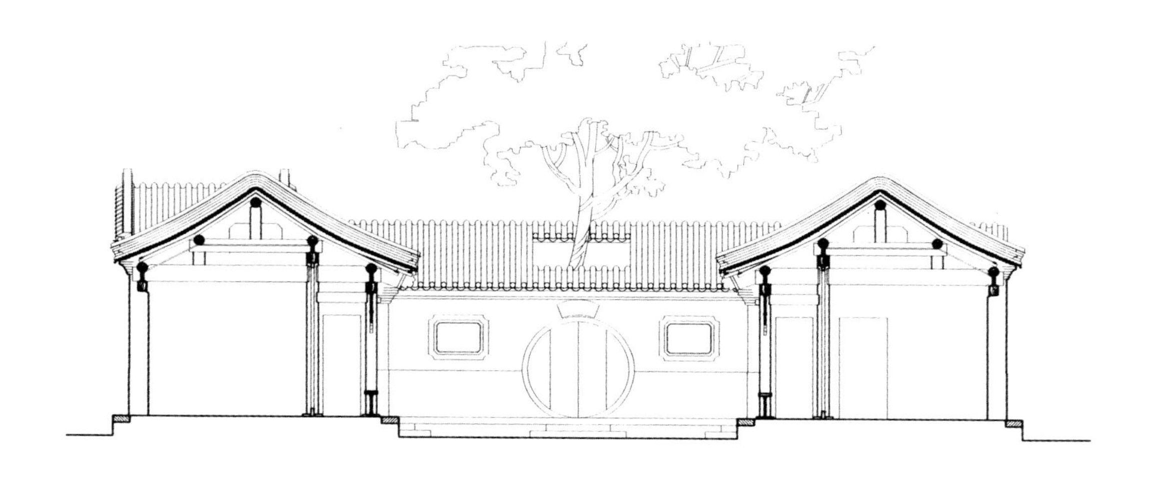

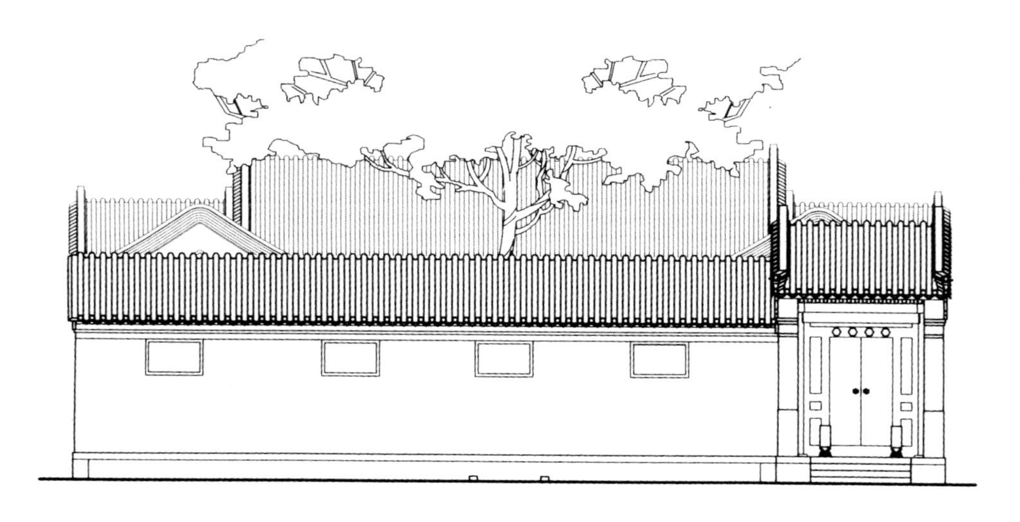

图8—1—2．1　东四八条某号立面图、剖面图

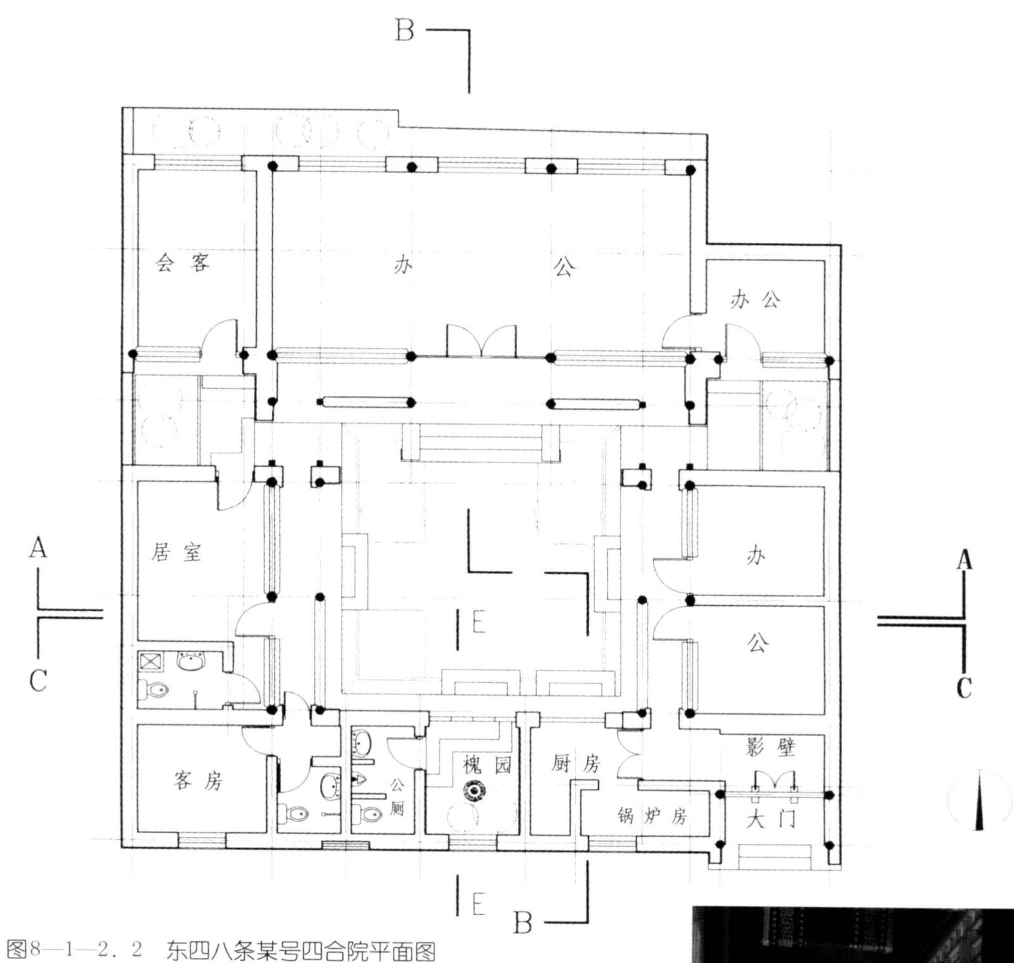

图8—1—2．2 东四八条某号四合院平面图

图8—1—2．3
从西房廊下看北房

图8—1—2．4
从东南角看北房
和西厢房

图8—1—2．5
从西北角看东厢房及槐园

图8—1—2．6
从东北角看西厢房及倒座房

2. 一种风格，多种形式——四合院外形设计的原则

四合院是老北京人世代居住的传统民居。老北京人由各个阶层组成。他们之中有官宦之家，有商贾之家，有书香门第，有市井平民。各人的社会地位不同，贫富程度不等，爱好不一，修养各异。他们对住宅的审美观念和功能要求亦千差万别，这就从客观上造成了北京四合院在总体风格一致前提下的差异。加上东西南北城工匠的师承不同、做法有别这些因素，使北京四合院形式变化多样。仅以院门为例，就有达官贵人惯用的广亮大门和金柱大门，有不留前廊的蛮子门，有雕饰华丽、内涵丰富的如意门，有与西洋建筑融合而形成的西洋门，有体量虽小，但仍不失华丽的小门楼，还有朴实简易的随墙门……同样形式的门，在不同城区还有不同风格，同样一种形式又会派生出几种不同做法。比如，如意门就有冰盘檐栏板望柱式、须弥座栏板望柱式、冰盘檐花瓦式、冰盘檐花板式、花活、素活等多种不同做法。这些门的差别不仅标志着房主人不同的社会地位、经济实力，而且表现出他们不同的志趣爱好和修养程度。因此，我们做四合院设计，必须在风格统一的前提下，力求形式的多样性，不仅宅门要形式各异、繁简不同、色彩有别，符合房主人的身份，而且在处理其他房间的檐口、脊饰、花纹等细部时，也要注意强调个性，保留差别。

3. 尺度亲切，体量宜人——四合院尺度确定的原则

四合院是居住建筑，其尺度一定要符合居住功能的要求，人身居其中要感到建筑体量适中，房间亲切宜人。如果将住宅搞得像王府、宫殿那样尺度高大、院落空旷、建筑森严，则是最大的失败。近年来，在四合院改建翻建中，有一种倾向值得注意，就是台基越做越高，房子越盖越大。出现这种情况，有房主人求吉利、摆阔气的因素；也有设计施工人员投其所好，无原则地加大尺度、提高台基、增加柱高、加大进深的因素。这样，就使得本应是亲切宜人的住宅失去了传统民居平易近人的风格。

4. 清新淡雅，华而不奢——四合院装饰色彩设计的原则

清新淡雅，华而不奢，是从事四合院装饰、色彩设计应遵循的原则。我们已经在油饰彩绘一节中谈到，传统四合院油饰彩画的应用是讲制度、分等级的。在色彩油饰以及彩画装饰的应用方面，除王府一类等级很高的建筑外，作为民间的住宅是极少遍施彩绘的，油饰的颜色也分为若干等级，有严格区别。近年来搞的四合院建筑一概朱红大门，大面积彩画，大面积贴金，金辉耀目，光怪陆离，显然是不符合传统的。

四合院民居在用料方面也应以朴素、淡雅为主调。如石料多用小青石或青白石，而不应当使用洁白的汉白玉；瓦宜使用灰色板瓦，而不能用琉璃瓦；至于街门上面的金属饰件，也应以铁制、黑色为主调，不宜过多采用镀金铜板一类炫耀显赫的装饰材料；在雕刻或彩画的纹饰当中，更不宜用龙的图案。

5. 室外保留传统，室内风格多样——四合院室内设计的原则

随着人们生活水平的日益提高和新型建筑材料、装饰材料的飞速发展，人们对室内的装饰水平和装饰风格提出了越来越多的要求。尽管传统的室内装修和陈设——碧纱橱、花罩、多宝格以及明清家具、文玩字画等，永远都会是人们玩赏不尽的艺术珍宝，但毕竟人们兴趣爱好不同、审美观念不同、艺术追求不同，如果一律是中国传统的室内装修和陈设，就难以满足不同人的不同要求。因此，室外保留传统形式，室内采取多种风格，是现在和今后四合院设计，尤其是室内设计的一个重要指导思想。各国的传统文化中，都有不少值得吸取的精华。西方室内家具的安逸舒适，洁净明快，也是我们应当吸收的。尤其是现代的室内设施——暖气、空调、卫生设备以及家用电器已成为居家的必需品。这些现代设施的采用，也不可避免地给室内带来现代生活气息。总而言之，做传统的四合院建筑设计，外部应保持传统，以保留四合院的格局形式和建筑风格，室内装修则应按房主人的要求装饰，应求实用、舒适，切忌千篇一律。

6. 利用原有构（饰）件，保留文物价值——四合院翻建改建的原则

传统四合院中，有许多具有文物价值和艺术价值的构件、饰件，如抱鼓石、砖雕戗檐、垫花、透风、上马石、拴马桩、木雕饰件、室内碧纱橱、花罩等。这些构件、饰件，记载着百年历史，积淀着古代艺术文化，铭刻着传统文明。精心保留这些构件、饰件是非常有意义的。如果能在四合院的翻建改建中有意识地保留并在建设中用上这些构件，那将更有意义。将雕刻精美的砖石构件镶嵌在新建的四合院的相应部位，继续实现它们的价值，不仅能使新建院落增添艺术感染力和历史沧桑感，为人们留下一段古老而神秘的话题，还能杜绝那些粗劣俗气的“作品”出现在仿古四合院中，实为一举两得。

在进行四合院翻建改建的时候，应当特别注意保护那些保存尚好且具有艺术价值和观赏价值的构配件，切不可将这些宝贝作为垃圾扔掉。即使有些构件已无法用在建筑上，也可以有意识地将其保护陈列起来，供人们观赏。

7. 结合主题思想，营造文化氛围——四合院及其园林设计的灵魂

四合院不仅是一个具有居住功能的实体，而且还有它的主题思想。四合院的主题思想是同宅主人的理想、追求、治家思想有着直接关系，而这些又往往通过门联、楹联、雕刻等形式表现出来。如“忠厚传家久　诗书继世长”，是我们见的最多的门联之一，它表达出宅主人以忠厚传家、诗书继世的处世原则，表现出中国人民忠厚、诚信、崇尚诗礼、追求文明的优秀品质。又如“行义致多福　积善有余庆”，这幅门联所表达的是“行义为先、积善为本，行义行善即可积福积庆”的主题。再如“忠诚济世　恕道存心”，表达的是以忠诚和恕道对待人和事的主题思想。这些都是中国人民几千年形成的传统美德和处世原则，将它们作为主题思想反映在宅院的醒目部位是很有意义的。

传统宅院如此，新建宅院依然有其主题。如商人追求财源茂盛，百姓向往吉祥平安，官员祈求仕途坦荡，文人标榜出污不染。当宅主人对其宅子表达出明确的主题后，我们在从事装饰设计时，就应紧紧扣住主题思想，通过砖雕、木刻、彩绘、匾联等各种艺术手段，去表现这个主题，营造文化氛围，而不应无目的无主题地进行装饰。

营造私家园林，更应当有鲜明的主题思想，要表达明确的思想内容，而不仅仅是掇山、理水、种树而已。

中国是文明古国，处处都有文化的积淀，一座宅院，一处园林，如果没有主题思想，没有文化内涵，就如同只有躯壳没有灵魂，是不可取的（图8—1—3）。

图8—1—3　新四合院设计一例——国子监街 43 号院

图8—1—3．1　门楣砖雕

图8—1—3．2
宅门内的坐山影壁及盆栽植物

图8—1—3．3
东厢房

图8—1—3．4
正房的彩绘装饰

图8—1—3．5　北房及外廊一角

图8—1—3．6　廊内彩画装饰

四、新材料、新工艺、新技术、新内容在设计中的体现

传统的四合院建筑以木结构为骨架。木结构承重，砖墙只起围护分隔作用。砖墙的砌筑是以整砖砌外皮，碎砖背里，屋面钉木望板、苫泥背、灰背、宽瓦。随着时代的发展，有些传统材料逐渐被新材料所代替。除去砖墙屋面还继续采用传统材料和做法之外，其余都可采用新材料、新工艺。最为突出的是以钢筋混凝土构架代替木构架，已成为不可逆转的趋势。随着结构材料的变更，相关工艺技术也都发生了根本变化，传统的一麻五灰地仗几乎全被单披灰所代替，木质槛框的安装也由榫卯连接变为钢件连接。但不管如何变化，只要是仿古四合院建筑就应当仿得真、仿得像，要做到以假乱真，才能保持四合院的神韵。

随着居住要求的提高，现代的仿古四合院还增加了许多新内容，如汽车库代替了马号，暖气代替了煤火炉子，增添了供暖用的锅炉房。其他诸如空调、电视、电话、网络等现代生活设施，均是四合院中出现的新事物。对这些，要在设计中予以充分考虑，同时又不能因这些新内容的出现而冲淡了四合院古朴典雅的气氛。尤其是汽车库这种现代设施，很容易影响传统的街面效果。所以，在设计时应格外小心，应当使它的外形、色彩尽量与传统建筑相协调，而不要搞得明晃耀目，格格不入。只有将新材料、新工艺、新技术、新内容与传统的建筑形式风格特点统一起来，才能创造出成功的作品。

第二节　四合院的施工

传统民居四合院是中国古建筑的重要组成部分。它的施工程序和工艺技术，与一般古建筑的施工程序和工艺技术基本相同。它们主要包括基础与台基工程、大木构件预制加工、石活加工、灰浆制备和砖料加工、大木构件安装、墙体工程、屋面工程、地面工程、木装修工程、墙面抹灰工程、地仗油漆工程、彩绘工程。现按一般顺序，将各工程部位的施工程序和工艺技术简述如下。

一、基础与台基工程

1. 放线、挖槽

在施工场地做到三通一平之后，首先要根据设计图纸确定的建筑位置、墙体、柱网轴线尺寸放线，并定出标高，以便根据轴线和标高确定墙体位置和基槽宽度、深度，然后挖槽。

一般四合院属民间工程，小式做法，基槽多采取沟槽形式，沟槽一般为墙体宽的2倍以上。基槽深度应按设计要求，并应在冰冻线以下。

2. 常见的基础做法及施工程序

四合院常见的基础做法及施工程序如下。

（1）*素土夯实*　传统做法是用大硪拍底1～2遍，现代做法是用机械夯实。

（2）*打灰土*　一般民居的基础灰土比例为3:7，即三成白灰，七成黄土，搅拌均匀，在槽内虚铺7寸（约22cm），耙平，先用人工踩1～2遍，称为“纳虚踩盘”，然后用夯筑打。传统筑打程序有“行头夯”（又称“冲海窝”）、“行二夯”（又称“筑银锭”）、“行余夯”（又称“充沟”或“剁梗”）、“掖边”（冲打沟槽边角处），然后用铁锹铲平。这种夯打称为“旱活”，可重复1～3次。为使灰土密实，在“旱活”之后还要“落水”，又称“漫水活”，即用水将灰土洇湿，水量控制在将最底层的灰土洇湿为度，判断方法是“冬见霜”“夏看帮”，即冬天看灰土表层结霜，夏天槽帮侧面洇湿高度相当于灰土厚的2～3倍。“落水”一般在晚上进行。第二天再筑打之前为防灰土粘夯底，应先撒砖面灰一层，称为“撒渣子”，然后再进一步夯打密实。基础灰土一般为1～3步，每步均按以上程序进行。现代做法也常在基槽夯实后打灰土，或打素混凝土垫层。

（3）*砌磉墩掐栏土*　传统基础多为独立基础，支撑柱顶的独立基础称为“磉墩”。磉墩之间的墙称为“栏土”。它是为栏档回填土用的，一般不与磉墩连接。

（4）*摆放柱顶石*　磉墩砌至一定高度（室内地平高度减去柱顶石鼓镜以下部分），即可在上面摆放柱顶石。摆放柱顶石时要注意，柱顶石上面的十字中线要与柱网轴线相对，外圈柱子一定要加出侧脚尺寸，柱顶石顶面要平。

（5）*包砌台明*　房屋台基露出地面部分称为台明。台明四周应用砖石砌筑，包砌台明可与砌磉墩、掐栏土同时进行，也可滞后进行，需根据具体工程情况而定（图8–2–1）。

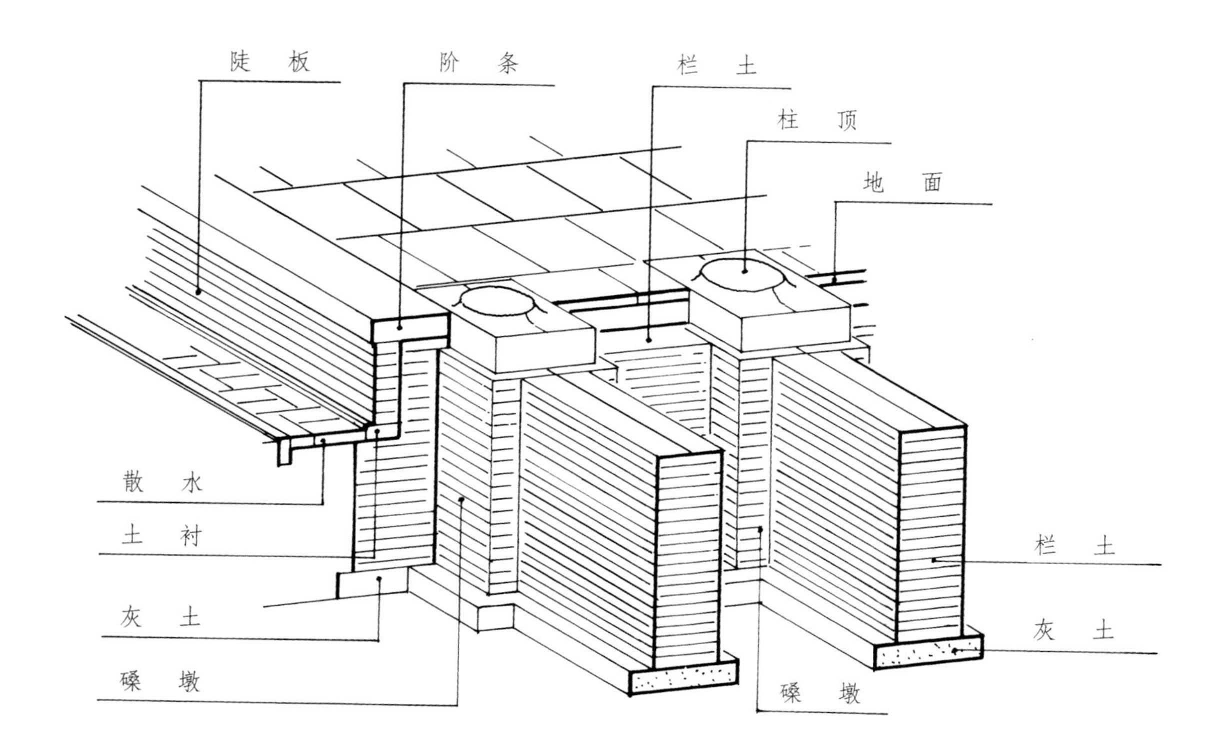

图8—2—1　基础、台基构造示意

二、大木构件预制加工

古建筑中将柱、梁、枋、檩等木构件称为大木。传统四合院建筑是以木构为骨架。木构架由柱、梁、枋、檩、垫板等预制构件组装而成。因此，木构件的预制加工工作应率先进行，以保证在基础工程完成后即能进行组装。

古建木构架是凭借榫卯结合在一起的。大木构件预制加工就是按尺度和构造要求做出构件及其榫卯。大木构件预制加工之前应做好以下准备工作。

(1) 备料　按设计要求，以幢号为单位开出料单。备料时要考虑“加荒”，材料的长度及截面尺寸都要留出供加工的余量。

(2) 验料　根据工程对木材质量的要求，检验有无腐朽、虫蛀、劈裂、空心以及节疤、裂缝、含水率等疵病程度，不合质量要求的木材不能使用。

(3) 材料初步加工　将荒料加工成制作木构件所需要的规格材。如柱、檩等圆构件的砍圆刨光，梁枋等方形构件的砍刨平直，以备画线制作。

(4) 排丈杆　丈杆是古建筑大木制作和安装时使用的一种既有度量功能、又有施工图作用的特殊工具，用优质干燥木材制成，有总丈杆和分丈杆，分别在上面标注柱、梁、枋、檩等构件的实际长度和榫卯位置、尺寸。排丈杆是一项非常严格细致的工作，绝对不能出差错，一般都由技术最高的工匠师傅或工地技术负责人进行。丈杆排出后至少需经两人严格检查，确认无误后方可使用（图 8–2–2）。

大木构件制作的首要工作是画线。大木画线的工具除丈杆之外还有弯尺、墨斗，画檩碗用的样板，画榫头用的样板，岔活用的岔子板等等（图 8–2–3）。大木制作的传统工具有锯、锛子、刨子、斧子、扁铲、凿子等。

大木画线有一套传统的、独特的符号，分别用来表示中线、升线、截线、断肩线、透眼、半眼、大进小出眼、枋子榫、正确线、错误线等，至今仍在工程中承传应用（图 8–2–4）。

由于一幢建筑的木构架是由千百件木构单件所组成的，为使这些构件在安装时有条不紊，各有各的位置，在木构件制作完成后需标注它的具体位置。大木位置号的标写有一套传统方法。以柱子的位置号为例，通常要写明所在幢号、在明间的哪一侧、前或后檐、什么柱、所标的位置号朝哪个方向等。这就如同我们寄信写地址那样，要注明×省、×市、×区、×街、×号，住楼房还要注明×楼×门×室。国际间还要写明×国，以确保位置准确无误。图8–2–5是四合院最常见的柱位平面。假定该平面为北房三间，那么，①号柱的位置号应为北房明间东一缝前檐柱向北；②号柱位置号应为北房明间西一缝前檐金柱向北；③号柱位置号应为北房明间西二缝后檐金柱向南，依此类推。其余构件，如梁、枋、檩等，也都有具体标注方法。这些方法至今仍在施工中沿用。

大木构件分为柱类、梁类、枋类、檩类、板类以及椽子、连檐、望板等不同类别，分别用不同的丈杆画线，然后按线制作。木构件制作的成品应妥善保管，不可日晒雨淋，碰撞损伤，以备顺利安装。

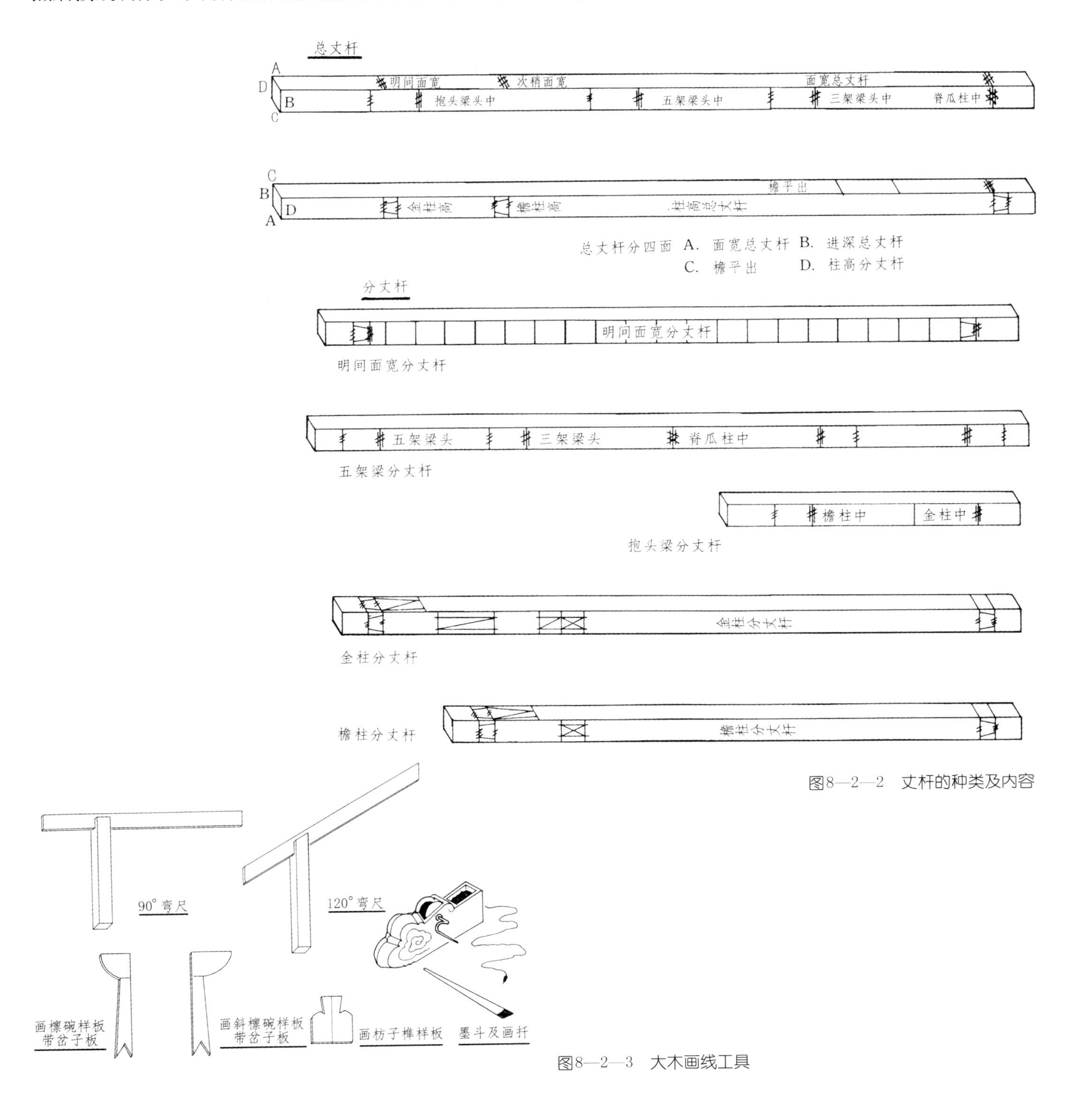

图8—2—2 丈杆的种类及内容

图8—2—3 大木画线工具

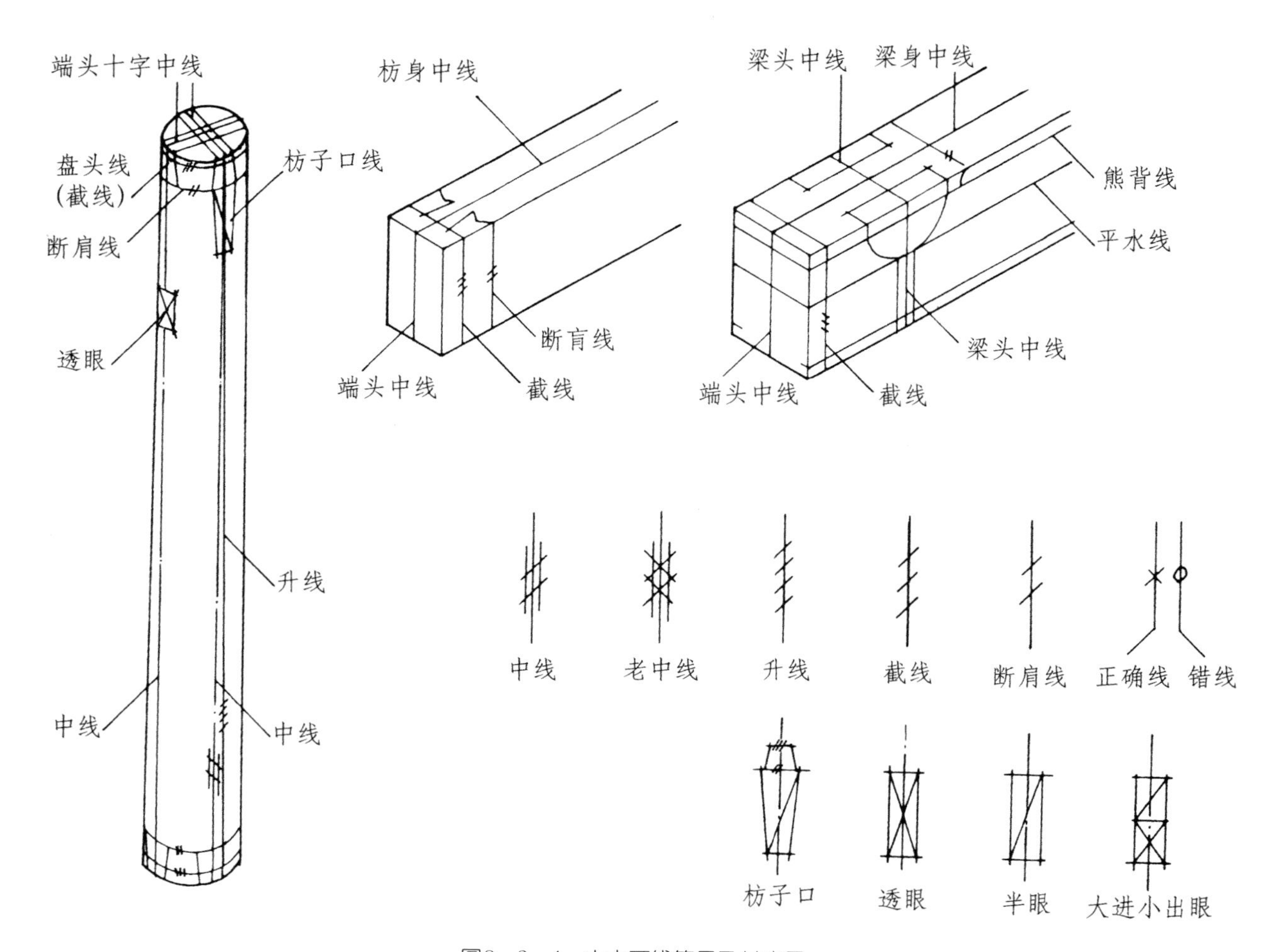

图8—2—4 大木画线符号及其应用

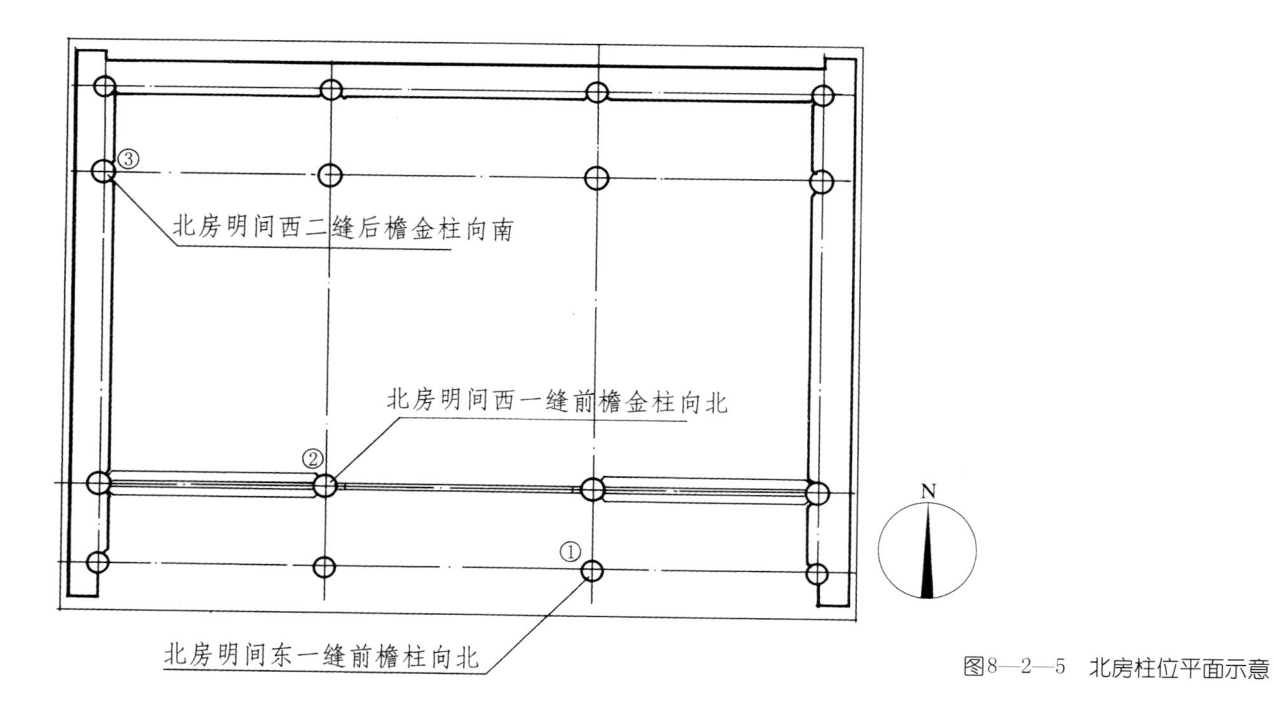

图8—2—5 北房柱位平面示意

三、石活加工

用于四合院的石构件主要有阶条、土衬、埋头、垂带、踏跺、柱顶、角柱、腰线、挑檐石等。其中，阶条、土衬、埋头等用于台基部分，角柱、腰线、挑檐石用于墙身部分。由于石活要在台基、墙体施工时使用，所以也需要事先进行加工。

四合院建筑所用石料多为长方形，属一般材料。这种一般石料的加工程序主要有：选定荒料、打荒、打大底（即打出大面）、弹线打小面、砍口齐边、刺点或打道（找平）、截头、砸花锤、剁斧（通常剁三遍）、打细道等，需要进行雕刻的石构件还要做石雕（石雕工艺及程序参见第六章第一节）。现代石料加工多采用机械，程序要简化得多。

四、砖料加工和灰浆制备

（一）砖料加工

中国传统建筑所用的砖瓦材料的形成和发展历史悠久、品种繁多。这些材料多为手工制作，经砖瓦窑焙烧而成，外形比较粗糙。但传统建筑的墙体摆砌却十分考究，有干摆、丝缝等多种，对砖料的精度要求很高。为适应墙体摆砌的需要，要对砖料预先进行加工。

四合院常用的砖料有停泥砖（分大、小停泥砖）、方砖（有尺二、尺四、尺七等不同规格）、开条砖、四丁砖等等。需加工的种类主要有摆砌墙身用的停泥砖，墁地用的方砖，做盘头、博缝、戗檐用的檐料砖，屋脊上用的脊料砖，以及影壁、檐口、须弥座等处用的杂料砖等。

砖料加工是凭砍、磨等手段，将糙砖加工成符合尺度和造型要求的细料砖。现以干摆、丝缝墙所用的砖料为例简要介绍如下：首先根据墙体尺寸和做法（如墀头宽度、山墙进深等），定出所需砖料的尺寸（长短薄厚都应小于糙砖尺寸），砍出“官砖”（标准砖），并按“官砖”尺寸定出制子（确定砖尺寸的简易度量工具）。砍砖程序：①先将看面用刨子铲平并用磨头磨平；②沿一侧画出直边，用扁子和敲手将多余部分铲去，称为“打扁”；③在打扁的基础上用斧子进一步进行劈砍，此工序称“过肋”，过肋时要在侧面留出“包灰”；④以砍磨过的直边为准，用制子、平尺在另一侧画线、打直、打扁、过肋，并在后口留出包灰；⑤截一头，在砖的一头用90°角尺画出直线，用扁子和敲手打去多余部分，然后用斧子劈砍，过肋并留出包灰；⑥以截好的头为准，用制子和角尺画线，截另一头，并进行同样加工。经过这一系列程序，一块用于干摆（或丝缝）的砖料即加工完毕，因这种砖料的加工须动五个面，所以又称“五扒皮”（图8–2–6）。

砖加工的内容很庞杂，用于不同部位的砖料，其加工程序和方法均不相同。关于这方面的内容，《古建园林技术》杂志总第1、2期刊有程万里的文章《古建砖料及加工技术》，刘大可《中国古建筑瓦石营法》一书中亦有详细介绍，在此不复赘述。

（二）灰浆调制

传统古建筑瓦石工程所用灰浆种类繁多，有“九浆十八灰”之说。按灰的泡制方法分，有泼浆灰（经水泼过的生石灰过细筛后用青灰浆分层泼洒，闷15天后使用）、煮浆灰（即石灰膏，用生石灰加水煮后过滤而成）、老浆灰（青浆、生石灰过细筛后共同发涨而成）。按灰内掺和麻刀的程度分，则有素灰（灰内无麻刀）、大麻刀灰（灰与麻刀重量比为100：5）、中麻刀灰（灰与麻刀重量比为100：4）、小麻刀灰（灰与麻刀重量比为100：3,且麻刀较短）。按灰的颜色分则有纯白灰、月白灰（泼浆灰加水或加青浆搅拌，必要时加麻刀）、葡萄灰（即红灰，泼灰加红土或氧化铁红）、黄灰（泼灰加包金土或地板黄）。按用途分，则可有驮背灰、扎缝灰、抱头灰、节子灰、熊头灰、花灰、护板灰、夹垄灰、裹垄灰等。因用途不同，灰浆中还可加添加剂，调出江米灰、油灰、纸筋灰、砖面灰、青浆、桃花浆、烟子浆、红土浆、包金土浆、江米浆等。这些灰浆，要根据不同部位的不同用途，事先进行调制。

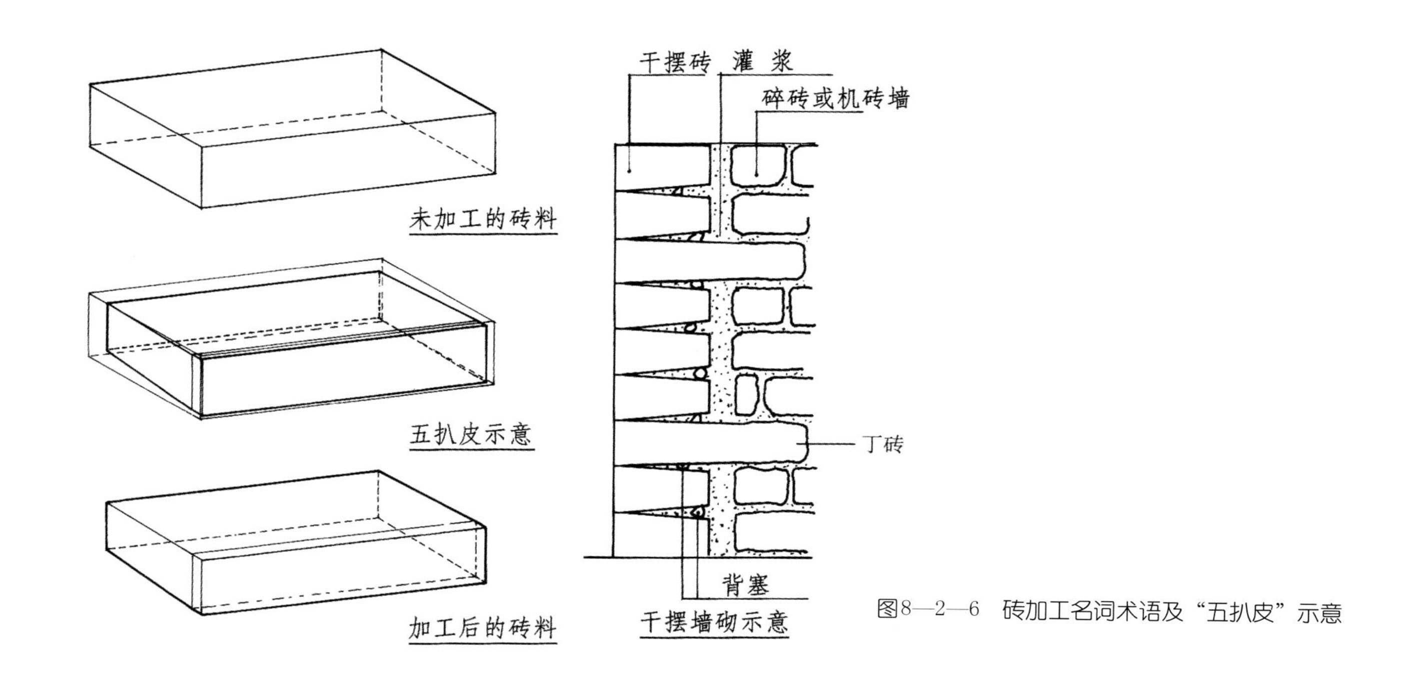

图8—2—6 砖加工名词术语及“五扒皮”示意

五、大木构件安装

大木构件安装是在基础和台基工程完成之后的工序，大木安装又称“立架”，即立木构架。

大木安装之前要对预制加工的木构件进行一次尺寸和数量的全面核对工作，同时，还要对柱顶石操作的摆放质量进行认真检查。应重点检查有无偏离轴线，有无加出侧脚，有无侧偏不平。大木安装之前还要做好操作人员的组织分工和必要的物质准备。

大木安装的一般程序和注意事项可以概括为这样几句话：“对号入座、切记勿忘；先内后外，先下后上；下架装齐，验核丈量，吊直拨正，牢固支戗；上架构件，顺序安装，中线相对，勤校勤量；大木装齐，再装椽望；瓦作完工，方可撤戗。”

其中，“对号入座，切记勿忘”是说必须按木构件上标写的位置号来进行安装，不得以任何理由调换构件的位置，更不能安错位置。“先内后外，先下后上”是讲要按照先内后外、先下后上的程序进行安装，一幢建筑不论有多少间，应先从明间安起，明间应先从内檐柱安起，逐步向外发展，不能违背规律。

“下架装齐，验核丈量，吊直拨正，牢固支戗”，是讲大木以柱头为界，分为下架和上架两部分。当安装至柱头部位时，应当对尺寸进行一次严格的校核，以防闯退中线（实际尺寸大于或小于图纸轴线要求的尺寸）。尺寸验核完毕后应将下架榫卯及构件固定，这就是榫卯处掩卡口（背楔子）和支戗杆，完成这些工作以后才能继续向上安装。“上架构件，顺序安装，中线相对，勤校勤量”，是讲上架构件的安装，也要遵循由内向外、由下向上的顺序，在安装过程中要不断验核尺寸，以确保安装质量。

“大木装齐，再装椽望；瓦作完工，方可撤戗”，是讲大木和椽子、望板安装的顺序。特别强调了墙身、屋面工程完工以后才能撤掉戗杆。这64字要诀，是在总结前人的施工经验和技术的基础上提出来的，按照这些要诀去做，就能保证大木安装工程的顺利进行。

六、墙体工程

在木构架安装工程完成以后，紧接着就要进行墙体工程的施工。四合院建筑的墙体种类很多，有山墙、槛墙、檐墙、廊心墙、院墙以及影壁等等。墙体的砌筑大多数比较讲究，其中干摆和丝缝的砌法最具代表性。

干摆即所谓“磨砖对缝”做法。它是用经过砍磨加工的砖料摆砌墙体的外皮，用碎砖衬里，里外皮砖体凭丁砖拉接。干摆砖之间不坐灰，因而无缝隙，里口有包灰，凭灰浆（一般是用白灰和黄土调成的桃花浆）灌筑

成为一体。干摆墙每摆一层即需灌浆一次，并且要将不平之处磨去，以求上口平齐，称为“刹趟”。每摆三层抹一次线，五层以上应放置一段时间，待灰浆初凝后再继续作业，称为“一层一灌，三层一抹，五层一蹲”。摆砌完成以后还要对墙面进行打点修理，主要工序有：墁干活，将砖接缝突出之处磨平；打点，用砖面灰将残缺部分和砖上面的砂眼勾抹填平；墁水活，用磨头沾水将打点过的地方以及砖接缝处磨平，并将整个墙面通磨一遍；最后，通过冲水将墙面洗净使墙体完全现出砖的本色。

丝缝是与干摆相配合采用的另一种讲究砌法，一般常将墙体下碱做干摆，上身做丝缝。丝缝即细缝的意思，砖与砖之间留有2～4mm的细砖缝。砌筑丝缝墙时，要在砖棱处用老浆灰打灰条，在里口打两个灰墩（称为瓜子灰），然后进行砌筑。丝缝墙也要在里口灌浆，凭灰浆筑成套体。丝缝墙砌完后也要进行打点、墁干活、水活，还要进行耕缝，以使墙面美观。

除以上两种讲究的砌法之外，还有淌白、糙砌等不同做法，分别用在不同部位。

传统四合院建筑墙面除砌筑讲究之外，还常采用许多艺术手段使灰色的墙面显出活泼变化的效果，常见的有落膛做法、砖圈做法、五出五进做法、圈三套五做法、砖池子做法、方砖陡砌、人字纹砌法、砖墙花砌、花瓦墙帽等等（参见图5—3—7）。

七、屋面工程

传统四合院民居的屋面常见有两种做法，一种是合瓦屋面，一种是筒瓦屋面。一般民居多采用合瓦屋面。高等级四合院及王府级建筑常采用筒瓦屋面，其他还有棋盘心屋面、干搓瓦屋面、仰瓦灰梗屋面等，是合瓦屋面的简易做法，多用于低等级的民宅。

1. 四合院建筑屋面的分层做法

北京四合院建筑的屋面基层多数为木椽子、木望板。有些低等级的建筑采用苇箔和荆笆做基层，但不具代表性。屋面分层做法，自下至上依次为护板灰、滑秸泥背1～2层、青灰背、瓦泥、瓦面。其中护板灰采用月白色大麻刀灰，厚度1～2cm,起保护望板和找平层作用。苫背1～2层，每层不超过5cm,七八成干时用杏儿拍子拍打密实。泥背之上苫青灰背，采用大麻刀灰，反复刷青浆赶轧，称三浆三轧，务求密实，以防渗漏。

2.屋面瓷瓦

屋面瓷瓦也有十分严格的操作规程。瓷瓦一般用掺灰泥，厚4cm左右，底瓦要三搭头，称压七露三或压六露四。背瓦翅子（将底瓦两侧的灰泥用瓦刀抹严抹齐）要严实，扎缝（两趟底瓷瓦之间的当子用大麻刀灰塞实）灰要能盖住瓦翅子。瓷盖瓦时，盖瓦泥要实，熊头灰要足，捉节要严，夹垄要实。总之，要使瓦面起到有效防水层的作用，以防屋面渗漏。

八、地面工程

四合院的地面工程包括室内地面和室外地面。室外地面又包括甬路、散水和海墁地面。地面做法有细墁地面、淌白地面、糙墁地面几种。细墁地面所用砖料需经过砍磨加工，统一规格，使表面光平，棱角完整。细墁地面的灰缝细，洁净美观，表面还要经桐油浸泡，常用于讲究的室内。淌白地面不如细墁地面讲究，砖料加工简单，多用于一般建筑室内。糙墁地面用砖不需砍磨加工，因砖缝较大，一般用于室外。

四合院的地面做法中，细墁地面是较有代表性的。它的施工程序大致如下：①素土或灰土夯实作为垫层；②按设计的标高要求抄平，室外地面、廊内地面还要找出泛水，以保证排水通畅；③冲趟，在两端拴好线各墁一趟标准砖，必要时还需居中墁一趟砖；④样趟，以冲趟为基准拴线墁砖，泥先不要打得太足，以留浇浆之余量；⑤揭趟浇浆，将预铺的砖揭下来，浇上白灰浆；⑥上缝，在砖楞角小面抹上油灰，将砖按原位重新墁好，并用墩锤振打严实；⑦铲齿缝，用竹片将挤出的多余油灰铲掉，然后用磨头将接缝不平之处铲平；⑧刹趟，以卧线为准检查砖棱，将侧面突出的砖棱磨平；⑨打点，用砖药将表面残缺砂眼之处勾抹严实；⑩墁水活，用磨头沾水将凸出之处进一步磨平，并将地面普遍磨一遍后擦净，露出真砖实缝；⑪钻生，待地面完全干透后，在地面上倒上 3cm 左右厚的生桐油浸泡，令桐油浸入砖内，待油不再继续渗入时，起出余油，将生石灰面加青灰面拌和成青砖色灰撒在地面上，厚约 3cm，2 ～ 3 天后将灰起掉，扫净，并用布揉擦即成。以生桐油钻过的地面不仅光洁亮泽，有较好的防潮性能，而且坚固耐磨。讲究的室内地面均采用此种做法。

九、木装修工程

木装修工程又称木装修安装工程，是将预先做好的槛框、门窗、栏杆、楣子及室内碧纱橱、花罩等木装修安装就位的工程。木装修工程还包括天花、壁板、护墙板、木地板等室内装修的施工和安装。外檐隔扇、风门、帘架、支摘窗的安装，是木装修工程的主要内容之一。门窗安装，首先是槛框、榻板的安装，槛框是门窗的外框，相当于现代建筑的门窗口。它是由单件组成，凭榫卯连接，附着在柱枋之间。槛框、榻板的安装要求平、直、方正，如门窗安于檐柱间，要随柱升线，因为升线是垂直于地面的线，如果随中线（有侧脚的柱子中线与地面不垂直），那么，门窗开启时就会走扇。抱框与柱子结合面应当有抱豁，以保证牢固严实。隔扇安装时，扇与扇之间要留缝路，并应留出地仗油漆所占余量，以保证开启自如。外檐倒挂楣子安装，应保证各间之间高低出入平齐跟线，以求整齐美观。坐凳楣子安装除应平齐之外还须坚固耐用，以供人凭坐休息。

内檐碧纱橱、花罩等木装修安装，其原理与外檐装修并无差异，但要求更加严格细致。内檐装修外表不做油灰地仗，不施混油刷饰。要讲实用功能，装饰功能也很重要，因此，从制作到安装都应精益求精，不得敷衍应付。同时，为了拆装移动方便，所有节点都应凭榫卯木梢结合，不得用铁钉组装。

十、地仗油漆工程

在木作、瓦作、石作诸项工程都相继完成之后，就要进行地仗油漆工程了。

在木构件表层披麻挂灰，形成灰壳以保护构件免受风雨侵蚀，是中国传统建筑独特的做法。传统四合院建筑，凡比较讲究者，都要在木构表面做地仗。

古建油漆地仗分单披灰地仗和麻灰地仗两种，单披灰地仗只挂灰不披麻，代表性做法有二道灰、三道灰、四道灰。麻灰地仗有一麻五灰、一布五灰、二麻六灰、一麻一布六灰、二麻一布七灰等，四合院建筑常采用且最具代表性的做法是一麻五灰地仗。

现以一麻五灰地仗工艺做法为例，对古建油漆地仗施工做法略加说明。

（1）斩砍见木　这是做地仗的第一道工序。对于旧构件，要去掉旧灰皮，砍出新木槎，要求砍净挠白，但不要伤及木骨。对于新构件，则要用剁斧剁出深 1 ～ 3mm 的斧迹，以创造在木件表面挂灰的条件，称为“剁斧迹”。

（2）撕缝、揎缝、下竹钉　古建木构无论新老构件，其表面都有裂缝。在做地仗之前，大缝内需嵌入木条，并用胶粘住，称为揎缝；较小的缝子需将表面修出八字口；较长的缝子还要向缝内下竹钉。这些措施都是为防止因木件裂缝膨胀收缩导致地仗开裂，使地仗灰与缝子结合牢固。

（3）汁浆　将用血料、满（满的成分是白面加灰油、石灰水调成的灰褐色糊状物，具有较好的粘接作用）和清水调成的稀浆涂于木构件表面，这道工序相当于新建中木件表层的操底油。

（4）捉缝灰（第一道灰）　用籽灰（成颗粒的砖灰）加细砖灰，加血料，加满调成稠糊状，用以捉填构件表面的大小缝隙，找补缺棱短角处以及不圆、不平、不直的部分，填补木构件表面的缺陷。

（5）通灰（第二道灰）　这道工序是在构件表面满做灰。所用材料与捉缝灰基本相同，目的是通过满做灰使构件表面达到平整、浑圆、直顺的效果。

（6）使麻　垂直于木纹方向缠绕麻丝纤维，用浆粘在地仗表面。浆的成份是血料和满。程序有开浆（用棕刷刷浆）、粘麻、砸干轧（用轧子将麻与灰压实，底浆要透过麻层）、稍生（个别不到之处添浆再轧）、整理（找补欠缺之处）。

（7）压麻灰（第三道灰）　将灰满刮于麻层表面，方法同通灰。

（8）中灰（第四道灰）　材料调制同通灰，但灰要细得多，灰中颗粒不明显，满的比例亦适当减少。其作用在于使构件表面灰层细腻化。但灰不宜厚。

（9）细灰（第五道灰）　细砖灰加血料加光油，材料中不再用满。在构件表层满做，厚度略厚于中灰，目的在于完善构件外形并进一步使之细腻化。

地仗各道灰的用油（灰油）量是，底层用油比例最大，然后逐层减少，面层最少。具体比例须根据不同建筑等级而定。

（10）磨细灰、钻生油　待灰干透后即用砂纸修磨表层，以去掉缺陷，达到外观要求。要求磨断斑，即要把表层的浆皮全部磨开。磨完细灰后要立即将生桐油刷在构件表面，称为钻生。一般要求至少钻两遍，以保证桐油浸至中灰层，但表面不得挂甲（不得有浮油及结皮）。

以上程序，再加上每层灰干后的打磨工作，共十三道工序，俗称“十三太保”。这是颇具代表性的古建地仗做法，在四合院建筑施工中经常采用。

待生油干后，即可在表层涂刷油漆或绘制彩画。

古建油饰工程施工，要设专门的材料房，由经验丰富的高级匠师负责统一进行调灰、拌料、调油等关键性工作。这一点在油漆地仗工程中是至关重要的。

十一、彩绘工程

在梁枋构件表面绘制彩画以装点建筑，是沿袭几千年的传统做法，宫殿、庙宇、园林、住宅概莫能外。这些，我们已在前文作了详细阐述。关于古建彩画的施工，用于不同等级建筑的彩画制度不同，做法也略有差别。现以北京四合院中应用最多的苏式彩画为例，略述其施工过程 。

（1）磨生过水　首先，对要做彩画的构件表面磨生过水，通过用砂纸打磨及过水等工序，去掉地仗面层的油痕、浮灰，为彩画创造良好的作业条件。

（2）分中　中国传统建筑彩画的图案一般都是以中线为准、左右对称，因此，在进行绘画之前首先要找到构件的中线，以便在1/2构件的范围内布置纹饰（起谱子）。

（3）起谱子、扎谱子　在厚纸（一般用比较结实的牛皮纸）上按构件实际尺寸画出彩画的线描图，称为起谱子。画谱的图案要准确、清晰。然后，沿图案线条用大针扎出均匀的针孔，称为扎谱子。

（4）拍谱子　拍谱子又称打谱子，是将扎好的彩画谱子覆于构件表面，用白粉包沿谱子拍打，使白粉透过谱子上的针孔印在构件之上。拍出的画谱应准确、清晰，花纹连贯不走样。

（5）沥大、小粉　沥粉是通过沥粉工具和材料使彩画图案线条成为凸起的立体线条，固结在构件上，其目的是为强调彩画主线的立体效果和贴金箔后的光泽效果。沥粉材料主要由土粉、青粉、胶液、少量光油和水合成，工具有粉袋和粉尖子。沥粉的程序应先沥大粉后沥小粉。大粉用来表现彩画中起主体轮廓作用的线条，如箍头线、方心线等；小粉用来表现细部纹饰线条。沥粉应严格按谱子进行，准确表现纹饰图案。要达到粉条饱满，图案对称、端正，线条流畅，具有连贯性，且要求粉条的粗细高低一致。

（6）刷色　刷色包括刷大色、抹小色、剔填色、掏刷色。刷色应先刷大色（如大青、大绿色），后刷各种小色。无论涂刷何种颜色，都应按彩画施色制度进行。要求涂刷均匀、整洁、不虚不花、不掉色。

（7）接天地　苏式彩画的白活（用白色做衬底的绘画内容称为白活），如线法山水、洋抹自然过渡的画面底色。一般应将浅蓝色涂于上方，谓之“接天”，下方谓之“接地”。这是画白活之前的一项重要工作，它的主要作用是创造出置身于自然天地间的画面效果。

（8）包黄胶　将画面中要贴金的部位涂上黄颜色或黄色油。这种黄色起着标示贴金范围和衬托其上的金胶油不被地仗吸收的作用。包黄胶要求涂刷的范围准确、齐整，不能有多出和落掉的地方。

（9）拉晕色、拉大粉　晕色是表现彩画色彩层次的一种手段，它通过色阶的过渡，达到由青至白、由绿至白或由其他颜色（如紫、红、黑等）至白的晕染效果，使颜色间过渡自然柔和。其施工程序应是先拉晕色后拉大粉。拉晕色是用捻子（彩画中专用的一种刷色工具）沿大线的轮廓画出（要求画浅于大色的二色、三色）。拉大粉即画最浅的一道白色。晕色的色度要准确，色阶要匀，无论晕色或大粉，都应直顺、均匀、不虚不花、整齐美观。

（10）画白活　白活包含彩画中各种绘画内容，如翎毛、花卉、山水、人物等。白活多画在包袱、枋心、聚饰、池子内及廊心等处，有“硬抹实开”“落墨搭色”“洋抹”“拆垛”等各种不同制度和做法，须严格按这些做法进行，才能达到各自的制度要求和艺术水准。

（11）攒退活　攒退活包含两个内容，其一为“攒活”，泛指一般的工细图案的装色。其中运用同一色相但分为不同色度的颜色，须分层次施色，使图案装点成有层次感的晕染效果。其二为“退活”，一般特指退烟云，即包袱边框、方心岔口等处，用同一颜色由浅至深分道摹画，以便产生强烈的立体效果。无论攒活还是退活，其色度应用都应准确，色阶层次自然分明，无骤深骤浅，不虚不花，洁净美观。

（12）刷老箍头、拉黑掏、压黑老　这三项都是用黑颜色完成的工序。“刷老箍头”是用黑色刷构件最端头的部分；“拉黑掏”是用黑色拉饰两个构件相交的秧角部分，如檩与垫板、檩与随檩枋的相交处，还有某些金线老的外圈等部位，可起到齐色或齐金的作用。“压黑老”用于彩画的某些特殊部位，如斗拱、角梁、霸王拳等处。这项工艺，起着对彩画某些部位的强调、突出、衬托和齐界的作用。

（13）打点活　这是彩画的最后一道工序，即用颜色对已完成的彩画部位进行全面的检查、修饰、校正，使之达到尽善尽美的程度。

古建彩画是不同于其他传统绘画艺术的一种艺术形式。它专门用于装饰建筑，为我们的生活环境增添了迷人的色彩，其高雅的艺术形式和丰厚的文化底蕴，值得认真继承和弘扬。

第三节　四合院的保护与修缮

修缮工作，是对四合院进行有效保护的重要手段。任何建筑物，在经历过一定的年代之后，都需要进行修缮保养，这样才能使建筑物完好无损。这对于以木结构为主体的四合院建筑来说更是如此。如今保留下来的四合院建筑，大多数建于清代，有少数建于明代。这些建筑能较好地保留下来，都因经历过多次不同规模的修缮。可见，搞好修缮工作对四合院的保护至关重要。四合院的保护修缮，主要应做好以下几项工作。

一、进行认真勘察，确定修缮方案

勘察是对四合院建筑的损坏情况进行详细的、普遍的检查。勘察工作是一项非常细致的工作，要有专业工程技术人员和木作、瓦作、石作、油漆、彩画等各工种中有经验的工匠共同参与。

勘察工作的着眼点，首先应当是建筑物主要结构的状况，如整体构架是否歪闪，柱子是否糟朽，屋顶是否漏雨，主要承重梁架檩木是否劈裂、下垂、折断，檐头是否下垂、漏雨，基础是否下沉，墙身是否歪闪、裂缝等。对于油漆彩画状况等，也要予以关注。

勘察工作应以院为单位，逐幢逐间地按序进行，发现问题随时记录下来，必要时还需要进行拍照、测绘，留下详尽的资料。

在经过详细勘察掌握了建筑物损坏情况之后，要进一步分析，找出损坏的主要原因，有针对性地制定出行之有效的修缮方案。

四合院建筑的保护修缮，主要有以下措施和手段。

（1）木作方面　有墩接柱子、抽换柱子或辅柱、更换椽子望板、揭宼檐头、打牮拨正、归安、拆安、落地重建。

（2）瓦石作方面　有墙体修缮，包括拆砌或摘砌山墙、槛墙、挖补酥碱砖块；有屋面修缮，包括揭宼檐头、抽换底瓦、更换盖瓦、捉节夹垄、局部挑顶、局部添配瓦件；有石活修缮，包括石活归安、石活添配、表面见新、剔凿控补、补抹粘接、灌浆加固等。

（3）油漆彩画方面　有重做油漆彩画、下架油饰见新、局部地仗找补等。

二、主要的修缮手段和技术措施

1. 墩接柱根

木构建筑的柱根常受风雨侵蚀，最易糟朽，尤其是掩砌在墙体内的柱子，更容易受潮腐烂。因此，解决柱根糟朽问题便成为四合院建筑的主要修缮内容之一。柱根糟朽程度较轻的可以采取包镶的办法，除去朽木后，在表面包上新木，并用铁箍缠箍结实。如果柱根糟朽严重（糟朽面积占截面面积1/2以上，或有柱心糟朽现象）时，则应采用墩接的方法，这种方法是将糟朽部分截掉，换上新料，新旧料搭接长度一般应为柱径的1～1.5倍，并在端头作榫。为确保坚固，还应在外面打上铁箍。柱子的墩接高度，如为露明柱，应不超过柱高的1/5,如是砌在墙内的柱子，应不超过1/3。

2. 抽换柱子及辅柱

如果木柱严重腐朽、高位腐朽或折断，不能采用墩接的方法进行修缮时，可以采用抽换柱子或辅柱的方法来解决。

抽换柱子即人们通常所说的“偷梁换柱”，是在不拆除与柱子相关的其他构件的前提下，用牮杆和千斤顶将梁枋支顶起来，撤下朽柱，换上新柱。抽换柱子是有条件的，并不是所有木柱都能抽换，如中柱、山柱就不能抽换。如果这种柱子发生高位腐朽或折断时，可采取辅柱的方法解决。辅柱即在柱的两侧（或三面、或四面）辅以枋木构件，并用铁箍将辅柱与原柱箍牢，使之形成整体，如同骨伤打夹板那样。

3. 辅柁、辅檩

柁（又称梁）和檩是四合院建筑的水平承重构件，长期受重力，容易下垂弯曲，如再遇屋面漏雨等不利因素作用，还会糟朽折断。如一幢建筑物中大部构架尚完好，仅有个别柁檩出现以上问题时，可以采用在柁或檩下辅上一根木件的方法将其托住，以免发生危险。辅柁可于两端加抱柱支撑，辅檩可加斜撑支于梁上。辅件要与原件用铁件连接为一体，以防外力作用时脱落。

4. 揭宼檐头

传统建筑檐头最易漏雨，即使问题出在屋脊或屋面中部，渗水也会汇集于檐头部分，导致望板、椽子、连檐糟朽。揭宼檐头是将檐头部分的瓦面、灰背、泥背等拆除，将糟朽的望板、飞椽、连檐等拆掉，换上新的飞椽、望板、连檐并重新苫背宼瓦。揭宼檐头的瓦、木工程除一般技术要求外，还应特别注意新旧屋面接茬的处理。要用水将旧茬处洇湿洇透，与新苫灰泥背形成整体。瓦面接茬也要整齐一致。

5. 打牮拨正

打牮拨正是在木构架并无严重损坏腐朽，只是整体歪闪严重时采取的一种修缮措施。它的主要程序和技术措施是：①先在歪闪严重的建筑支保上戗杆，以防其在修缮过程中进一步歪闪倾圮；②揭去瓦面，铲除泥背、灰背，拆除山墙、檐墙及其他支顶物；③将木构架榫卯处的涨眼料（即木楔子）去掉，将加固的铁件松开；④在柱子四面弹上中线、升线，如旧线清晰可辨，也可利用旧线；⑤向构架歪闪的反向支顶戗杆，并吊直拨正使歪闪的构架归正；⑥稳住戗杆，在节点处重新掩上卡口，塞住涨眼，加固铁件，并砌山墙、槛墙、后檐墙，苫背宽瓦。待全部完成后撤掉戗杆，打牮拨正工作即告完成。

打牮拨正的工作往往是与拆砌山墙、槛墙、檐墙、重新宽瓦等工作结合在一起进行的。这些工作的内容与程序均与房屋新建时的工序相同，只不过多用旧料，故此不复赘述。

6. 大木归安、拆安

归安是针对大木构件严重拔榫而采取的一种修缮措施。它是采取支撑、打缥紧固的方法使拔榫的构件重新归位。归安不必拆散构架。

拆安是将损坏的构件拆下，或将全部木构件落地，对损坏严重、不能再用者进行更换，对损坏轻微者进行整修后重新安装。拆安时要在构件上编号，以保证按原位组装。编号的方法可以采取大木位置号的标写方法，如大木构件原有编号清晰，也可沿用原有编号。大木拆安又称落架大修，是修缮工程中规模最大的一种，木、瓦、油漆彩画各作都要涉及，石活修缮也可同时进行。

三、经常性维修是保护四合院的有效措施

经常性维修即小型维修，是一种将隐患消灭在萌芽中的修缮方法。四合院建筑的损坏，主要源于基础和屋面。地基返潮容易造成柱根糟朽、墙身酥碱，屋面漏雨则会导致椽望糟朽、大木腐烂乃至墙体潮湿鼓闪。因此，及时发现和排除基础和屋面出现的返潮、渗漏等现象，是保护四合院建筑使之延年益寿的有效方法。

造成地基返潮除先天的原因之外，主要有院内积水不能及时排出、下水堵塞造成污水漫延、柱根墙脚长期不能通风等。这就需要及时发现并排除这些隐患，疏通下水系统，排出淤积的雨水及使阴暗潮湿的墙脚通风见干。如果发现因潮湿造成柱子糟朽，应及时剔去糟朽部分，补上新木并涂刷防腐剂。对墙体上的透风等设施也应加以保护，以发挥其效用。墙体，尤其是下碱部分如发生砖块酥碱应及时进行剔凿挖补，以防其继续损坏。其他诸如及时归回移位的台明、阶条，及时勾抹打点阶条石的缝隙，进行灌浆处理，都是防止雨水渗入台基的有效措施。

屋面漏雨的原因主要有底盖瓦碎裂、松动、残破，夹腮灰剥落，局部灰背损坏，脊根檐头渗漏，屋面长草、长树，瓦垄内积存杂物，屋面局部坑洼造成流水不畅等。解决的方法是及时拔除屋面滋生的杂草，保持瓦垄内的清洁及走水通畅，保证底盖瓦完好无损，捉节灰、夹垄灰剥落及时打点，及时查补雨漏部位等。只要发现小问题及时解决，就可以防止出现严重损坏。

此外，地仗损坏及时找补修整，每隔十数年就重刷一次油漆，都是确保木构件完好的有效措施。以木结构为骨架的四合院建筑，只要保证木架完好，整座建筑就能健康延年，长时间地发挥它的作用，不断实现历史的、文物的、艺术的和实用的价值。

第九章
北京老城保护与四合院的恢复性修建

《北京城市总体规划（2016—2035）》确定了首都的功能定位，即：政治中心、文化中心、国际交往中心和科技创新中心。“第四章　加强历史文化名城保护，强化首都风范、古都风韵、时代风貌的城市特色”当中，特别强调要“加强老城整体保护”，要“保护北京特有的胡同、四合院”，“将核心区内具有历史价值的地区规划纳入历史文化街区名单，通过腾退，恢复性修建，做到应保尽保，最大限度留存有价值的历史信息……强化文化展示与继承……重新唤起对老北京的文化记忆，保持历史文化街区的生活延续性”。其中关于对老城区及胡同、四合院进行“恢复性修建”的要求至关重要，这个提法准确、具体，具有很强的指导性。

究竟怎样做才是“恢复性修建”？“恢复性修建”以恢复到哪个历史时期为宜？根据目前老北京胡同、四合院的现状，需要采取哪些措施，走出哪些误区，防止哪些倾向才能保证“恢复性修建”的顺利实施？能否准确回答这些问题，关系到老城区保护和胡同四合院“恢复性修建”的成败，至关重要。

一、“恢复性修建”以恢复到哪个历史时期为宜

北京老城区历史文化街区的“恢复性修建”以恢复到哪个历史阶段为宜？对这个问题，有不同的主张。一种倾向性的意见是，主张按清乾隆时期的京城图作为依据来恢复。诚然“乾隆盛世”是北京城最为兴盛的时期，但毕竟已经过去近280年。280年中，北京城经历了巨大变化，除去过去的九坛八庙（天坛、地坛、朝日坛、夕月坛、太岁坛、先农坛、社稷坛、先蚕坛、祈谷坛诸坛以及太庙、奉先殿、传心殿、寿皇殿、雍和宫、堂子、历代帝王庙、孔庙等）以及现存比较完整的王府还基本（或大部）保留着原始形态，还可以按照乾隆图进行修复之外，其余的寺庙、府邸、宅院、街巷等已经发生了很大的改变。这些变化了的部分，除文物保护单位外，按照乾隆京城图去进行恢复已不可能。因此，笔者主张，可以参照民国末期或新中国成立初期老北京的风貌进行修复。那时北京还没有开始大规模改造整饬，还比较完整地保留着清代的街巷、府邸、寺庙、宅院，且有较丰富的影像资料可以供参考。清末至民国时期建造的近代建筑，尤其是那些具有西洋风格的建筑，已经成为历史建筑的重要组成部分，成为老城保护的对象。所以按照1949年前后北京的街巷、胡同、院落遗存作为风貌依据进行恢复性修建是比较现实的。

二、恢复性修建要适当引入文物保护修缮的理念和做法

《中华人民共和国文物保护法》规定，文物古建筑保护原则是“不改变文物原状”，或者叫保持文物建筑的“真实性”“原真性”。具体讲，就是在修缮时要保持文物建筑的“原形制、原结构、原材料、原工艺”，不能有任何改变。

老城历史文化保护区主要是保护历史建筑的传统风貌，其中除去文物保护单位外，其余大部分都是民居。所以对这些历史建筑的保护，主要是保护它的外表。可以具体化为五个立面，即前、后、左、右、屋顶，另外再加上庭院（环境），也就是老建筑的表皮。对它的外表要做到“原形制、原材料、原工艺”（如果需要，可以改变结构，但外形不能变）。

室内不影响风貌的地方可以改造，添加现代设施、设备，提升居民生活水平，使住在平房区的居民过上与楼房区居民同等的生活。

这就是所谓“老胡同，现代生活”。

这是北京老城风貌保护应遵循的理念和做法，也是“恢复性修建”应当遵循的原则。

三、恢复性修建要做的具体工作

根据当前老城区胡同四合院的现状，恢复性修建应当抓好以下主要环节。

1. 恢复胡同内的传统建筑要素及环境

展现在胡同里的传统建筑要素主要有屋面、墙体及檐口、各种不同形态的宅门、沿街的窗，以及附属于这些要素的上马石、拴马桩（环）、抱鼓石（门墩）以及宅门墀头及门楣上面的砖雕及其他装饰。除这些建筑要素外，还有胡同里的绿化。

（1）*关于屋面*　胡同里能看到的屋面主要有两种，一种是筒瓦屋面，一种是合瓦屋面。其他如棋盘心屋面、干碴瓦屋面、仰瓦灰梗屋面、灰平台屋面等，可以看做是合瓦屋面的衍生品种。

筒瓦屋面，在清代是只有寺庙或王府才能采用的屋面，与筒瓦屋面配套的还有吻兽、跑兽以及大式做法的正脊、垂脊等。平民百姓和过去一般的官员都是不能用筒瓦屋面的。这些规定作为一种制度、一种文化，一直沿袭下来。这是我们在“恢复性修建”时应当特别注意的。

但是，近二三十年来，一些先富起来的人在建新四合院时往往选用筒瓦而很少用合瓦。这是一种不合规制的做法，对“恢复性修建”有负面作用。今后，规划部门应严格把关，凡不是历史上留存至今的王府、公主府或寺庙，一律不得采用筒瓦屋面。对已经采用筒瓦屋面和吻兽、大脊的新建四合院民居或其他民用建筑，应当在本次“恢复性修建”中进行整改。（图9—1）

合瓦屋面，在老城区胡同四合院中占绝大多数，这是传统四合院的主要屋面做法。在过

图 9—1　民居滥用筒瓦屋面和脊饰吻兽的例子

去，百姓在翻建房屋时，由于财力不足或其他原因，在合瓦屋面的基础上又衍生出多种不同的屋面做法，如棋盘心屋面、干碴瓦屋面、仰瓦灰梗屋面、灰平台屋面等，使胡同民居屋面做法呈现出多样性。这种屋面做法的多样性，是北京四合院在历史发展中自然形成的，是四合院传统风貌的重要组成部分，应当予以保留。但是近几十年来，经房管部门修缮、改造之后，棋盘心、干碴瓦、仰瓦灰梗以及灰平台几乎见不到了，统统改成了合瓦屋面。在此次“恢复性修建”中，在有条件和征得居民同意的前提下，应适当恢复一些棋盘心、仰瓦灰梗等屋面的做法，以保持风貌的真实性。（图9—2）

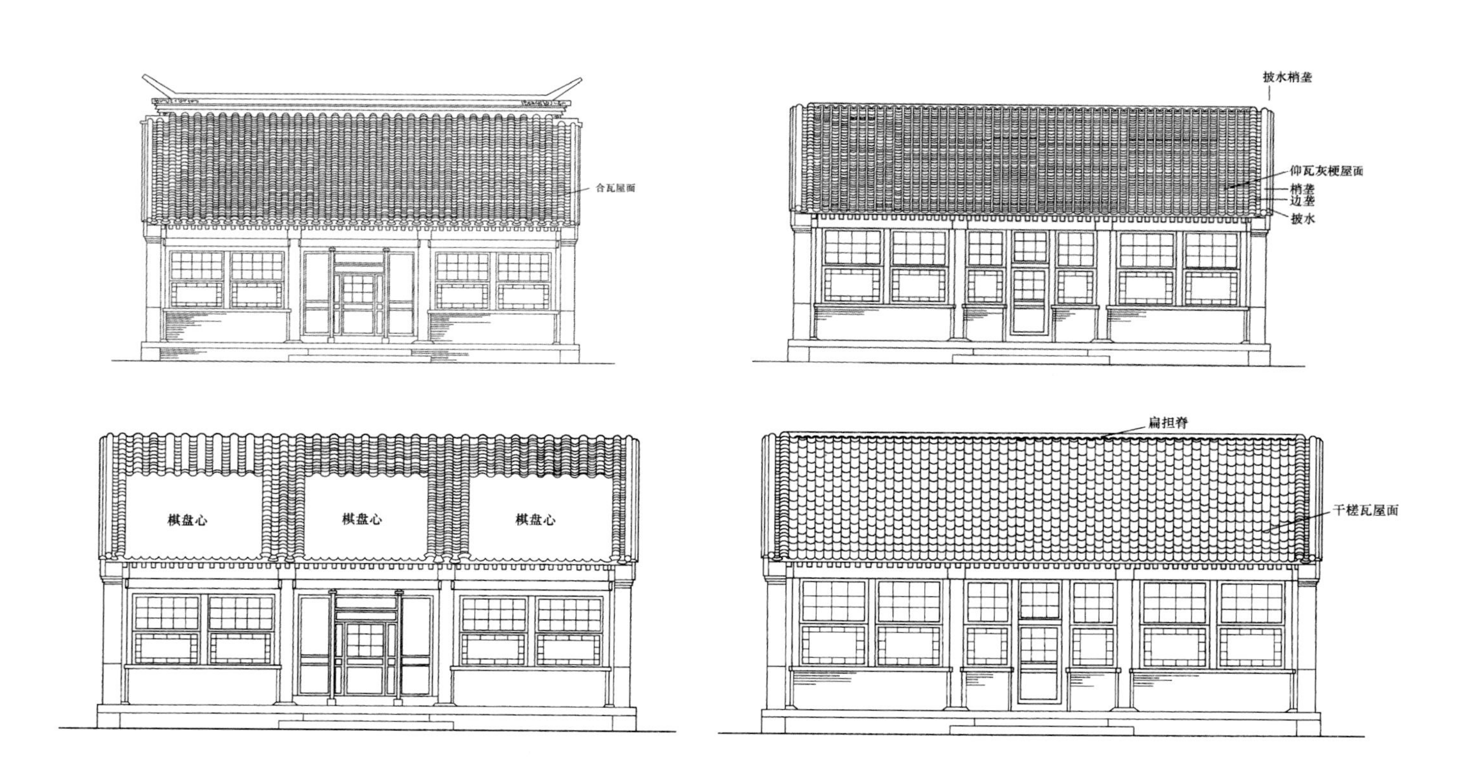

图 9—2 合瓦屋面、棋盘心屋面、仰瓦灰梗屋面等

（2）关于墙体 胡同里见到的墙体，主要是四合院的后檐墙或山墙。这些墙体的做法各有不同，有的差别很大。这主要是由于宅主人的社会地位、经济条件等差别造成的。

过去，居住在同一条胡同里的，有官员、富商，有文人、雅士，也有一般的平民百姓。人们的社会地位不同，经济实力不同，价值追求不同，所建的宅子也必然会有很多差别。经济实力雄厚的人家，建造的房子豪华、讲究；读书人清高，建宅子追求清新脱俗；普通百姓人家，经济实力不强，精神上亦无过高追求，建房时用工、用料标准也一般；而生活较为拮据的人家，甚至会采用简陋粗糙的材料和工艺，为的是省工省料，减少花费。这些差别，造成胡同里不同宅院的墙体存在多种不同的做法。如：有的墙身通体干摆丝缝，檐口线脚式样讲究；有的墙体四角干摆，内心采用淌白砌法，稍微节省成本仍效果不差；有的四角采用淌白砌法，墙心则采用碎砖砌筑，外表抹灰；有的采用糙砌做法，简陋实用。这种自然形成的不同等级做法，不仅呈现出胡同风貌的多姿多彩，而且形成了丰富的胡同建筑文化。这些都是“恢复性修建”时应当格外注意的。（图9—3）

图9—3．1 丝缝墙

图9—3．2 外干摆内淌白墙

图9—3．3 干摆墙

(3) 关于宅门　北京胡同四合院的宅门是最为讲究的部分。四合院的宅门主要有广亮门、金柱门、蛮子门、如意门、西洋门、随墙门（小门楼）等数种。（图9—4）

其中，广亮门和金柱门在过去是官员居所的宅门，大门外面有较大的空间，檐柱与檐枋交角处安装有雀替。整樘大门为木质，门口较宽，用四支门簪，门两侧有余塞，体量较

图 9—4. 1　金柱门

图 9—4. 2　广亮门

图 9—4. 3　如意门

图 9—4. 4　蛮子门

图 9—4．5 随墙门

图 9—4．6 西洋门

大，气派显赫。蛮子门系由官宅演变而来。从蛮子门的体量构造看，应为由广亮门或金柱门改造而成：将广亮门或金柱门的门扉开启位置移至外檐柱间，便成了蛮子门。据说清末及民国时期福建、广东一带北上经商的有钱人买下官宅后，为防止贼人夜间隐藏在门洞里伺机偷盗，便将门扉移至前檐柱间，去除了贼人藏身的空间。于是这类大门便有了“蛮子门”之称。由此可见，蛮子门应属民宅的宅门。如意门、西洋门、随墙门等，则是民宅常采用的宅门形式。

在封建社会，对门第的规定是非常严格的，不仅宅门的型制不同，而且对门上附属的构件、门的油饰色彩也有严格规定。《大明会典》卷六十二“房屋器用等第”记载：“洪武二十六年定：官员盖造房屋，并不许……绘画藻井。公侯……梁枋斗拱檐角用彩色绘饰……门用金漆。窗枋柱用金漆或黑油饰。一品、二品……梁栋斗拱檐角用青碧绘饰……门用绿油……三品至五品……梁栋檐角用青碧绘饰……门用黑油……六品至九品……梁栋止用土黄刷饰……黑门……庶民所居房舍……不许用斗拱及彩色妆饰。”

《大清会典》规定：“亲王府正门、殿寝凡有正屋正楼门柱均红青油饰，梁栋贴金，绘五爪金龙及各色花草。凡房屋楼层均丹楹朱户，其府库廪厨及祗侯各执事房屋，随宜建置，门柱黑油。”“公侯以下官民房屋梁栋许画五彩杂花，柱用素油，门用黑饰。官员住屋，中梁贴金，余不得擅用。”

根据以上规定，清代公侯可用旋子彩画，红绿油饰，普通官民住宅可做中低等苏画，素油或黑油，一般做黑红净。民国时期民宅一般用掐箍头搭包袱的简易苏画做法，用素油或黑油，多做黑红净。

在中国传统五行中，黑代表水。中国人把水看做是财富，所谓“肥水不外流”“四水归堂”，即是这种观念。同时，黑（水）还有尅火的含义，所以一般官民都乐于用黑色油饰。黑色的运用，从秦汉一直传承到清代。清灭亡之后，西方文化传入中国，传统文化受到冲击，有

些规定或讲究在人们心目中逐渐淡化或失传。在当下的胡同四合院恢复性修建中，这些文化传统应当得到恢复。（图9—5）

图 9—5　黑红净油饰（本次修建中恢复的民宅宅门）

（4）关于上马石、拴马桩（环）、石敢当、抱鼓石等附属物或建筑小品　上马石、拴马桩（环）、石敢当（泰山石）、抱鼓石等四合院宅门外或墙面上的这些附属物件或小品，是传统胡同四合院中不可缺少的文化元素。它承载着过去老北京人的民俗民风和某些生活内容，我们可以从中读出很多故事和历史。保存好这些构件，可以使老北京文化内容更丰富。（图9—6）

图 9—6．1　上马石

图 9—6．2　拴马环

图9—6．3 石敢当一

图9—6．4 石敢当二

图9—6．5 抱鼓石

（5）关于门楣及墀头砖雕 四合院宅门的门楣砖雕（用在如意门上）以及大门两侧墀头上端的砖雕花饰是展示宅门文化的重要内容，这些砖雕各有其吉祥寓意和主题，如：岁寒三友（松竹梅）、喜鹊登梅、居家欢乐、富贵白头、连升三级、太平有象、玉堂富贵等。不同的房主人有不同的追求，会在门楣上选择不同的雕刻题材，从而展示出丰富多彩的形式和内容。这些雕刻，都是非常精致的，许多都是民间艺人花费毕生心血的经典之作，非常珍贵。（图9—7）

图9—7．2 如意门门楣砖雕

图9—7．1 墀头砖雕

与历史遗留下来的这些经典艺术品形成强烈对比的，是近些年在进行四合院修缮或复建时加上去的一些十分拙劣丑陋的所谓“砖雕”或“石雕”。在对精美的砖石雕刻进行保护的同时，必须将这些丑陋拙劣的东西予以清除，替换上能与那些精美的砖石雕刻相匹配的作品（图9—8）。

图 9—8 丑陋“砖雕”与精美砖雕对比

(6) 关于胡同绿化 四合院的绿化主要在庭院当中，胡同里的绿化相对简单，主要是在胡同一侧（一般是北侧）种一排树木，树种以槐树为多。槐树是北京地区的主要树种之一，树形优美，树冠覆盖面大，遮阴效果好，在入夏时节槐花盛开时散发出浓郁的香气，沁人心脾。伏天知了（蝉）的鸣叫，更营造出一种特有的环境气氛，深受老北京人喜欢。过去北京胡同里种植花卉的情况较为少见，主要是因为胡同尺度较窄，主要功能是交通，不能随意占用。偶尔有一些小面积空地，也可种植一些花草，如月季、茉莉、指甲草、西番莲、大丽花、夹竹桃等一些为老北京人喜欢的草花或盆栽，营造出浓郁的生活气息。随着时间的推移和居民的改变，这些花草有些现在已经见不到了。要恢复胡同的传统元素，绿化形式和内容也是非常重要的。

2. 拆除杂院内的违章建筑，恢复庭院环境，提升居民生活品质

四合院，顾名思义，四面是房，中间是院。院落是四合院的中心，是人们活动和交流的场所。没有院落就不能称之为四合院。

在清代及以前，四合院是以一家一户居住的形式存在的，叫“独门独院”。一个院子只能由一家人居住，这一家可以是两代人、三代人，甚至是四代人、五代人（叫四世同堂、五世同堂），但必须是一家人，不能是两家、三家，更不能多家共同居住。这种独门独院的居住方式与现在的单元楼房是一个概念。一个单元只能住一家人，这一家人可以是两代人、三代人，但不应该是互相之间没有血缘关系的两家人或多家人。

四合院由一家人居住的独门独院逐步演变成多家共同居住的杂院（现在叫“共生院”，即多户人家共同生活之意），这种变化始于清朝灭亡之后，是社会变动的结果。清朝灭亡之后，原来的王公贵族断了俸禄，没了生活来源。这些长期“吃皇粮”的贵族缺乏基本生存技能，只能靠“吃家底”维持生活。开始是凭借变卖文玩字画、器皿家具度日，可动产卖光了，就开始变卖或出租房屋。这样，本来的独门院落，便迁进了其他房客。独门独院开始变为多家居住的杂院。这种现象随着社会进一步动荡——外敌入侵、军阀混战、抗日战争、连年内战而愈

演愈烈。

1949年新中国成立，所有制变更又使房屋产权关系发生了变化，公房分配、私房转租等现象使更多的四合院变为杂院。在民国时期和新中国成立初期，尽管存在很多“杂院”，但在院内私搭乱建的现象没有出现。那时的杂院，各家都住着自己固有的房子，对院内的公共区域，或植树木，或种花草，大家都不去侵占。小孩子在院里玩耍，成年人在院里乘凉聊天，邻里关系大多数融洽和谐，庭院环境也不错。

在院里违章私搭乱建，始于上世纪60年代，主要原因是院内居民的孩子逐渐长大了，需要分开房间；年轻人结婚，没有房子，只能在院里挤占空间搭建临时房屋。这样就出现了在原有房子外面搭建小厨房或婚房的“贫嘴张大民”现象。这些现象，是从个别住户开始的，但住房拥挤问题是普遍的。由于这种现象的出现是源于老百姓住房的实际困难，政府不好干预，采取了放任政策，于是又出现了仿效效应。其他住户看个别人私建房子政府不管，也纷纷仿效，院落被蚕食的情况就出现了。尤其到了“文化大革命”无政府主义泛滥时，这种现象更加普遍。

杂院里公共空间被挤占，还有一个重要原因就是地震。1965 年发生了邢台地震，1976 年发生了唐山地震，都波及北京。唐山地震对北京影响很大。老百姓为抗震，纷纷在院子里搭建“抗震棚”；地震过后，有些“抗震棚”不拆除，四合院空间进一步被挤占。随着大杂院居住环境逐步变差，一些有条件的原住居民逐步搬离了大杂院住进了楼房。腾出的旧房子进行转租。由于旧平房生活设施不齐全，水暖电设施不配套，私搭乱建的违建设施更差，因此租金相对便宜。外地来京打工的不少人都租住这种简陋房屋。四合院走了老住户，住进了外来户，杂乱、拥挤状况更加恶化，居民成分也发生了相当大的变化，本来是很有文化内涵和京城特色的四合院，现在已经面目全非，成了环境极其杂乱的“贫民窟”。（图9—9）

图 9—9 杂院

北京四合院从由一家一户居住的独门独院，变成由多家居住的杂院（“共生院”），经历了近百年时间。实践证明，只要加强管理，居者自觉遵守公德，彼此和谐共处，这种多户同住一院的居住形式还是能为大家所接受的，传统四合院的居住文化大多还能得以保留。

在院里私搭乱建，使杂院逐渐变得杂乱无章，大约经历了五六十年时间。这几十年间，中国最有文化品味、最有特色的四合院变成了环境极其恶劣、极其脏乱差的“贫民窟”，这是历史的严重倒退，与日益强大富裕的中国的身份，与北京作为历史文化名城和现代世界大都市的身份极不相符。这种状况绝不能再继续下去，必须在尽可能短的时间内彻底改变，使北京传统民居四合院恢复到原有状态。要做到这一点，就要按《北京城市总体规划（2016—2035）》的要求，通过对非首都功能的疏解和腾退，拆除违章建筑，对四合院进行恢复性修建。

拆除违章建筑是一项十分艰难的任务，它涉及居民的切身利益，需要政府投入大量资金，花费大量精力，做深入细致的工作，才能达到预期目的。四合院之所以有魅力，是因为它有一个环境优美的院落。把院内私搭乱建的破棚屋拆除，恢复庭院环境，是提升居民生活品质的重要步骤。拆除违章建筑的工作完成以后，必须要对四合院原有建筑进行修缮。

四合院主要是木构建筑，建筑年限至少已有百八十年，有的已经近二百年，木构件（主要是墙内的柱子和屋顶的椽子、望板）糟朽是普遍问题，有相当一部分建筑已经是危房，进行大规模修缮（即对老房子落地翻建）是此次修建的主要方式。

通过恢复性修建，恢复老北京的传统风貌，必须在修建中适当引入文物修缮的理念和做法，尤其是对建筑外表的屋面、墙体、门窗、色彩，都要按原型制、原材料、原风格进行修缮，不能随意更改。

北京的老旧房子，有相当一部分在过去的几十年中陆续进行过翻改建。当时的翻改建，是以解决安全问题为主要目的，并没有把风貌保护问题放在重要地位。前些年翻改建的房子，不管原来墙体是干摆丝缝做法，还是四角硬软心做法，经翻建后一律改为兰机砖墙，虽然仍然保持了灰色，但是材料做法变了，质感变了，外观古朴精细的老墙不见了，古建筑的感觉不在了，对传统风貌而言无疑是一个重大损失。（图9—10）

图 9—10　兰机砖墙改变了古建筑的风貌

屋面也是如此。前些年对老旧房进行的翻改建，虽然保持了青砖灰瓦的色彩，但瓦屋面的做法却由多种形式变成了合瓦屋面一种形式，再也难以找到棋盘心屋面、仰瓦灰梗屋面等多种屋面形式。

门窗同样存在类似问题。经翻改建以后的房子，仍保留传统的步步锦、龟背锦、灯笼锦棂条形式的窗格已经少之又少，大部分改成了现代塑钢窗。

以上这些做法都不同程度“蚕食”了老北京的传统风貌，这也正是本次恢复性修建应引为教训的。

四合院在这次的大修、翻建中如何严格遵循“恢复性修建”原则，认真做好风貌保护，是个重要课题。前面，我们已经就此次恢复性修建中，如何保持胡同风貌问题作了详尽的阐述。对于庭院的修建，与修建胡同的原则和方法是相同的。有人认为，胡同保持风貌就行了，院子里可以放松一些。这是不对的。《北京城市总体规划（2016—2035）》要求保护老城的传统风貌，要“最大限度地留存有价值的历史信息……强化文化展示……重新唤起对老北京的文化记忆，保护历史文化街区的历史延续”。这就必须要在建筑五个立面的形式、特点、做法、色彩的恢复上面做足文章，并恢复庭院种植和绿化。至于室内装修、设施这些不影响风貌的部分，则应尽可能改进，尽可能现代化，尽最大努力予以提升和改善。要让住在胡同四合院平房区的居民，过上与住在楼房里的居民一样的现代生活。

3. 恢复性修建要解决“低洼院”问题

在对北京老城区胡同四合院进行“恢复性修建”的过程中，还有一个问题是必须要解决的，这就是关于“低洼院”问题。

低洼院的形成，主要是胡同（包括主干道）不断修路造成的，每修一次即垫高一次，久而久之，路面就高过了院内地坪。越是年代久远的院子，院内地面低于胡同的情况越严重。现在，北京胡同里的低洼院，院落地坪比胡同地面低30 cm 以上的不在少数，有的甚至低70～80 cm，居民进院如同跳坑。

院落低洼，不仅造成雨季院内积水，严重时积水还会漫到居民室内。由于胡同地面升高，雨水也会直接侵蚀沿街房屋的后墙，造成墙体和室内潮湿，使掩埋在墙内的木柱糟朽加剧，给建筑的安全带来更大威胁。因此，在恢复性修建时，低洼院是必须要解决的问题。（图9—11）

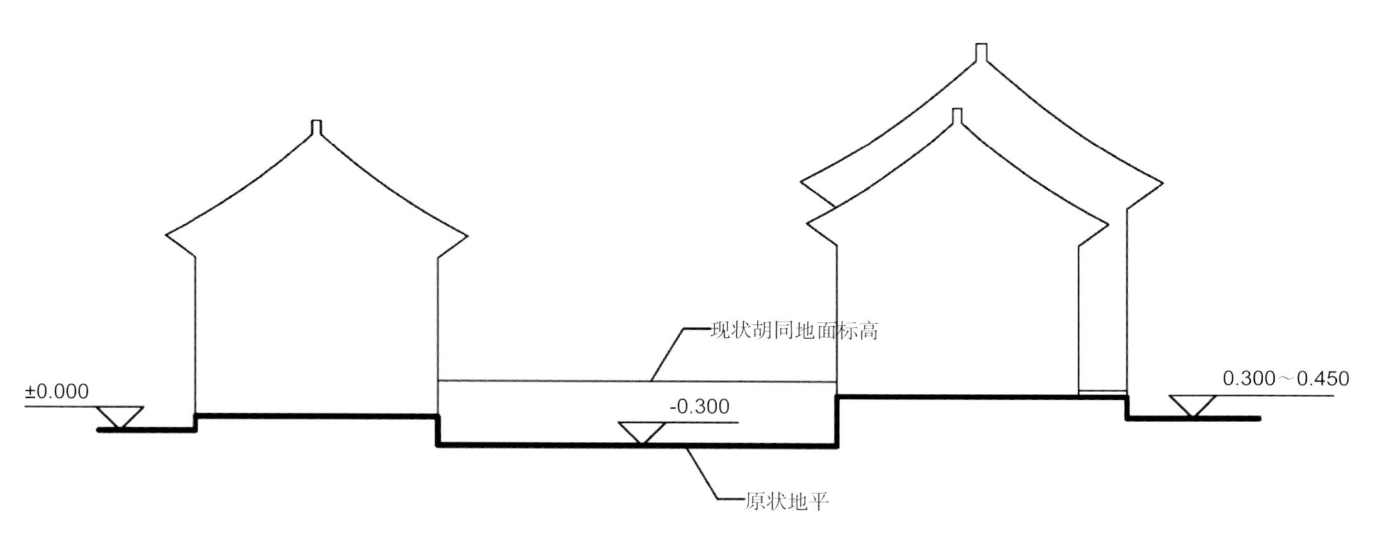

图 9—11 低洼院情况示意

解决低洼院问题，无非是两个办法：一是升院子，二是降路面。究竟哪个办法更便于实施，应当由相关部门（如房管部门、市政部门、规划部门等）进行充分论证，找到一个妥善的、相对简单的办法，并制定出相应的政策。

四、恢复性修建要走出误区，纠正乱象

通过这两年胡同整治和“恢复性修建”的实践，笔者发现一些认识误区，需要认真纠正。

误区之一：墙面及沿街整饰追求整齐划一、高大上

当前，参加胡同、四合院整饰修建的大部分设计人员对北京胡同四合院的认识和理解局限在青砖灰瓦、干摆丝缝这些简单概念上，缺乏对老北京建筑材料做法多样性的了解和认识。甚至有人把南方民居的粉墙黛瓦与北京四合院的青砖灰瓦相混淆。他们在整饰设计中，往往把整条胡同的瓦面、檐口一律做成黑色，把墙面一律按丝缝墙的样子贴上砖片（图9—12），把沿街门窗一律油成红色（图9—13.1）。这显然是错误的，必须加以纠正。

图 9—12　被错误刷成黑色的博缝、檐口、屋面和墙面贴砖

误区之二：滥用“中国红”、汉白玉、龙凤图案及墙面装饰

近年来，在对胡同、四合院整饰修建中，普遍存在着滥用“中国红”、汉白玉、龙凤图案及墙面装饰等现象。

先谈颜色，在中国传统文化和典章制度中，对颜色的运用是非常严格的。在明清两代，朱红色和明黄色是皇家专用的颜色，一般官员、庶民都不得使用。历史上，秦代崇尚黑色，汉代崇尚黑与红的搭配，俗称“黑红净”，这种传统一直延续到近代。可是在那个特殊的年代，“黑”成了反动阶级的代表色，“红”成了革命阶级的代表色。这种将颜色贴上政治标签的做法，颠覆了中国几千年的文化传统。这种影响至今还在一些人的头脑中作祟，以致阻碍了正确的色彩运用。（图9—13）

图 9—13 油漆做法正误对比

图 9—13．1 错误的油漆做法

图 9—13．2 传统的黑红净油漆做法

图 9—13．3 被改造过的油漆做法

滥用汉白玉，也是近些年在修建四合院时出现的一种乱象。汉白玉这种高档石料十分稀缺，过去只在皇家宫殿、坛庙的重要部位（如栏板、望柱）使用。其上施以精美雕刻，颇具观赏性。而皇宫里的阶条、踏跺等易损部位，也不用汉白玉，而用青白石。可见皇家建筑在用料上也是很节制的。寻常百姓住宅用的石构件，一般都是小青石，其颜色灰中透绿，十分雅致，与青砖灰瓦搭配协调，浑然一体，透出一种质朴之美。但是，近些年来，有钱人建四合院，盲目追求高档，阶条石、门墩、腰线石、角柱石均采用汉白玉石料，与青砖灰瓦色调不搭配，显得格外刺眼。（图9—14）

滥施彩绘、滥用龙凤图案等，也是近些年出现的乱象。和玺彩画、龙凤题材，在传统建筑中都是皇家专用的形式和题材，近些年在四合院民居中也时有出现。有些地方还在胡同街巷中滥加墙面雕饰和彩画，内容五花八门，有些粗俗丑陋，难以入目。这种种现象除去反映当事人（包括建设单位、设计方、施工方）不懂规矩，缺乏文化素养之外，更多的是透露出房主人的浅薄与无知。（图9—15）

图 9—14　四合院住宅滥用汉白玉情况举例

图 9—15．1　滥施墙面装饰情况举例

图 9—15. 2　滥施彩画情况举例

误区之三："内衣外穿"，南装北饰，门窗及装饰元素应用的误区

"内衣外穿"、南装北饰，主要是指在胡同四合院整饰中，将用于室内的木装修（古建筑室内用的木装修包括碧纱橱、落地罩、多宝格、圆光罩、八角罩等细木装饰）上面的棂条、纹样搬到室外来用；将多见于江、浙、赣、皖等地区的窗格纹样用于北京胡同四合院。这显然是不妥当的。

北京四合院用于室外和室内的窗格棂条，在截面尺寸、空当大小、纹样选择、内涵讲究上是有很大差别的。一般来说，用于室外的窗格棂条断面较大，看面大约6 分（折合约19 mm），进深大约8 分（折合约25 mm），棂条之间的空当为3～3.5倍棂条宽（称为一空三棂，折合60～70 mm）。室外用的窗（门）有一定防护作用，棂条比较结实，用料一般不太讲究。用于室内的窗格棂条看面大约4 分（折合约13 mm），进深6 分（折合约19 mm），相对纤细，且用料考究（多用楠木、楸木甚至花梨、红木）。室内装修常与硬木家具相互映衬，具有很强的艺术价值和观赏价值。（图9—16）

图 9—16. 1　内檐木装修

图 9—16. 2 外檐木装修

室外窗格棂条重在实用，因此比较粗壮，室内窗格棂条除实用外，主要是观赏，将二者混淆是不妥的。

南方与北方的窗格，在棂条形式、内容上有很大差异。一般来说，南方窗格细密繁缛，北方窗格疏朗大气，这是地方文化差别。这些年北京有些地方将南方建筑及其窗格引入北京。在胡同四合院整饰中也将其到处乱用，其结果是改变了北京胡同四合院的风貌，也是不恰当的。（图9—17）

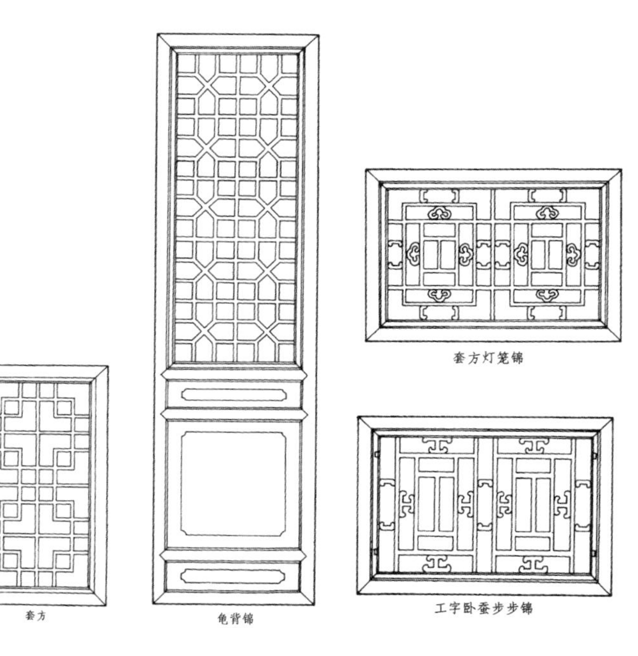

图 9—17.1 北方风格棂条

图 9—17.2 用于北京某沿街店面的南方风格棂条

这里还要顺便提及：近年做的软心墙（墙体四角硬中心抹灰），大多数将抹灰墙面做成白色，将瓦面刷成黑色，这也是不对的。这种“粉墙黛瓦”的做法是南方地区建筑常用的色彩搭配方法，与环境气候和民俗文化有直接关系，是地域建筑的特征。北京地区凡软心墙的抹灰墙面都是月白色（即浅灰色），是与青砖灰瓦相协调的颜色。将“粉墙黛瓦”搬到北京也属于南装北饰，应当加以纠正。（图9—18）

解决上述问题的唯一办法，是要认真学习、了解、熟悉北京胡同四合院的建筑和文化特色，分清哪些是北京本地的，哪些是外地移植进来的，哪些是原有的，哪些是后改的。要对设计方案进行严格审查，施工过程中发现问题应及时纠正。

图9—18 北方青砖灰瓦与南方粉墙黛瓦比较

纠正乱象：开办传统建筑讲座，普及四合院文化与技术知识，是搞好“恢复性修建”的关键

以上种种问题和乱象的出现，是因为从事胡同四合院保护、修建的相关人员，包括主管部门、主管领导、设计人员、施工人员、街道干部，乃至居民百姓不懂得、不了解北京四合院的相关知识，或者对这些知识一知半解。根本原因则在于多年来对传统文化，尤其是传统建筑文化不重视、不学习、不宣传、不普及，而建筑院校缺乏进行传统建筑知识教育则是造成这种结果的直接原因。

为及时纠正这些乱象，必须进行补课。具体办法是开展传统四合院知识的普及和教育。要组织参与北京老城区四合院保护整治的主管部门、主管领导、规划设计人员、施工及管理人员，甚至包括参与审计监察的人员，还包括街区骨干和群众代表，系统学习了解与北京四合院有关的历史、文化、建筑、装饰、艺术、技术等方面知识，通过学习武装头脑，填补知识空白，才能做好工作。

四合院恢复性修建是涉及面很广、知识面很宽的一项长期、细致的工作，没有相应的知识技能做支撑，是注定做不好的。

五、禁止继续在历史文化保护区内“做实验”，禁止对胡同四合院进行“转基因”改造

对老北京胡同四合院进行风貌保护，有一件事必须予以制止，这就是：在历史文化保护区内“搞创新”“做实验”，对胡同四合院进行“转基因”改造。

近些年来，一些不了解北京传统文化和建筑知识的建筑师为了进行“旧城更新，院落改

造”，在历史文化保护区内做起了“实验”，搞起了“创新”。这些缺乏文化根基的“实验”和“创新”，究竟效果如何？对老北京的传统风貌保护起到了什么作用呢？且举几个例子看。

一个例子是，几年前在北京东城区西打磨厂整治提升中，引进了一些国内外“大师”来搞“创新”。其中一处是打磨厂街220号，腾退后由某位外籍建筑师进行“改造提升”。他将四合院原来很完整、很规则的前后檐墙和窗加以“非标准式改造”，把整齐的墙体做成“断壁残垣”的效果，在“残缺”处安上大玻璃，在外面附上一层类似渔网的金属装饰，把好端端的四合院搞得不伦不类。另一处，銮庆胡同37号，由一位国内建筑大师主刀，把一个很规则的庭院用竹钢制品分割成不规则的空间，改变了四合院的空间形态。还有一位西方建筑师，在还未读懂中国建筑的情况下，对四合院过厅的隔扇门滥施“手术”，将灵活实用、开启自如的传统木制隔扇门改为金属转门，以追求中门大开、一眼看到底的视觉效果，不仅忽视了中国人千百年形成的生活习惯，而且对原有建筑造成了破坏。（图9—19）

图 9—19　北京东城区西打磨厂“改造”实例

另一个例子是2017年春季的“白塔寺院落更新”。此项目由国内某著名高校国际班的研究生和某设计机构共同做了一个“学术性建筑实践项目”，试图通过建筑设计“应对中国快速发展的社会变革”，以“改变胡同的刻板印象”，“给传统文化带来新的三维视角”。具体内容

是在院落里建一座“天桥”，使人可以“漫步在天桥上”，“看胡同建筑的兴衰变化”，“寻找日落的最佳视点”。这个实践活动还将四合院空间分割成若干空间，并命名为“四分院”。（图9—20）

图9—20　“白塔寺院落更新”项目

再一个例子是对北京东城区草场某宅的改造，旨在“让原本普通的合院住宅经改造后继续着平凡的使命”，以改善“由于人口极度膨胀和长期私搭乱建”而消失的“基本生活空间”。建筑师“以满足住户基本生活需求为前提，力图在极度拥挤的环境中实现传统四合院建筑的空间格局”。建筑师的用心是好的，但改造的结果是，人们已经很难再找到老北京四合院的文化特点与空间感受（图9—21）。

图 9—21　东城区草场胡同某宅改造

还有一个例子，也是某设计单位为探讨“新中式”建筑对老城区传统建筑的改造提升的效果和作用而做的。建筑师的出发点应该是好的，但可惜的是，他们没搞清楚不同国家、不同民族的居住习惯及生活方式对居住建筑风格的影响，错把日本、韩国等民居常用的落地玻璃、格子推拉门以及枯山水引进了北京，混淆了“北京”与“东京”的差别……（图9—22）

图 9—22　东四雨儿胡同某宅的改造实验

新悦派建筑师总乐于把落地玻璃门或窗搬到老北京四合院，尤其是大杂院的建筑改造中。理由是落地玻璃门窗时尚，代表新潮流，采光好，住用舒适（图9—23）。这似乎是个很站得住脚的理由，也容易被许多人接受。但是，他们忘记了两点：第一，落地玻璃门窗最早是源自于进门席地而坐的国家和民族，降低门窗透明部分的高度是出自他们特殊生活方式的视觉需要；第二，凡装落地玻璃门窗的，大多数是一家人居住的独门独院，对于多户杂居的杂院（共生院）是不适宜的，因为它缺乏最起码的私密性。

还可以举出一些例子，但以上几例已很有代表性。

图 9—23 北京大院胡同 28 号四合院改造

通过上述案例可以反映出两种不同情况。

一是有些人对我们祖先传承了几千年的建筑文化和历史建筑缺少起码的尊重和敬畏之心。他们不认为历史建筑是古人劳动和智慧的结晶，只是一些没什么价值的破房子。在他们心目中建筑师的天职就是“创作”，没有创作就不是称职的建筑师。当他们面对历史建筑和文物建筑的时候，不是甘当翻译家——尊重原著，而是要当作家——要对历史名著进行修改，以展示他们的创作才华。这是由他们的价值观决定的。

二是由于长期的文化缺失、教育缺失造成许多建筑师对中国传统建筑文化和四合院建筑缺乏基本的了解和认知，他们分不清地域建筑及其文化的差别。具体地说，分不清江浙、皖南民居与北京传统四合院的区别，甚至分不清日本、韩国民居与中国传统民居的区别，以致误将北京四合院改造成江浙、皖南，甚至日本、韩国的民居建筑。

不论是上述哪种情况，对于老北京胡同四合院建筑文化的保护都是极为不利的，甚至有意或无意地起到了歪曲和破坏作用。如果这种现象继续存在下去，北京的胡同四合院将面临被“转基因”的危险。

上述这些“创新”“实验”是在《北京城市总体规划（2016—2035）》出台前做的，当时的指导思想是“旧城更新，院落改造”。在这种思想指导下，作这样的探索改造无可厚非。《北京城市总体规划（2016—2035）》出台以后，指导思想发生了根本性的变化，对北京老城确定的原则是：要“加强老城整体保护”，要“保护北京特有的胡同、四合院”，“通过腾退，恢复性修建，做到应保尽保，最大限度留存有价值的历史信息”，“强化文化展示与继承”，“重新唤起对老北京的文化记忆，保持历史文化街区的历史延续”。

有鉴于此，老北京不能再搞“旧城更新、院落改造”，而是要尽最大努力保住胡同四合院的传统文化基因，为子孙后代留下一份原汁原味的历史文化遗产。

参考书目

1 刘致平．中国居住建筑简史．王其明，增补．北京：中国建筑工业出版社，1990．

2 刘敦桢．中国古代建筑史．2版．北京：中国建筑工业出版社，1984．

3 林洙．北京的四合院——建筑叙述．汉声杂志，1987．

4 程敬琪，杨玲玉．北京传统街坊的保护刍议——南锣鼓巷四合院街坊．中国建筑科学院建筑情报研究所建筑理论与历史研究室．建筑历史研究第二辑，1983．

5 北京市宣武区建设委员会，北京市古代建筑研究所合编．王世仁主编．宣南鸿雪图志．北京：中国建筑工业出版社，1997．

6 陆翔，王其明．北京四合院．北京：中国建筑工业出版社，1996．

7 王其亨．风水理论研究．天津：天津大学出版社，1992．

8 一丁，雨露，洪涌．中国古代风水与建筑选址．石家庄：河北科学技术出版社，1996．

9 何俊寿．大游年辨析．古建园林技术杂志，1996(4):22—31．

10 何俊寿．“门光尺”析证．古建园林技术杂志，1994(3):23—32．

11 刘大可．中国古建筑瓦石营法．北京：中国建筑工业出版社，1993．

12 马炳坚．中国古建筑木作营造技术．北京：科学出版社，1991．

13 王世襄．明式家具珍赏．香港：三联书店(香港)有限公司，北京：文物出版社，1985．

14 杨耀．明式家具研究．北京：中国建筑工业出版社，1986．

15 刘策．中国古代苑囿．银川：宁夏人民出版社，1979．

16 曲居仁．清代五彩、洋彩和珐琅彩的发展演变．陶瓷研究，1998(2):37—43．

17 胡德生．明清家具的种类及风格．收藏家，1997(5):56—62．

18 彭一刚．中国古典园林分析．北京：中国建筑工业出版社，1986．

19 冯钟平．中国园林建筑．北京：清华大学出版社，1988．

20 陈植，张公驰．中国历代名园记选注．合肥：安徽科学技术出版社，1983．

21 (日)冈大路．中国宫苑园林史考．常瀛生，译北京：农业出版社，1988．

22 宗白华．中国园林艺术概观．南京：江苏人民出版社，1987．

23 北京市社会科学研究所《北京史苑》编辑部．北京史苑．北京：北京出版社，1983．

24 邓云乡．北京四合院．北京：人民日报出版社，1990．

后记

经过半年多的紧张工作，《北京四合院建筑》一书终于脱稿了。为此，我们感到十分欣慰。

我在本书的前言中曾经提到，这本书既是为继承、弘扬中华传统建筑文化而作，又是为庆祝第二十届世界建筑师大会在北京召开而作，其意义十分深远。我和我的同事们都为能在此时此刻奉献这样一部著作而引以为自豪。

我们能在短时间内完成这部书，与有关领导、先哲和各界朋友的支持是分不开的。在这里，我首先要感谢一贯支持我们事业的北京市房屋土地管理局和我的上级单位——北京市第二房屋修建工程公司（北京古代建筑工程公司）的领导，是他们在关键的时候给予了我们有力的支持，使我们获得了研究这一课题的客观条件。

我还要感谢天津大学出版社的同志们，是他们为我们提供了一个极好的机会，使我们能将多年的研究成果发表出来，把它献给祖国，献给世界。

我更要感谢十几年如一日，一贯热心支持我们事业的建筑界、文物界的专家、前辈和先哲。当古建文物界的老专家、德高望重的罗哲文先生得知《北京四合院建筑》即将出版的消息后，不胜欢喜，挥笔写下热情洋溢的题词。年近八旬的高履泰教授满腔热情地支持本书的编著，并为本书的英译部分作认真细致的审校工作，其情切切，感人至深。

我还要感谢正业集团董事长韩真发先生和北京市华鼎四合院房屋经营中心总经理周文水先生，他们对本书的出版十分关注，并给予了道义上和财力上的支持，使我们对本书的出版发行更加信心十足。

本书是集体劳动的成果。在成书过程中，除本部的蒋广全、相炳哲、阎静明、蒋涛、唐婧持诸同志全部投入工作之外，还得到北京市房屋土地管理局职工大学校长王希富教授、故宫博物院古建部高级工程师王仲杰先生、北京市房屋设计院周仁杰先生、北京房屋土地管理局工会张永新、仇超以及夫人高玉芳同志的大力支持。为本书的出版做出贡献的，还有高选、郭凌玺等同志，在此一并致谢！

祝愿《北京四合院建筑》在弘扬中华传统建筑文化、促进中外建筑文化交流、保护古都风貌和促进北京城市建设方面发挥出应有的作用！

马炳坚

1999年4月

于营宸斋

再版后记

经过数日的忙碌，北京四合院建筑一书的修订终于完成了。

这次修订，是在《北京城市总体规划（2016—2035）》批复，北京市落实新总规提出的各项任务，加强历史文化名城保护，对老城区胡同、四合院进行恢复性修建的背景下进行的。这次修订将原第九章“北京四合院前景展望”一章删除，改为“北京老城保护与四合院的恢复性修建”，使本书在全面介绍北京四合院历史、文化、格局、风水、空间、构造、装修、装饰、设计、施工的基础上，深度介入风貌保护和恢复性修建这些敏感而现实的问题，使之更具有实用性和现实意义。

此次修订工作十分紧张。由于平时诸事繁忙，只好利用五一小长假花了不到一周时间完成了修订工作。

本次修订，得到了北京市古代建筑设计研究所张越同志、《古建园林技术》杂志社梅笑妍同志的帮助，得到家人的理解和支持，在此一并致谢！

马炳坚

2019年5月5日

于北京营宸斋